3D Ultrasound

3D ultrasound techniques have been increasingly used in diagnosis, minimally invasive image-guided interventions, and intraoperative surgical use. Today, most ultrasound system manufacturers provide 3D imaging capability as part of the systems. This availability has stimulated researchers to develop various machine learning tools to automatically detect and diagnose diseases, such as cancer; monitor the progression and regression of diseases, such as carotid atherosclerosis; and guide and track tools being introduced into the body, such as brachytherapy and biopsy needles.

This edited book is divided into three sections covering 3D ultrasound devices, 3D ultrasound applications, and machine learning tools using 3D ultrasound imaging. It is written for physicians, engineers, and advanced graduate students.

Aaron Fenster is a Scientist at the Robarts Research Institute and the founder and past Director of the Imaging Research Laboratories at the Robarts Research Institute. He is also a Professor and Chair of the Division of Imaging Sciences of the Department of Medical Imaging at the Western University. In addition, he is the founder and past Director of the graduate program in Biomedical Engineering and past Director of the Biomedical Imaging Research Centre at Western University.

Imaging in Medical Diagnosis and Therapy

Series Editors: Bruce R. Thomadsen, David W. Jordan

Beam's Eye View Imaging in Radiation Oncology
Ross I. Berbeco, Ph.D.

Principles and Practice of Image-Guided Radiation Therapy of Lung Cancer
Jing Cai, Joe Y. Chang, Fang-Fang Yin

Radiochromic Film
Role and Applications in Radiation Dosimetry
Indra J. Das

Clinical 3D Dosimetry in Modern Radiation Therapy
Ben Mijnheer

Hybrid Imaging in Cardiovascular Medicine
Yi-Hwa Liu, Albert J. Sinusas

Observer Performance Methods for Diagnostic Imaging
Foundations, Modeling, and Applications with R-Based Examples
Dev P. Chakraborty

Ultrasound Imaging and Therapy
Aaron Fenster, James C. Lacefield

Dose, Benefit, and Risk in Medical Imaging
Lawrence T. Dauer, Bae P. Chu, Pat B. Zanzonico

Big Data in Radiation Oncology
Jun Deng, Lei Xing

Monte Carlo Techniques in Radiation Therapy
Introduction, Source Modelling, and Patient Dose Calculations, Second Edition
Frank Verhaegen, Joao Seco

Monte Carlo Techniques in Radiation Therapy
Applications to Dosimetry, Imaging, and Preclinical Radiotherapy, Second Edition
Joao Seco, Frank Verhaegen

Introductory Biomedical Imaging
Principles and Practice from Microscopy to MRI
Bethe A. Scalettar, James R. Abney

Medical Image Synthesis
Methods and Clinical Applications
Xiaofeng Yang

Artificial Intelligence in Radiation Oncology and Biomedical Physics
Gilmer Valdes, Lei Xing

3D Ultrasound
Devices, Applications, and Algorithms
Aaron Fenster

For more information about this series, please visit: https://www.routledge.com/Imaging-in-Medical-Diagnosis-and-Therapy/book-series/CRCIMAINMED

3D Ultrasound

Devices, Applications, and Algorithms

Edited by
Aaron Fenster

CRC Press
Taylor & Francis Group
Boca Raton London New York

CRC Press is an imprint of the
Taylor & Francis Group, an **informa** business

Designed cover image: shutterstock_677354815

First edition published 2024
by CRC Press
2385 NW Executive Center Drive, Suite 320, Boca Raton FL 33431

and by CRC Press
4 Park Square, Milton Park, Abingdon, Oxon, OX14 4RN

CRC Press is an imprint of Taylor & Francis Group, LLC

ISBN: 9781032288192 (hbk)
ISBN: 9781032289816 (pbk)
ISBN: 9781003299462 (ebk)

DOI: 10.1201/9781003299462

Typeset in Times LT Std
by KnowledgeWorks Global Ltd.

Contents

SECTION III   3D Ultrasound Algorithms

Contributors

Joeana Cambranis-Romero
Robarts Research Institute
Western University
London, Canada

Patrick K. Carnahan
Robarts Research Institute
Western University
London, Canada

Elvis C. S. Chen
Robarts Research Institute
Western University
London, Canada

Xueli Chen
Department of Electrical Engineering
City University of Hong Kong
Kowloon, Hong Kong

Bernard Chiu
Department of Electrical Engineering
City University of Hong Kong
Kowloon, Hong Kong

Robert Dima
Departments of Medical Biophysics and
    Mechanical Engineering
Western University
London, Canada

Houran Dou
Centre for Computational Imaging and
    Simulation Technologies in Biomedicine
    (CISTIB)
School of Computing
University of Leeds
Leeds, West Yorkshire, England

Carla du Toit
Departments of Medical Biophysics and
    Mechanical Engineering
Western University
London, Canada

Aaron Fenster
 Departments of Medical Imaging and Medical
    Biophysics
Imaging Research Laboratories & Robarts
    Research Institute
Western University
London, Canada

Rory Geoghegan
Department of Bioengineering
University of California
Los Angeles, California, USA

Derek J. Gillies
Department of Medical Biophysics
Western University
London, Canada

Leah A. Groves
Robarts Research Institute
Western University
London, Canada

Xiaoqiong Huang
National-Regional Key Technology
    Engineering Laboratory for Medical
    Ultrasound
School of Biomedical Engineering
Shenzhen University Medical School
Shenzhen University
Shenzhen, China

Yuhao Huang
National-Regional Key Technology
    Engineering Laboratory for Medical
    Ultrasound
School of Biomedical Engineering
Shenzhen University Medical School
Shenzhen University
Shenzhen, China

Megan Hutter
Departments of Medical Biophysics and
    Mechanical Engineering
Western University
London, Canada

Emily Lalone
Departments of Medical Biophysics and
    Mechanical Engineering
Western University
London, Canada

Tian Liu
Department of Radiation Oncology and
    Winship Cancer Institute
Emory University
Atlanta, Georgia, USA

Mingyuan Luo
National-Regional Key Technology
    Engineering Laboratory for Medical
    Ultrasound
School of Biomedical Engineering
Shenzhen University Medical School
Shenzhen University
Shenzhen, China

Randa Mudathir
Departments of Medical Biophysics and
    Mechanical Engineering
Western University
London, Canada

Shyam Natarajan
Department of Urology
University of California
Los Angeles, California, USA

Dong Ni
Medical Ultrasound Image Computing
    (MUSIC) Lab
Shenzhen University
Shenzhen, China

Hareem Nisar
Robarts Research Institute
Western University
London, Canada

Samuel Papernick
Departments of Medical Biophysics and
    Mechanical Engineering
Western University
London, Canada

Claire Keun Sun Park
Department of Medical Biophysics
Schulich School of Medicine & Dentistry
Western University
London, Canada

Jake Pensa
Department of Bioengineering
University of California
Los Angeles, California, USA

Terry M. Peters
Robarts Research Institute
Western University
London, Canada

Jessica Robin Rodgers
Department of Physics and Astronomy
University of Manitoba
Winnipeg, Manitoba, Canada

Zachary Szentimrey
School of Engineering
University of Guelph
Guelph, Canada

Eranga Ukwatta
School of Engineering
University of Guelph
Guelph, Canada

Jing Wang
Department of Radiation Oncology
Winship Cancer Institute
Emory University
Atlanta, Georgia, USA

Tonghe Wang
Department of Medical Physics
Memorial Sloan Kettering Cancer Center
New York, New York, USA

Robert Wodnicki
University of Southern California
Los Angeles, California, USA

Shuwei Xing
School of Biomedical Engineering
Western University
London, Canada

Xiaofeng Yang
Department of Radiation Oncology
Winship Cancer Institute
Emory University
Atlanta, Georgia, USA

Xin Yang
National-Regional Key Technology
    Engineering Laboratory for Medical
    Ultrasound
School of Biomedical Engineering
Shenzhen University Medical School
Shenzhen University
Shenzhen, China

Jesse Yen
University of Southern California
Los Angles, California, USA

Ran Zhou
The School of Computer Science
Hubei University of Technology
Wuhan, Hubei, China

Yuxin Zou
National-Regional Key Technology
    Engineering Laboratory for Medical
    Ultrasound
School of Biomedical Engineering
Shenzhen University Medical School
Shenzhen University
Shenzhen, China

Section I

3D Ultrasound Devices and Methods

1 2D Arrays

Jesse Yen and Robert Wodnicki

BEYOND 1D ARRAYS

From a transducer perspective, three-dimensional (3D) ultrasound imaging can be accomplished in a number of ways. Perhaps the most straightforward method is to mechanically move a one-dimensional (1D) array in the elevation direction while acquiring multiple two-dimensional (2D) slices. Mechanical movement can be in a linear direction or rotational (wobbler) [1]. The 2D slices are then stacked in the elevation direction to form a 3D volume. The disadvantage of this approach is that it is generally slower than electronic scanning, and it relies on mechanical movement, translation or rotation, of 1D array transducers. In addition, there are issues with reliability due to the use of mechanical scanning related to the need for an acoustic coupling fluid that can leak over time and may be prone to air bubbles which degrade the imaging quality. There is also a lack of dynamic elevational focusing leading to worse lateral resolution in the elevation direction away from the predetermined focal depth of the acoustic lens. This lack of elevational focusing can be alleviated through the use of multi-row arrays. So-called 1.25D, 1.5D, and 1.75D arrays employ a varying number of rows and differ in the quality of their elevational focusing capabilities [2]. Figure 1.1 shows cross-sectional drawings of a 1D array (Figure 1.1a) along with these multi-row arrays (Figure 1.1b–d). A 1D array appears as a single transducer element in the elevation direction with a mechanical acoustic lens for fixed focusing at a predetermined depth. A 1.25D array (Figure 1.1b) has three rows where the outer two rows can be electrically connected or disconnected with the central row via a switch. This expanding aperture enables adjustment of the aperture size as a function of depth which in turn translates the axial focal point to improve contrast resolution uniformity with depth. A 1.5D array generally has more than three rows where the rows are symmetrically tied together about the centerline and connected to separate system channels. This allows for electronically controlled dynamic focusing in the elevation direction as a function of depth. Lastly, a 1.75D array will likely have more rows than 1.5D array. However, the number of rows is substantially smaller than the number of elements in the azimuthal direction. Each element is separately controlled to allow for dynamic focusing and some steering of the beam in the elevation direction.

With these arrays, the azimuthal element pitch is on the order of 0.5–1 wavelength while the elevational element pitch is usually larger on the order of 2–8 wavelengths. While strictly speaking, electronic focusing in elevation should obviate the need for a mechanical lens, in practice it can be still useful to include both (Figure 1.1), with the acoustic lens improving the focus near the center of the axial range over the weak electronic focus provided by large elements in 1.25D and 1.5D array implementations [2]. For 1.75D arrays where the elements are small in elevation, an acoustic lens is typically not necessary.

2D arrays provide the ability to steer and focus an ultrasound beam equally in both azimuthal and elevational directions. Assuming the 2D array is same size in both lateral directions, the lateral resolution will be same in these directions leading to uniform image quality. These two capabilities enable the potential for rapid scanning of a large volume without the need for mechanically moving components. Advanced imaging methods such as Doppler, elastography, and ultrafast imaging are more readily extended to 3D in real time with 2D arrays compared to mechanically translating 1D arrays. For noninvasive imaging, simple 2D arrays have on the order of 1000 elements (e.g., 32×32 array matrix), while more complex arrays for specialized applications (e.g., vascular) can have tens of thousands of elements. 2D arrays are also preferred if advanced imaging methods, such as 3D Doppler, are desired.

DOI: 10.1201/9781003299462-2

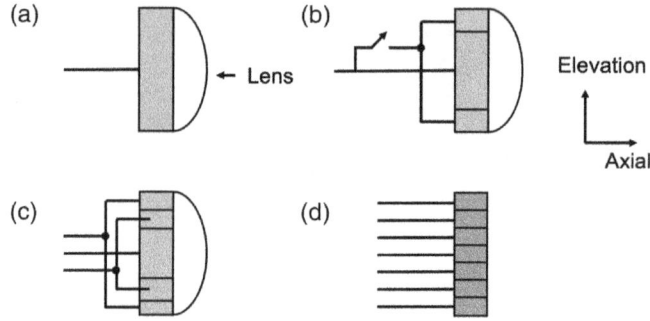

FIGURE 1.1   Side view of different array types. (a) 1D array, (b) 1.25D array, (c) 1.5D array, and (d) 1.75D array. (Adapted from Shung [3].)

TRANSDUCER AND ARRAY FUNDAMENTALS

We begin this section by summarizing basic ideas fundamental to transducer and array technology. These concepts are widely applicable to transducers in many applications but are also important to understanding the challenges associated with 2D arrays. At the time of this writing, most transducers use piezoelectric materials. Piezoelectric materials have the unique ability to convert electrical energy into mechanical energy and vice versa. Thus, analogous to audio speakers and microphones, the ultrasound transducer is capable of acting as both a transmitter and a receiver of acoustic signals. A typical transducer consists of a piezoelectric material, most commonly lead zirconate titanate (PZT), which acts as the physical transducing device; an electrical interconnect such as a flexible circuit which transmits the electrical signals between the imaging system and the piezo layer; a backing material which absorbs extraneous and unwanted backside echoes; at least one matching layer to improve coupling of sound from the piezo to the body; an electrical ground layer; and a protective outer layer to improve robustness and reliability for clinical use. Figure 1.2a shows a simple four-element 1D array with backing, flexible circuit interconnect, piezoelectric material, and matching layer. The ground and protective layer are not shown for purposes of clarity. The backing material provides mechanical support to the transducer or array while also damping out some of the backward propagating energy. In front of the transducer, one or more quarter-wave matching layers are used to improve the transfer efficiency as well as the bandwidth of the transducer. Piezoelectric ceramics normally have an acoustic impedance on the order of 30 MRayls making for a poor match to tissue which has an acoustic impedance of around 1.5 MRayls. If a single matching layer is used, it will have an intermediate acoustic impedance Z_m given by ref [4] where Z_p and Z_t are the acoustic impedances of the piezo and tissue respectively:

$$Z_m = \left(Z_p Z_t \right)^{1/3} \tag{1.1}$$

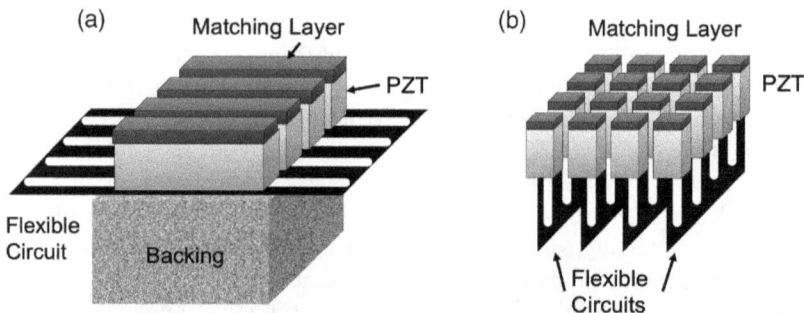

FIGURE 1.2   (a) Diagram of a four-element 1D array and (b) its 2D array equivalent.

In general, it is desirable to optimize the matching layer system to provide improved coupling of sound for greater transmit and receive sensitivity which increases the depth of penetration into the tissue for a deeper field of view. In addition, optimized matching layers can improve the fractional bandwidth of the acoustic response of the elements which leads to broadband excitation. This is important for harmonic imaging and contrast-enhanced ultrasound (CEUS), and it also improves axial resolution by reducing the length of the pulse in tissue.

For 2D arrays, the construction is similar conceptually, but is more challenging to realize in practice. Figure 1.2b is an illustration of 2D array, showing a 16-element 2D array with approximately the same overall aperture dimensions as the 4-element 1D array of Figure 1.2a. A number of factors in manufacturing conspire to reduce overall element yield which in turn reduces imaging performance [5]. These include the small element size which makes them more prone to yield loss during dicing, challenges in sensitivity due to the high electrical impedance of the smaller elements, issues with reliable and dense electrical interconnection to the large number of tightly pitched elements due to limitations of modern flex circuit manufacturing capabilities, as well as the complexity of accurate acoustic design of the 2D array elements themselves whose exact performance is best described using sophisticated finite element modeling techniques [6]. These design and implementation challenges have historically made the realization of high-quality 2D arrays mainly the purview of high-volume original equipment manufacturers (e.g., GE, Philips, Siemens) which have the resources to implement high-quality fabrication to resolve these issues systematically.

CHALLENGES FOR 2D ARRAY DEVELOPMENT

The development of 2D arrays is challenging on several fronts. These include the need for electrical connection to potentially thousands of elements, as well as the substantial electrical impedance mismatch between transducer elements and system electronics which leads to poor roundtrip sensitivity. A 2D array element made of conventional PZT can have an electrical impedance magnitude on the order of 1–10 kΩ whereas 1D array elements are usually on the order of 50–200 Ω. The higher electrical impedance relative to fixed 50 Ω ultrasound receive electronics leads to significant degradation in the receive signal. In addition, the small size of the elements reduces the absolute acoustic output pressure as compared to 1D array elements. Lastly, data acquisition and beamforming for thousands of elements also needs to be addressed. A primary goal of this chapter is to review the work done to address these challenges.

2D ARRAY MATERIALS: ADDRESSING THE IMPEDANCE PROBLEM

Compared to traditional 1D linear array elements which can be on the order of a few mm in height and 0.5 mm in width, the size of individual 2D array elements is significantly smaller (e.g., 0.2 × 0.2 mm). The reduction in cross-sectional area for 2D arrays leads to increased electrical impedance, lower power output, and concomitant mismatches with system electronics. 2D array elements directly connected to system channels via standard micro-coaxial cable will experience substantial losses in signal due to electrical loading when driving the large capacitance of (typically 3 m) long ultrasound cables. Strategies to minimize or overcome these losses include the use of novel transducer materials which include materials with high dielectric constants, multi-layer transducers, composites, and single-crystal materials.

The capacitance, C_0, of a transducer element away from resonance, is given by:

$$C_0 = \frac{\varepsilon_0 \varepsilon_r A}{t} \tag{1.2}$$

where, ε_0 is the permittivity of free space (8.85 × 10^{-12} F/m), ε_r is the relative dielectric constant, A is the cross-sectional area, and t is the thickness of the transducer material between the

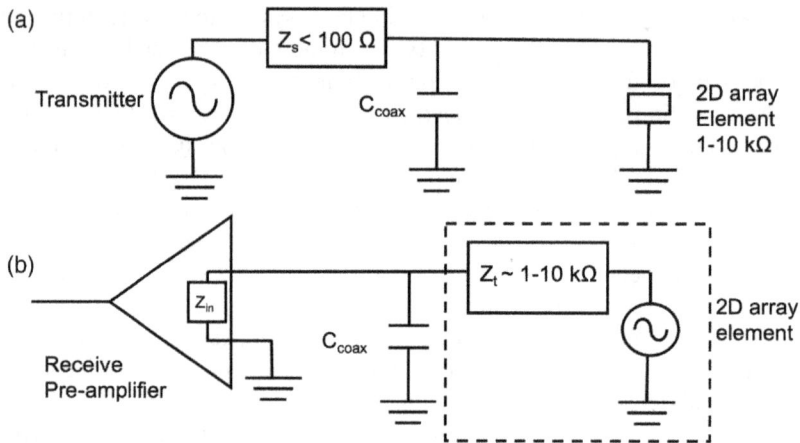

FIGURE 1.3 Simplified circuit diagrams of an ultrasound system front end connected to (a) transmitting channel and (b) a receive channel with 2D array element (dashed box).

plate electrodes. As can be appreciated from (1.2), as A decreases exponentially with an increasing number of 2D array elements in a given array footprint, the capacitance of the elements drops rapidly. The output impedance of the elements varies with the inverse of this capacitance and therefore increases significantly as the element pitch is reduced.

In early production arrays for clinical use, PZT-5H was the material of choice due to its wide use in 1D arrays and its reasonably high dielectric properties (ε_r ~3,400). The impedance of a 1D array element is on the order of 30–80 Ω representing a good match to system electronics which typically have 50 Ω receive impedance. However, when the same material is diced to form 2D array elements, the electrical impedance increases to the order of 1–10 kΩ, making it a very poor match to system electronics and coaxial cabling.

Figure 1.3 shows simplified circuit diagrams of an ultrasound system front end connected to a transmitting channel and a receiving channel. The output impedance of a transmitter Zs is oftentimes less than 100 Ω. A 2D array element having an impedance greater than 1 kΩ means that power transfer is inefficient. In receive mode, the 2D array element is modeled as an ideal voltage source with output impedance again on the order of 1–10 kΩ. A nominal coaxial cable will have a capacitance of around 200 pF or impedance of about 150 Ω depending on the ultrasound frequency. Because of the very high impedance of the array element, at least 80% of the signal is lost due the voltage divider formed between the coaxial cable impedance and the transducer element impedance. Signal-to-noise ratio (SNR) will be compromised assuming that noise from the pre-amplifier and other amplifiers is the dominant source of noise. Dielectric materials such as CTS 3203HD and HK1HD (TRS Technologies) can also improve the impedance issue to the degree that the dielectric constant can be increased, as this in turn will increase the capacitance of the elements. Multilayer approaches have also been pursued [7, 8]. Multilayer transducers consist of multiple layers of PZT arranged acoustically in series but electrically in parallel. If N is the number of layers, the effective area increases by a factor of N, while the thickness of each layer is reduced by a factor of N. Combined, this reduces the impedance by a factor of N^2 where N is the number of layers. A tradeoff is that the receive sensitivity of multilayer transducers may be reduced [8]. Figure 1.4 shows a drawing of a single-layer piezoelectric and a multilayer piezo with eight layers. For the multilayer, the poling direction alternates with each subsequent layer so that all layers expand and contract in unison. The overall thickness remains same as a single layer.

Single-crystal materials [9–11] have also been used in commercially available matrix arrays. Single crystals offer greater sensitivity and bandwidths compared to polycrystalline PZT due to their improved electromechanical coupling coefficient. However, fabrication of transducers with

(a) (b)

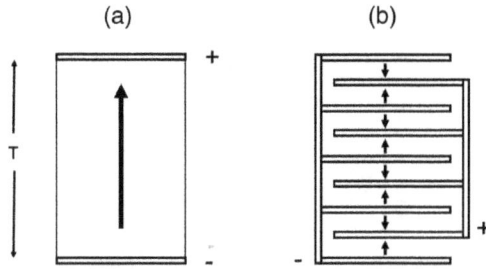

FIGURE 1.4 (a) A single-layer piezoelectric material and (b) an eight-layer multilayer piezoelectric material. (Adapted from Shung [3].)

single-crystal materials can be quite challenging because they are prone to fracture due to their fragile nature, and they also have a relatively low Curie temperature and are therefore more easily de-poled than other piezo materials. These issues require extensive optimization of fabrication process such as dicing to realize the full potential of their performance enhancement and have been implemented by most modern manufacturers of clinically useful probes.

Capacitive micromachined ultrasonic transducers (cMUTs) use micromaching and silicon fabrication methods to create ultrasonic transducers. These devices are an attractive solution for 2D arrays because they are amenable to high-volume electronic packaging methods which in turn make them much more readily coupled with local buffering electronics to help alleviate the impedance mismatches discussed above. The basic concept of cMUTs is the realization of evacuated cavities in silicon which have suspended above them a vibrating membrane constructed using materials which are standard in typical microelectromechanical system (MEMs) fabrication processes (e.g., Si, SiO$_2$ and silicon nitride). The cavity geometry and the membrane thickness of materials are optimized using sophisticated 3D finite element modeling to yield high sensitivity in the clinical range of ultrasound (1–20MHz). Given that they are essentially constructed using standard semiconductor fabrication techniques, they are also amenable to an array of standard and advanced semiconductor packaging technologies which makes them more easily integrated with front-end electronics. Furthermore, cMUTs are fabricated using wafer-scale processes which means that they benefit from the economies of scale that drive the relentless march of Moore's law and commercial electronics miniaturization. Once suitable designs for cMUTs and electronics can be established, fully sampled arrays can be more easily mass-produced compared to traditional piezo-based arrays whose fabrication has historically been very labor-intensive and time-consuming. cMUTs benefit from a wide acoustic bandwidth, performing comparably or better than single-crystal composite-based piezo arrays [12]. Due to their high source impedance, cMUTs require locally integrated buffering electronics or impedance matching to drive the cable to the system in order to improve their sensitivity and imaging penetration [12]. Figure 1.5 shows the basic construction of one cMUT cell.

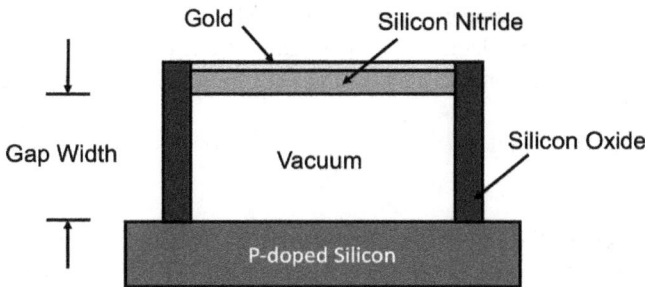

FIGURE 1.5 Basic construction of one cell of a capacitive micromachined ultrasonic transducer (cMUT). (Adapted from Shung [3].)

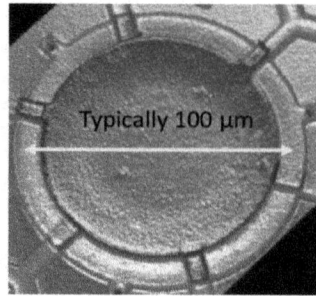

FIGURE 1.6   A single cMUT cell. (Courtesy of Philips.)

The basic operation of a cMUT, relative to Figure 1.5, is as follows. The transmit and receive sensitivity of the device are optimized by maintaining a very thin gap between the vibrating silicon nitride membrane and the p-doped silicon substrate which is the bulk of the device and shared by all elements of the array. To create this thin gap, a DC bias voltage is applied across the device terminals and this causes the membrane to be pulled down toward the substrate in the same way a violin string must be tightened to achieve high-quality resonance. Typical DC bias depends on the design of the devices and can be anywhere from 50 V to as much as 200 V. Superimposed on this DC voltage is an AC signal which is the transmitting signal for the device. This transmitting signal causes the membrane to vibrate at clinically relevant ultrasound frequencies (e.g., 1–20 MHz, depending on design). When the front face of the device which is the vibrating membrane is placed in contact with a propagating medium (e.g., water, oil, or human tissue), the vibration of the membrane causes sympathetic oscillations that effectively launch a wave of ultrasound forward into the medium. Similarly, for the receive case, vibrations in the medium cause the membrane to vibrate. These vibrations change the distance between the silicon nitride membrane and the silicon substrate which effectively creates an oscillating charge signal that appears across the electrical terminals of the device. This signal can be read by buffering electronics as the receive signal of the array. A photo of a single cMUT cell is shown in Figure 1.6. In typical applications, transducer elements are constructed by ganging together a large number of these individual unit cells. In this way, the acoustic parameters of the device (e.g., resonance frequency and bandwidth) can be independently optimized from the physical aperture size of the overall array elements.

Piezoelectric MUTs (pMUTs), much like cMUTs, also offer a potentially attractive solution for 2D arrays due the use of established MEMS fabrication techniques [13, 14]. With this technology, formulations of traditional piezo materials that have been optimized for integration with standard semiconductor processes flows are deposited directly on the surface of a silicon wafer. The backside of the silicon underneath the individual elements is typically etched to form cavities which improves the resonance performance of the transducer elements. pMUTs benefit from the same wafer-scale fabrication gains in yield and volume cost reduction as cMUTs and are also capable of being closely integrated with signal processing electronics. These gains make them attractive for portable systems which require low-cost and reasonable performance for high-volume clinical applications such as point of care US (POCUS).

SPARSE ARRAY DESIGNS

Sparse 2D arrays minimize the interconnect challenge by selecting a subset of the available elements of a 2D array [15]. Transmit and receive elements are judiciously chosen and several different approaches can be used to determine the most appropriate configuration for a given aperture size and the number of available transmit and receive channels. One key concept is the idea of the co-array or the effective aperture. These concepts were explored by Lockwood, Khuri-Yakub, and

Kassam [16–18] to develop novel sparse array designs. The idea of the co-array begins with the fact that a transmit aperture $a_t(x_0)$ and its beam in the far field $\phi_t(x)$ are Fourier transform pairs where x_0 is the azimuthal coordinate at the transducer surface and x is the azimuthal coordinate in the far field. In (1.3), $\mathcal{F}$ indicates the Fourier transform operation. Likewise, the receive aperture $a_r(x_0)$ and the receive beam $\phi_r(x)$ are also Fourier transform pairs. The variable x_0 is the azimuthal coordinate in the aperture plane and x is the azimuthal coordinate at the focal depth.

$$\phi_t(x) = \mathcal{F}\left[a_t(x_0)\right] \tag{1.3}$$

$$\phi_r(x) = \mathcal{F}\left[a_r(x_0)\right] \tag{1.4}$$

The transmit-receive beam pattern, $\psi(x)$, is the product of the transmit and receive beams. This product is also the Fourier transform of the convolution of the two aperture functions as indicated in (1.5) where * indicates convolution.

$$\psi(x) = \phi_t(x)\phi_r(x) = \mathcal{F}\left[a_t(x_0) * a_r(x_0)\right] \tag{1.5}$$

Therefore, the co-array or the effective aperture concept states that the co-array or effective aperture is the convolution of the transmit and receive apertures. The above equations use only the azimuthal coordinate, but the elevational coordinate can also be added to analyze 2D arrays. The effective aperture can be found by performing 2D convolution of the transmit and receive apertures.

MILLS CROSS ARRAY

A Mills Cross array (Figure 1.7) [19] has a column of transmit elements and an orthogonal row of receive elements. Transmit focusing is only achieved in the elevation direction. The transmit beam is wide in the azimuth direction. In receive elements, focusing is achieved in the azimuthal direction and the beam is wide in elevation. Due to the multiplicative pulse-echo process, the point spread function is given by $\mathrm{sinc}(x)*\mathrm{sinc}(y)$. Likewise, a 2D convolution of the transmit aperture and the receive aperture will show a flat response over the central portion of the co-array. To perform rapid 3D scanning, a transmit beam focused in elevation and wide in azimuth is emitted. Parallel receive focusing in the selected azimuthal plane is performed to make a single image slice. A new elevation coordinate is selected and the process is repeated.

ROW-COLUMN ARRAY

A variation of the Mills Cross array, known as the row-column array, has gained attention recently (Figure 1.8) [20–27]. Using a matrix of array elements, the top side of the matrix array has an electrode pattern resembling a 1D array along the elevation direction. On the bottom side, the electrode

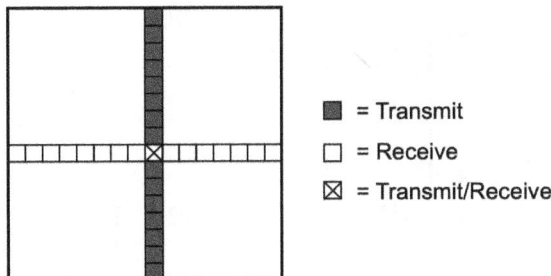

FIGURE 1.7   Mills Cross array.

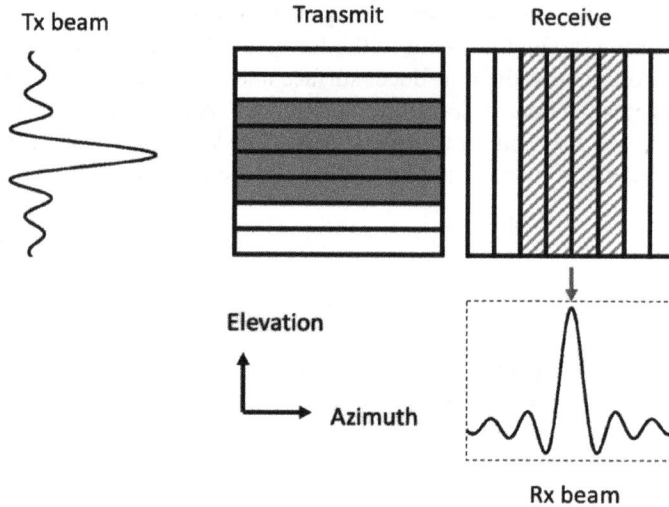

FIGURE 1.8   Row-column array.

pattern resembles a 1D array along the azimuth direction. In transmit, the top electrodes are active while the receive electrodes are grounded. This forms a 1D array along the elevation direction allowing for focusing in this direction. Elements shaded in gray indicate the active transmit elements. In receive, the roles of the electrodes are switched: the top electrodes are now grounded and the bottom electrodes are active. The active receive elements are indicated by the hatched pattern. This effectively creates a 1D array with elements along the azimuthal direction. Similar to the Mills Cross array, 3D scanning occurs on a slice-by-slice basis. Different slices can be selected by activating a different subset of elements in transmit much like a moving sub-aperture for a 1D array. Because the element areas are much larger than the area of 2D array elements in a Mill Cross array, the element impedance is much lower and is therefore a better match to system electronics. Large 256×256 row-column arrays have been fabricated with PZT as well as cMUTs versions [22, 24]. In addition, at least one manufacturer is currently marketing such a device for research use (Verasonics/Vermon) [28].

VERNIER ARRAY

A periodic Vernier array [29] uses every Nth element in transmit and every Mth element in receive where N and M are different integers. The periodic spacing gives rise to grating lobes. An example of a Vernier array is shown in Figure 1.9. For clarity, the transmit and receive elements are shown

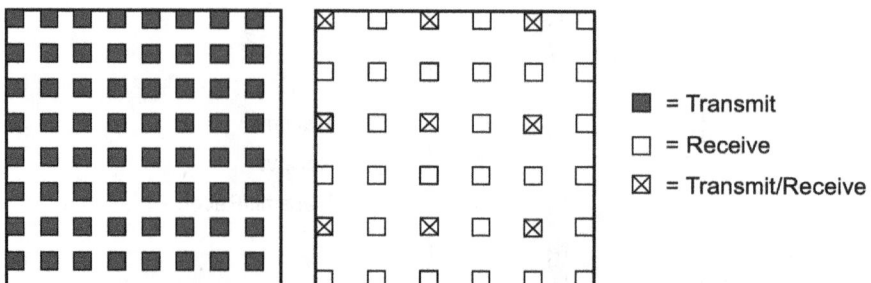

FIGURE 1.9   Vernier array with transmit and receive elements shown separately.

separately. In transmit, every other element is used in both azimuth and elevation directions. In receive, every third element is used. Thus, every sixth element will be used as both a transmitter and receiver indicated by the squares with an "*x*". The grating lobe positions for transmit and receive will occur at different locations since the location of the *n*th grating lobe is given by the following equation:

$$\theta_n = sin^{-1}\left(\frac{n\lambda}{d}\right)$$

(1.6)

where *n* is the location of the *n*th grating lobe, λ is the ultrasound wavelength in tissue, and *d* is the center-to-center distance between active elements. Continuing with the Vernier example and assuming a phased array with $\lambda/2$ pitch using every other element in transmit would have the first pairs of grating lobes located at +/− 90° and the receive grating lobes would be located at +/− 42°. Due to the multiplicative process of transmit and receive and the difference in transmit and receive grating lobe locations, the pulse-echo point spread function sees a reduction in grating lobe levels since the transmit and receive grating lobes are in separate locations. A common configuration for the Vernier array is to use every *N*th element in receive and *N*−1th element in transmit. Typical values of *N* may range from 3 to 6. Increasing the value of *N*, and therefore the sparseness, will see diminishing results since the transmit and receive grating lobe locations will begin to overlap.

RANDOM ARRAY

As its name suggests, transmit and receive elements in a random array [30] are chosen in a random fashion. Having randomly selected elements helps reduce the magnitude of grating lobes. The random distribution of elements can be realized in a uniformly distributed manner, for example, using a Gaussian distribution of the element locations or something similar. Hundreds, if not thousands, of simulations can be run to determine the best performing random array for a given set of constraints such as array size and number of transmit and receive channels. Rather than having distinct grating lobes, a pedestal of acoustic clutter is created and this level is approximately given by [30]:

$$20log_{10}\left(\frac{1}{\sqrt{N_t N_r}}\right)$$

(1.7)

where N_t and N_r are the number of transmit and receive elements, respectively. This expression indicates that the magnitude of the pedestal decreases with increasing numbers of transmitters and receivers as one might expect. Figure 1.10 is an example of a 16×16 array with 256 active elements randomly distributed throughout the active aperture. To help ensure uniformity, the 16×16 array was broken up into sixteen 4×4 subarrays. Four elements are selected for each subarray. Numerous permutations exist for a given array size and number of active elements. Computer simulations should be performed to determine the most desirable random array layout.

BOUNDARY ARRAYS

As their name suggests boundary arrays use only the outermost elements of a 2D array aperture. Rectangular boundary arrays (RBAs) have been proposed where the top and bottom rows are used in transmit and the left and right columns are used in receive [17, 18, 31]. Using the effective aperture concept, one can see that the size of the effective aperture is the same size as that of a fully sampled array. Therefore, it is theoretically possible to produce images of similar resolution in the elevation and azimuthal dimensions as compared to a fully sampled 2D array. However, since only the outermost elements are used, these arrays will have low SNR. RBAs may also be more

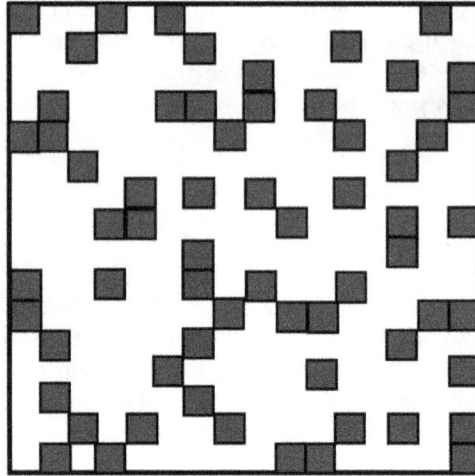

FIGURE 1.10   Example of a random array.

sensitive to loss of individual elements. In addition to rectangles or squares, other shapes are theoretically possible. Circular boundary arrays or ring arrays have been developed for miniaturized arrays needed for forward-looking intravascular ultrasound (IVUS) [32].

In summary, there are numerous strategies for designing sparse arrays, and researchers continue to develop different strategies for the design of sparse arrays. These include the Fermat spiral [33], density tapered arrays [34], and simulated annealing approaches [35]. In all cases, the design of the arrays can be guided by fundamental concepts such as the Fourier transform relationship between the aperture and the beam as well as the co-array concept. Greater numbers of active elements will always improve SNR, and improvement in the system point spread function in terms of reduced sidelobes, grating lobes, and clutter is highly likely. However this comes at the expense of additional system complexity.

FULLY SAMPLED ARRAYS WITH ASICs

Realization of fully sampled clinically useful 2D arrays requires miniaturization and placement of much of the front-end electronics in an ultrasound system adjacent to the fully sampled array. Otherwise, the transducer would require thousands of cables which would become expensive and unwieldy in a clinical setting. A very important advance in the past 20 years for production such fully sampled arrays is the use of application specific integrated circuits (ASICs) located directly in the probe handle either immediately below the transducer array or integrated adjacent to the arrays using a multi-chip module, high-density flex, or rigid organic build-up substrate [36–44]. These ASICs implement various signal processing functions including local multiplexing for scanning an active window across the array for imaging, as well as so-called sub-aperture beamforming which essentially groups analog signals from small sub-regions of elements to greatly reduce the number of system channels needed to form an image [43]. This reduction in signal density is a very important consideration for 2D arrays in particular since the element count grows exponentially with the size of the aperture. The large number of signal channels (e.g., anywhere from 256 to 10,000 or more) requires a large number of micro-coaxial cables to bring the signals back to the system for processing. A large aggregate cable with thousands of micro-coaxes presents challenges to the sonographer including the prevalence of occupational repetitive motion disorders due to the sheer weight of the combined imaging apparatus. In addition, a large number of signal processing channels presents challenges to the ultrasound system design itself in terms of the cost, energy consumption, and physical form factor of the processing electronics. The use of local ASICs in the probe handle has helped to mitigate these issues by greatly reducing the number of required micro-coax cables and in turn also reducing the beamforming requirements at the system end of the cable.

Starting in the early 2000s, fully sampled arrays were developed that began to take advantage of this new concept of ASICs in the probe handle [36, 38, 42, 43, 45–47]. These ASICs are placed adjacent or near the array. They are the bridge between the transducer and the system. ASIC-based imaging systems utilize highly integrated circuits which contain much of the needed front-end hardware which includes some combination of pulsers, pre-amplifiers, multiplexers, delay lines, and switches [37, 39, 41, 44]. Most ASIC architectures funnel echo signals from thousands of elements into a more manageable 64–256 channel ultrasound system.

The principal benefit of integrating microelectronics locally with the elements is a solution to the signal routing bottleneck that exists with 2D array fabrication. With a 1D array of aperture having N elements, there are N system channels, and therefore only N signal lines which need to be routed on the flex circuit that interfaces the system cables to the piezo array elements. The pitch of these lines must be fundamentally matched to the physics-based pitch of the acoustic elements which is commonly fixed at λ for a linear array and $\lambda/2$ in a phased array with beam-steering for greater than +/−30°. For imaging in tissue at 5 MHz, λ is 300 μm as an example. With flex circuit fabrication, 50 μm trace with 50 μm space lines are more or less standard, while 25 μm trace/space fabrication capability is considered state of the art. At 50 μm trace/space, in a single routing channel we may realize an element pitch of 100 μm (50 μm trace + 50 μm space). This implies that fabricating a linear array with standard technology is limited to 15 MHz operation for linear arrays, and 7.5 MHz for a phased array. The situation for 2D arrays however is quite different. In this case, we require $N \times N$ signal connections for a fully sampled array. This can be realized either by increasing the number of lines in every routing channel in a column or by increasing the number of signal routing layers or some combination of both. A further consideration is the requirement to use vias between layers if multiple layers are used and these are often relatively large (100–200 μm). Therefore, utilizing standard flex technology, creating a 2D array that is fully sampled is fairly limited to small array sizes, and depending on the frequency of operation it can be challenging if not impossible.

Utilizing locally integrated ASICs with direct assembly of the transducer array to the ASIC surface removes this bottleneck [36, 37, 44, 48]. Signal routing on the surface of an ASIC can be 100–1,000 times denser than state of the art flex. Moreover, locally integrated beamforming operations essentially reduce the routing requirements significantly by combining channels locally and yielding a much smaller number of signals that need to be routed off-ASIC and back to the system [36, 37, 44].

Further advantages of these ASIC implementations accrue due to improvements in signal processing capability [25, 37, 43, 44]. The ability to instantaneously operate all the elements in the array simultaneously both on transmit and receive provides significant advantage over multiplexed, synthetic aperture, and sparse array implementations. On transmit, utilizing the entire array provides improvements in SNR and penetration over synthetic aperture, or sparse array approaches since the maximum transmit pressure can be developed at the focus when all elements are active and electronically focused. On receive, the SNR improves as $\sqrt{N}$, where N is the number of active receive channels. Therefore, when the entire array can be sampled at the same time for a given beam formed in the image, optimal sensitivity and penetration depth will be achieved. Further advantages realized with ASICs integrated locally at the probe include the ability to pre-amplify and buffer the receive signals so that signal loss due to cable loading can be minimized, which serves to further improve penetration. As opposed to sparse arrays, fully sampled arrays also benefit from reduced beamforming clutter due to a larger number of simultaneously contributing system channels and elements. In addition, when compared to synthetic aperture methods, fully sampled arrays benefit from increased volume acquisition rate. While multiplexed realizations must scan through a product of $N \times N$ transmit/receive operations to create an image, fully sampled arrays can perform a single flash transmit and then implement multiple-aperture acquisition by simply rearranging the beamforming coefficients while performing dynamic receive focusing.

To realize these ASIC-based microbeamformer systems, each channel from the ultrasound system is connected to a patch or cluster of adjacent elements often referred to as a sub-array [43, 44]. The delay lines and switches perform microbeamforming of the patch. These beamforming delays are limited to a small number of nanoseconds which can be realized using analog delay lines. Having longer delays is not necessary because the maximum delay difference within a patch is limited to a few nanoseconds even when beam-steering to large angles. Final beamforming for all elements is done in the ultrasound system where much longer delays are possible due to the use of digital signal processing. This trade-off is necessary because with current technology it is difficult to integrate a large number of A/D converters in the probe handle. Short analog delays can be implemented locally with enough precision for acceptable image quality for volumetric imaging, and therefore splitting the beamforming operation between the probe and the system is a reasonable compromise which has made possible these fully sampled 2D array implementations. Most commercially available matrix arrays employ ASICs allowing for excellent volumetric image quality.

While ASICs are an attractive solution to the development of fully sampled matrix arrays, they are not without substantial challenges. One challenge is to remove the heat generated by the ASIC [49]. While passive solutions are preferred, active cooling mechanisms may be needed depending on the ability of the probe shell and cable to efficiently remove heat away from the skin surface that is being imaged. Even with cooling solutions in place, there may be limitations on the length of use in certain applications such as transesophageal echocardiography. There may also be limitations on the use of more power-intensive applications such as color Doppler and elastography. Compromises such as the *Tx* voltage may also be needed depending on the voltage stand-off supported by the particular ASIC fabrication process used. To ensure optimal performance of these tightly coupled systems, designers must adhere to strict power budgets for allocating the amount of power that can be dissipated in the analog and digital electronics as well as in the transducer elements themselves.

Another potential issue with the use of locally integrated ASICs is that the pitch-matched architecture essentially requires a completely new ASIC design process for each application that must be addressed. This is because the ASIC itself must conform to the specific requirements of the transducer array in terms of the pitch in X and Y dimensions as well as the number of elements in the overall footprint of the array. Some of the basic electronic building blocks can be reused in some instances, however, these will have to be further optimized to maximize performance due to different transmit frequencies, different impedance of the elements, etc. Therefore, the implementation of an ASIC-based 2D fully sampled transducer array can often take years to realize at a cost of millions of dollars. These designs require close coupling between transducer design engineers and ultrasound systems engineers, and require large teams to work together very efficiently to achieve the desired results. Often there are multiple design runs of the ASICs required before all performance parameters can be optimized in these highly complex systems. These iterations are costly and are beyond the scope of niche applications. Therefore, the decision to undertake an ASIC-based probe design is a significant commitment where the cost in resources and funding must be judiciously balanced against the expected revenue over the life of the product as well as maximum societal benefits to the clinicians and their patients.

FUTURE OUTLOOK AND APPLICATIONS OF 2D ARRAYS

At present, 2D arrays are primarily used in the mainstays of ultrasound—obstetrics and cardiology—with some increasing use in newer applications such as vascular. These applications are generally lower in frequency, making design requirements of 2D array less strict. In the future, one can envision continued innovation with ASIC designs to add more electronics to the transducer handle. ASIC designs of higher frequency will be developed. Larger arrays will also be developed to accommodate scanning of superficial targets near the transducer such as the breast and carotid artery. A particularly exciting line of research is the implementation of highly power efficient and advanced figure of merit A/D converters directly behind the elements [50, 51]. Such ADC-based microbeamformer arrays are the "holy grail" of 2D array imaging as they will eliminate the need for digital processing in a separated system console. Essentially, all of the beamforming operations could be performed digitally behind the array itself with expected improvements in image quality and speed of operation. Since digital circuits are constantly reducing in size and power requirements with the march of Moore's law, these "digital probe" systems would benefit considerably with reduction in power and system cost, with the elimination of the probe cable, and the loading for transmitting a large number of analog signals back to the processing system. In this case, the digital probe simply transmits fully formed images wirelessly to an image display. Alternately, researchers are currently working on patch-based ultrasound for 24-hour ambulatory monitoring of patients [52], and the digital probe with ultra-low power and high volume fabrication would be ideally suited to this implementation as well.

In research, implementation of advanced beamforming and imaging methods with fully sampled 2D arrays will be pursued. A particularly interesting application would be phase aberration correction which was initially pursued a couple of decades ago with 1D or multi-row arrays [53–55]. However, since the aberration is usually dimensional in shape, correction methods with 2D arrays should see substantial benefit. Such techniques include ultrafast imaging [56, 57], coherence-based beamforming [58, 59], and super-resolution imaging [60, 61]. These and other new applications will continue to advance the state of the art for 2D arrays in the coming years.

REFERENCES

1. A. Fenster, and D. B. Downey, "Three-dimensional ultrasound imaging," *Annual Review of Biomedical Engineering*, vol. 2, pp. 457–475, 2000. doi: 10.1146/annurev.bioeng.2.1.457.
2. D. G. Wildes, R. Y. Chiao, C. M. W. Daft, K. W. Rigby, L. S. Smith, and K. E. Thomenius, "Elevation performance of 1.25D and 1.5D transducer arrays," *IEEE Transactions on Ultrasonics Ferroelectrics and Frequency Control*, vol. 44, no. 5, pp. 1027–1037, 1997. doi: 10.1109/58.655628.
3. K. K. Shung, *Diagnostic Ultrasound: Imaging and Blood Flow Measurements*. Boca Raton, FL: Taylor & Francis, 2015.
4. C. S. Desilets, J. D. Fraser, and G. S. Kino, "Design of efficient broad-band piezoelectric transducers," *IEEE Transactions on Sonics and Ultrasonics*, vol. 25, no. 3, pp. 115–125, 1978. doi: 10.1109/t-su.1978.31001.
5. B. Weigang, G. W. Moore, J. Gessert, W. H. Phillips, and M. Schafer, "The methods and effects of transducer degradation on image quality and the clinical efficacy of diagnostic sonography," *Journal of Diagnostic Medical Sonography*, vol. 19, no. 1, pp. 3–13, 2003.
6. R. McKeighen, "Finite element simulation and modeling of 2-D arrays for 3-D ultrasonic imaging," *IEEE Transactions on Ultrasonics Ferroelectrics and Frequency Control*, vol. 48, no. 5, pp. 1395–1405, 2001. doi: 10.1109/58.949749.
7. R. L. Goldberg, and S. W. Smith, "Optimization of signal-to-noise ratio for multilayer PZT transducers," *Ultrasonic Imaging*, vol. 17, no. 2, pp. 95–113, 1995. doi: 10.1006/uimg.1995.1005.
8. R. L. Goldberg, C. D. Emery, and S. W. Smith, "Hybrid multi single layer array transducers for increased signal-to-noise ratio," *IEEE Transactions on Ultrasonics Ferroelectrics and Frequency Control*, vol. 44, no. 2, pp. 315–325, 1997. doi: 10.1109/58.585116.
9. Q. F. Zhou, K. H. Lam, H. R. Zheng, W. B. Qiu, and K. K. Shung, "Piezoelectric single crystal ultrasonic transducers for biomedical applications," *Progress in Materials Science*, vol. 66, pp. 87–111, 2014. doi: 10.1016/j.pmatsci.2014.06.001.
10. C. G. Oakley, and M. J. Zipparo, "Single crystal piezoelectrics: A revolutionary development for transducers," *2000 IEEE Ultrasonics Symposium Proceedings, Vols 1 and 2*, pp. 1157–1167, 2000.
11. S. E. Park, and T. R. Shrout, "Characteristics of relaxor-based piezoelectric single crystals for ultrasonic transducers," *IEEE Transactions on Ultrasonics Ferroelectrics and Frequency Control*, vol. 44, no. 5, pp. 1140–1147, 1997. doi: 10.1109/58.655639.
12. D. M. Mills, "Medical imaging with capacitive micromachined ultrasound transducer (cMUT) arrays," *2004 IEEE Ultrasonics Symposium, Vols 1-3*, pp. 384–390, 2004.
13. Y. Q. Qiu *et al.*, "Piezoelectric micromachined ultrasound transducer (PMUT) arrays for integrated sensing, actuation and imaging," *Sensors*, vol. 15, no. 4, pp. 8020–8041, Apr 2015, doi: 10.3390/s150408020.

14. J. Jung, W. Lee, W. Kang, E. Shin, J. Ryu, and H. Choi, "Review of piezoelectric micromachined ultra-sonic transducers and their applications," *Journal of Micromechanics and Microengineering*, vol. 27, no. 11, p. 113001, 2017. doi: 10.1088/1361-6439/aa851b.

15. A. Austeng, and S. Holm, "Sparse 2-d arrays for 3-D phased array imaging: experimental validation," *IEEE Transactions on Ultrasonics Ferroelectrics and Frequency Control*, vol. 49, no. 8, pp. 1087–1093, 2002. doi: 10.1109/tuffc.2002.1026020.

16. M. Karaman, I. O. Wygant, O. Oralkan, and B. T. Khuri-Yakub, "Minimally redundant 2-D array designs for 3-D medical ultrasound imaging," *IEEE Transactions on Medical Imaging*, vol. 28, no. 7, pp. 1051–1061, 2009. doi: 10.1109/tmi.2008.2010936.

17. R. J. Kozick, and S. A. Kassam, "Synthetic aperture pulse-echo imaging with rectangular boundary arrays," *IEEE Transactions on Image Processing*, vol. 2, no. 1, pp. 68–79, 1993. doi: 10.1109/83.210867.

18. R. T. Hoctor, and S. A. Kassam, "The unifying role of the coarray in aperture synthesis for coherent and incoherent imaging," *Proceedings of the IEEE*, vol. 78, no. 4, pp. 735–752, 1990. doi: 10.1109/5.54811.

19. J. T. Yen, and S. W. Smith, "Real-time rectilinear volumetric imaging," *IEEE Transactions on Ultrasonics Ferroelectrics and Frequency Control*, vol. 49, no. 1, pp. 114–124, 2002. doi: 10.1109/58.981389.

20. M. F. Rasmussen, and J. A. Jensen, and IEEE, "3-D Ultrasound Imaging Performance of a Row-Column Addressed 2-D Array Transducer: A Measurement Study," *2013 IEEE International Ultrasonics Symposium (IUS)*, pp. 1452–1455, 2013, doi: 10.1109/ultsym.2013.0370.

21. Y. L. Chen, M. Nguyen, and J. T. Yen, and IEEE, "Real-time Rectilinear Volumetric Acquisition with a 7.5 MHz Dual-layer Array Transducer: Data Acquisition and Signal Processing," *2011 IEEE International Ultrasonics Symposium (IUS)*, pp. 1759–1761, 2011, doi: 10.1109/ultsym.2011.0439.

22. C. H. Seo, and J. T. Yen, "A 256 x 256 2-D array transducer with row-column addressing for 3-D rectilinear imaging," *IEEE Transactions on Ultrasonics Ferroelectrics and Frequency Control*, vol. 56, no. 4, pp. 837–847, 2009. doi: 10.1109/tuffc.2009.1107.

23. K. Latham, C. Samson, C. Ceroici, R. J. Zemp, and J. A. Brown, and IEEE, "Fabrication and Performance of a 128-element Crossed-Electrode Relaxor Array, for a novel 3D Imaging Approach," *2017 IEEE International Ultrasonics Symposium (IUS)*, 2017.

24. A. Sampaleanu, P. Y. Zhang, A. Kshirsagar, W. Moussa, and R. J. Zemp, "Top-orthogonal-to-Bottom-electrode (TOBE) CMUT arrays for 3-D ultrasound imaging," *IEEE Transactions on Ultrasonics Ferroelectrics and Frequency Control*, vol. 61, no. 2, pp. 266–276, 2014. doi: 10.1109/tuffc.2014.6722612.

25. K. L. Chen, H. S. Lee, and C. G. Sodini, "A column-row-parallel ASIC architecture for 3-D portable medical ultrasonic imaging," *IEEE Journal of Solid-State Circuits*, vol. 51, no. 3, pp. 738–751, 2016. doi: 10.1109/jssc.2015.2505714.

26. C. E. M. Demore, A. W. Joyce, K. Wall, and G. R. Lockwood, "Real-time volume imaging using a crossed electrode array," *IEEE Transactions on Ultrasonics Ferroelectrics and Frequency Control*, vol. 56, no. 6, pp. 1252–1261, 2009. doi: 10.1109/tuffc.2009.1167.

27. A. I. H. Chen, L. L. P. Wong, S. Na, Z. H. Li, M. Macecek, and J. T. W. Yeow, "Fabrication of a curved row-column addressed capacitive micromachined ultrasonic transducer array," *Journal of Microelectromechanical Systems*, vol. 25, no. 4, pp. 675–682, 2016. doi: 10.1109/jmems.2016.2580152.

28. J. Hansen-Shearer, M. Lerendegui, M. Toulemonde, and M. X. Tang, "Ultrafast 3-D ultrasound imaging using row-column array-specific frame-multiply-and-sum beamforming," *IEEE Transactions on Ultrasonics Ferroelectrics and Frequency Control*, vol. 69, no. 2, pp. 480–488, 2022. doi: 10.1109/tuffc.2021.3122094.

29. G. R. Lockwood, J. R. Talman, and S. S. Brunke, "Real-time 3-D ultrasound imaging using sparse synthetic aperture beamforming," *IEEE Transactions on Ultrasonics Ferroelectrics and Frequency Control*, vol. 45, no. 4, pp. 980–988, 1998. doi: 10.1109/58.710573.

30. R. E. Davidsen, J. A. Jensen, and S. W. Smith, "2-dimensional random arrays for real-time volumetric imaging," *Ultrasonic Imaging*, vol. 16, no. 3, pp. 143–163, 1994. doi: 10.1006/uimg.1994.1009.

31. J. T. Yen, and E. Powis, and IEEE, "Boundary Array Transducer and Beamforming for Low-Cost Real-time 3D Imaging," in *Proceedings of the 2020 IEEE International Ultrasonics Symposium*, (IEEE International Ultrasonics Symposium), 2020.

32. F. L. Degertekin, R. O. Guldiken, and M. Karaman, "Annular-ring CMUT arrays for forward-looking IVUS: Transducer characterization and imaging," *IEEE Transactions on Ultrasonics Ferroelectrics and Frequency Control*, vol. 53, no. 2, pp. 474–482, 2006. doi: 10.1109/tuffc.2006.1593387.

33. O. Martinez-Graullera, C. J. Martin, G. Godoy, and L. G. Ullate, "2D array design based on Fermat spiral for ultrasound imaging," *Ultrasonics*, vol. 50, no. 2, pp. 280–289, 2010. doi: 10.1016/j.ultras.2009.09.010.

34. A. Ramalli, E. Boni, A. S. Savoia, and P. Tortoli, "Density-tapered spiral arrays for ultrasound 3-D imaging," *IEEE Transactions on Ultrasonics Ferroelectrics and Frequency Control*, vol. 62, no. 8, pp. 1580–1588, 2015. doi: 10.1109/tuffc.2015.007035.

35. A. Trucco, "Thinning and weighting of large planar arrays by simulated annealing," *IEEE Transactions on Ultrasonics Ferroelectrics and Frequency Control*, vol. 46, no. 2, pp. 347–355, 1999. doi: 10.1109/58.753023.
36. I. O. Wygant et al., "An integrated circuit with transmit beamforming flip-chip bonded to a 2-D CMUT array for 3-D ultrasound imaging," *IEEE Transactions on Ultrasonics Ferroelectrics and Frequency Control*, vol. 56, no. 10, pp. 2145–2156, Oct 2009, doi: 10.1109/tuffc.2009.1297.
37. C. Chen et al., "A front-end ASIC with receive sub-array beamforming integrated with a 32 × 32 PZT matrix transducer for 3-D transesophageal echocardiography," *IEEE Journal of Solid-State Circuits*, vol. 52, no. 4, pp. 994–1006, Apr 2017, doi: 10.1109/jssc.2016.2638433.
38. A. T. Fernandez, K. L. Gammelmark, J. J. Dahl, C. G. Keen, R. C. Gauss, and G. E. Trahey, "Synthetic elevation beamforming and image acquisition capabilities using an 8 × 128 1.75-D array," *IEEE Transactions on Ultrasonics Ferroelectrics and Frequency Control*, vol. 50, no. 1, pp. 40–57, 2003. doi: 10.1109/tuffc.2003.1176524.
39. G. Gurun, P. Hasler, and F. L. Degertekin, "Front-end receiver electronics for high-frequency monolithic CMUT-on-CMOS imaging arrays," *IEEE Transactions on Ultrasonics Ferroelectrics and Frequency Control*, vol. 58, no. 8, pp. 1658–1668, 2011. doi: 10.1109/tuffc.2011.1993.
40. E. Kang et al., "A reconfigurable ultrasound transceiver ASIC with 24 × 40 elements for 3-D carotid artery imaging," *IEEE Journal of Solid-State Circuits*, vol. 53, no. 7, pp. 2065–2075, Jul 2018, doi: 10.1109/jssc.2018.2820156.
41. W. *Lee* et al., "A Magnetic Resonance Compatible E4D Ultrasound Probe for Motion Management of Radiation Therapy," *2017 IEEE International Ultrasonics Symposium (IUS)*, 2017.
42. A. Nikoozadeh et al., "Forward-looking intracardiac ultrasound imaging using a 1-D CMUT array integrated with custom front-end electronics," *IEEE Transactions on Ultrasonics Ferroelectrics and Frequency Control*, vol. 55, no. 12, pp. 2651–2660, Dec 2008, doi: 10.1109/tuffc.2008.980.
43. B. Savord, and R. Solomon, "Fully sampled matrix transducer for real time 3D ultrasonic imaging," *2003 IEEE Ultrasonics Symposium Proceedings, Vols 1 and 2*, pp. 945–953, 2003.
44. D. Wildes *et al.*, "4D ICE: A 2D array transducer with integrated ASIC in a 10 fr catheter for real-time 3D intracardiac echocardiography," *IEEE Transactions on Ultrasonics Ferroelectrics and Frequency Control*, vol. 63, no. 12, pp. 2159–2173, 2016, doi: 10.1109/TUFFC.2016.2615602.
45. B. Dufort, T. Letavic, and S. Mukherjee, and IEEE, "Digitally Controlled High-Voltage Analog Switch Array for Medical Ultrasound Applications in Thin-Layer Silicon-on-Insulator Process," *2002 IEEE International Soi Conference, Proceedings*, pp. 78–79, 2002.
46. M. D. C. Eames, and J. A. Hossack, "Fabrication and evaluation of fully-sampled, two-dimensional transducer array for "sonic window" imaging system," *Ultrasonics*, vol. 48, no. 5, pp. 376–383, 2008. doi: 10.1016/j.ultras.2008.01.011.
47. K. Erikson, A. Hairston, A. Nicoli, J. Stockwell, and T. White, "A 128 × 128 ultrasonic Transducer Hybrid Array," *1997 IEEE Ultrasonics Symposium Proceedings, Vols 1 & 2*, pp. 1625–1629, 1997.
48. R. Wodnicki et al., "Co-integrated PIN-PMN-PT 2-D array and transceiver electronics by direct assembly using a 3-D printed interposer grid frame," *IEEE Transactions on Ultrasonics, Ferroelectrics and Frequency Control*, vol. 67, no. 2, pp. 387–401, Feb 2020, doi: 10.1109/tuffc.2019.2944668.
49. M. S. Richard Edward Davidsen, J. C. Taylor, A. L. Robinson, and W. Sudol, "Ultrasonic matrix array probe with thermally dissipating cable and backing block heat exchange," USA Patent 10178986B2, January 15, 2019, 2013.
50. Y. Hopf et al., "A pitch-matched ASIC with integrated 65V Tx and shared hybrid beamforming ADC for catheter-based high-frame-rate 3D ultrasound probes," presented at the 2022 IEEE International Solid-State Circuits Conference (ISSCC), 2022.
51. J. Lee et al., "A 5.37mW/channel pitch-matched ultrasound ASIC with dynamic-bit-shared SAR ADC and 13.2V charge-recycling TX in standard CMOS for intracardiac echocardiography," in IEEE International Solid-State Circuits Conference (ISSCC), San Francisco, CA, Feb 17–21 2019, vol. 62, in IEEE International Solid State Circuits Conference, 2019, pp. 190-U2139. [Online]. Available: <Go to ISI>://WOS:000463153600182
52. C. H. Wang et al., "Bioadhesive ultrasound for long-term continuous imaging of diverse organs," *Science*, vol. 377, no. 6605, pp. 517-+, 2022, doi: 10.1126/science.abo2542.
53. P. C. Li, and M. Odonnell, "Phase aberration correction on 2-dimensional conformal arrays," *IEEE Transactions on Ultrasonics, Ferroelectrics and Frequency Control*, vol. 42, no. 1, pp. 73–82, 1995.
54. B. D. Lindsey, H. A. Nicoletto, E. R. Bennett, D. T. Laskowitz, and S. W. Smith, "3-D transcranial ultrasound imaging with bilateral phase aberration correction of multiple isoplanatic patches: A pilot human study with microbubble contrast enhancement," *Ultrasound in Medicine and Biology*, vol. 40, no. 1, pp. 90–101, 2014. doi: 10.1016/j.ultrasmedbio.2013.09.006.

55. L. L. Ries, and S. W. Smith, "Phase aberration correction in two dimensions with an integrated deformable actuator/transducer," *IEEE Transactions on Ultrasonics, Ferroelectrics and Frequency Control*, vol. 44, no. 6, pp. 1366–1375, 1997. doi: 10.1109/58.656640.

56. J. Sauvage et al., "Ultrafast 4D Doppler Imaging of the Rat Brain with a Large Aperture Row Column Addressed Probe," *2018 IEEE International Ultrasonics Symposium (IUS)*, 2018.

57. M. Tanter, and M. Fink, "Ultrafast imaging in biomedical ultrasound," *IEEE Transactions on Ultrasonics, Ferroelectrics and Frequency Control*, vol. 61, no. 1, pp. 102–119, 2014. doi: 10.1109/tuffc.2014.2882.

58. D. Hyun, G. E. Trahey, M. Jakovljevic, and J. J. Dahl, "Short-lag spatial coherence imaging on matrix arrays, part I: Beamforming methods and simulation studies," *IEEE Transactions on Ultrasonics, Ferroelectrics and Frequency Control*, vol. 61, no. 7, pp. 1101–1112, 2014. doi: 10.1109/tuffc.2014.3010.

59. M. Jakovljevic, B. C. Byram, D. Hyun, J. J. Dahl, and G. E. Trahey, "Short-lag spatial coherence imaging on matrix arrays, part II: Phantom and in vivo experiments," *IEEE Transactions on Ultrasonics, Ferroelectrics and Frequency Control*, vol. 61, no. 7, pp. 1113–1122, 2014. doi: 10.1109/tuffc.2014.3011.

60. S. Harput *et al.*, "3-D super-resolution ultrasound imaging with a 2-D sparse array," *IEEE Transactions on Ultrasonics, Ferroelectrics and Frequency Control*, vol. 67, no. 2, pp. 269–277, Feb 2020. doi: 10.1109/tuffc.2019.2943646.

61. I. Ozdemir, K. Johnson, S. Mohr-Allen, K. E. Peak, V. Varner, and K. Hoyt, "Three-dimensional visualization and improved quantification with super-resolution ultrasound imaging-validation framework for analysis of microvascular morphology using a chicken embryo model," *Physics in Medicine and Biology*, vol. 66, no. 8, p. 085008, 2021. doi: 10.1088/1361-6560/abf203.

2 Mechanical 3D Ultrasound Scanning Devices

Aaron Fenster

ROLE OF MECHANICALLY SWEPT 3D US DEVICES

Commercial ultrasound vendors and researchers have developed a variety of three-dimensional (3D) ultrasound (US) imaging devices and capabilities that are now widely available and used routinely in many clinical applications. Some of these 3D US devices are based on free-hand 3D US imaging approaches and use electromagnetic or optical tracking approaches.[1-3] Some commercial systems are based on 2D piezoelectric arrays which produce real-time 3D US images (i.e., 4D US).[4,5] Although these systems produce excellent image quality, they have advantages that have been addressed by investigators and commercial companies by developing mechanically swept systems to meet unmet clinical needs.[6-8] Two approaches have been used in the development of mechanically swept systems by researchers and commercial companies: (1) integrated motorized ultrasound probes, in which the linkage is internal to the probe, and (2) external fixtures to the ultrasound probe that are motorized and able to control the motion of the ultrasound probe.

While systems that use 2D piezoelectric arrays and tracked free-hand scanning are highly useful, the ability to control the movement of ultrasound probe using internal or external fixtures has following limitations:

- An ability to control the probe's motion by motor allows generation of repeatable scans of the anatomy, removing much of human operator variability in generating useful 3D US images. The operator is still required to position the 3D US device over the anatomy, but the motion of the US probe is automated.
- A motorized scanning approach does not require additional tracking devices that add cost and complexity to the procedure, such as line of sight for optical tracking and avoidance of ferromagnetic materials near the site of imaging.
- Although new applications of 3D US imaging have been explored, we are still unable to use commercial 3D US systems because the available devices are not suitable for the application due to the US transmission frequency and/or the type of required motion (e.g., endocavity 3D US imaging).
- As deep learning algorithms are playing an increasingly important role in medical imaging (e.g., automated detection, segmentation, and classification), researchers are exploring the integration of these methods with 3D US acquisition systems. For researchers, this is currently accomplished on external computers connected to the ultrasound machines, requiring efficient methods to export the 3D US images.
- The use of external-fixture approaches that integrate commercial US probes that generate 2D US images can take advantage of improvements in the commercial US machines and generate improved 3D US image quality without extra time and cost investments in the improvements of the US imaging machine or probes. For example, the generation of 3D Doppler images is made possible by making changes to external acquisition software of the ultrasound machine.[9-11]

However, internal and external fixtures used for 3D US imaging result in bulkier devices as they must accommodate motorized mechanisms causing their handling by the operator more

DOI: 10.1201/9781003299462-3

uncomfortable compared to 3D US imaging that uses free-hand and 2D array approaches. In the following section, mechanically swept scanning approaches are described with specific examples of their applications.

EXTERNAL-FIXTURE-BASED MECHANICALLY SWEPT SYSTEMS

Mechanically swept 3D US scanning devices use motorized approaches to control the motion of conventional US probes and cause them to translate, tilt, rotate, or combine these motions. Since the type of motion and its extent is controlled by a motor and the mechanical linkage, the location and orientation of the US probe are either known through a preprogrammed motion control or controlled with feedback from the motor encoders. Since the scanning geometry in these mechanical 3D US systems is precisely controlled and predefined or monitored by position or rotary encoders, the relative position and orientation of the acquired individual 2D US images are known accurately and precisely. Thus, reconstruction of the 3D US image can be performed as the 2D US images are acquired to generate an optimized 3D US image and made available immediately after completion of the 3D US scan.

Since these mechanically swept 3D US scanning systems allow adjustment of the angular or spatial interval between the acquired 2D US images, the 3D US image quality and acquisition time of these systems can be optimized for the specific application. In general, the spatial interval between the acquired 2D US images can be adjusted based on the elevational resolution of the 2D US probe so that the scan time is minimized. The following sections describe the most commonly used external mechanical fixture devices used to generate 3D US images.

LINEAR MECHANICAL SCANNERS

The linear scanner mechanisms use an external motorized device to translate the conventional 2D probe over the patient's skin (see Figure 2.1). The 2D probe can be positioned in the external fixture to be perpendicular to the scan direction and the surface of the skin. The US probe can also be positioned at an angle to the scanning direction for acquiring 3D Doppler images of vessels. Since

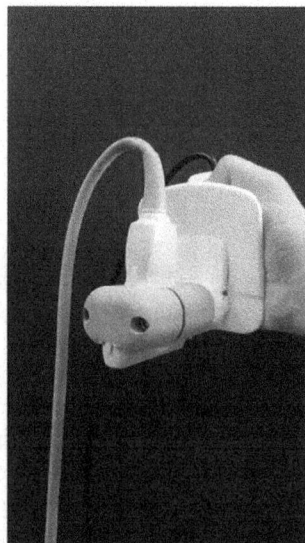

FIGURE 2.1 Photograph of a handheld motorized fixture with a conventional ultrasound transducer. By clicking the button on the top of the device, the motor is actuated and causes the transducer to translate over 5 cm.

the motion is translation, the acquired 2D US images will be parallel to each other. By adjusting the velocity of the US probe and the 2D US image acquisition rate, the spacing between the acquired 2D US images will be uniform and can be controlled by the user. The spacing between acquired frames should be one-half or smaller than the elevational resolution of the ultrasound probe to avoid undersampling of the anatomy in the scan direction. Since the elevational resolution (ultrasound beam width) will increase with depth, adjusting the spacing between the acquired 2D US images based on the elevational resolution in the focal zone will cause deeper parts of the anatomy to be oversampled. Nevertheless, the velocity of the US probe scanning is typically adjusted so that the temporal sampling interval can match the 2D US frame rate for the US machine.[12]

The linear scanning approach with external fixtures has been used in many vascular B-mode and Doppler imaging applications, particularly for imaging the carotid arteries and assessing carotid plaque burden in the management of the risk for stroke.[10,13–18] In this application, 3D US images of the carotid arteries are used to analyze and quantify the plaque burden, which is used to assess and monitor carotid atherosclerosis and its changes in response to therapy. Atherosclerosis is an inflammatory disease in which the inner layer of the arteries progressively accumulates low-density lipoproteins and macrophages over several decades forming plaques.[19] The carotid arteries are major sites for developing plaques, which may become unstable and suddenly rupture forming a thrombus that can travel to the brain causing a transient ischemic attack or an ischemic stroke by blocking the oxygenated blood supply to parts of the brain. The carotid arteries are superficial structures, thus can be easily imaged with higher frequency US probes (8–12 MHz) to generate high-quality 3D US images of the arteries and plaque burden.

To generate a 3D US image of the carotid arteries, the handheld fixture is positioned over the carotid arteries and the conventional US probe is translated over the carotid arteries generating a 3D US image that is approximately centered over the bifurcation and extends about 3 cm above and below the bifurcation. The 3D image can then be analyzed using conventional or deep learning software tools to generate a vessel wall volume as well as the volume of the plaques.[20–26] Figure 2.2 shows two examples of linearly scanned 3D US images of the carotid arteries made with an external fixture.

Since carotid atherosclerosis can be treated effectively using medical therapies (e.g., statins) that regress the plaques and inflammation, 3D US is used to assess the treatments as changes in the plaque burden can be directly measured, and changes in plaque morphology and/or composition can also be monitored.[13,20]

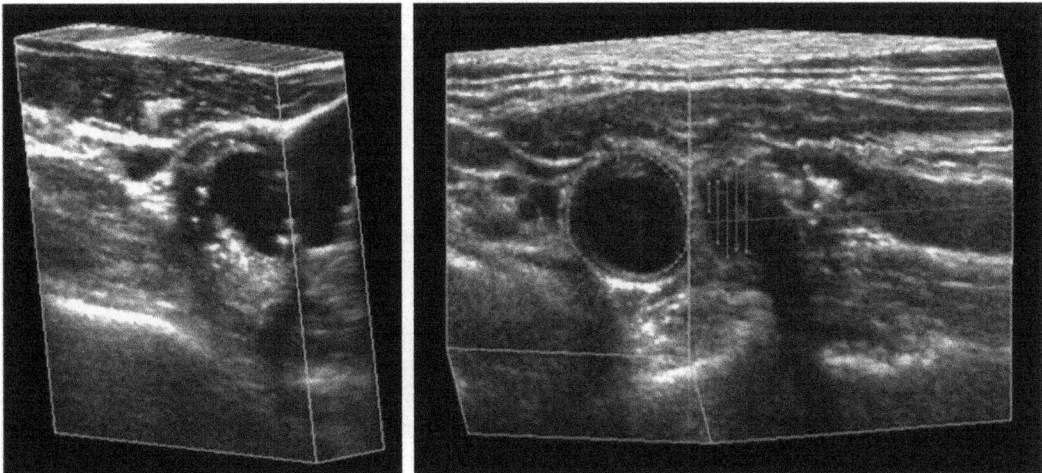

FIGURE 2.2 Two examples of 3D ultrasound images of the carotid arteries. The images are displaced using a cube view using multi-planar rendering and have been sliced to reveal the artery in two planes.

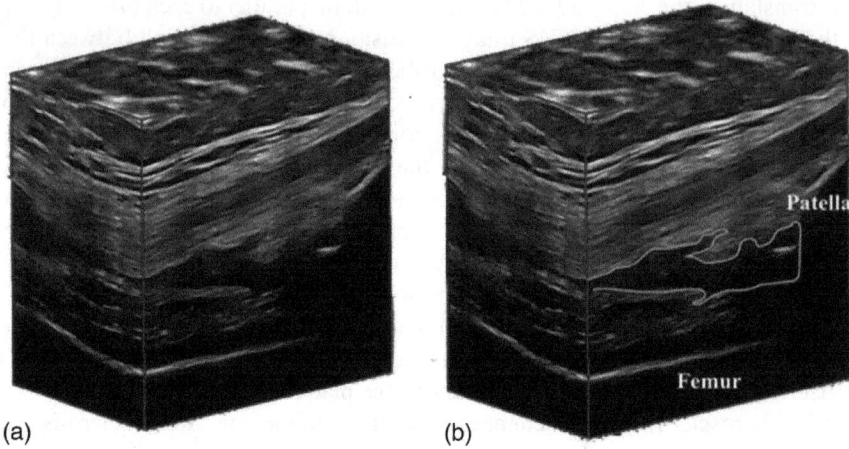

FIGURE 2.3 (a) 3D ultrasound image of an osteoarthritic patient's knee showing the high resolution achievable with the mechanical scanning approach. (b) Same image as in (a) but the effusion in the knee has been segmented and used to monitor the regression of the effusion.

Conventional 2D US imaging is increasingly being used for assessment of osteoarthritic and rheumatologic diseases.[27,28] However, to obtain more reliable and quantitative information on the examined joints, such as the knee, magnetic resonance imaging (MRI) is used. Since MRI systems are expensive and not readily available in many parts of the world, Papernick et al.[29] used a 3D US approach with a linear scanning device to image knee osteoarthritis. In this approach, a 2D US probe was mounted to a handheld motorized device that was custom 3D-printed, as shown in Figure 2.2. The 3D US linear scanning device contained a motorized drive mechanism that linearly translated the probe over 4.0 cm along the patient's skin. As with the carotid 3D US scanning approach, 2D US images are continually acquired at regular spatial intervals, which are reconstructed into a 3D image simultaneously. This approach yielded excellent 3D US images of the knee cartilage but could not image the complete knee cartilage as some parts are obscured by the patella and bones in the joint (see Figure 2.3).

Other applications include imaging of tumor vascularization[11,30–32] and methods for whole breast imaging of women with dense breasts, which are reviewed in another chapter.

A linear scanning system that combines some of the advantages of freehand scanning and a constrained linear scanning path was described by Huang et al.[33] Their system consists of a sliding track on which any conventional ultrasound probe is mounted (see Figure 2.4). The user can slide the probe along the track, while the location of the probe is read using a specially designed module. The position information is communicated via a Bluetooth module to a computer where the acquired images are reconstructed into a 3D US image. This approach provides accurate 3D US images, but its clinical implementation is problematic as placement of the track limits its application.

One of the constraints of linear scanning is that it can scan only a flat part of the anatomy. As most parts of the surface of the body are not flat, linear scanning over more than 5 cm will result in loss of coupling. To eliminate this issue, Du Toit et al. described a linear scanning mechanism for imaging the fingers and wrist, which have complex surfaces. In this method, a motorized linear scanning mechanism was submerged in a tank filled with a 7.25% isopropyl alcohol–water solution to match the speed of the sound of tissue. Since the US probe is mounted on the submerged scanning mechanism and positioned above the wrist or fingers, the anatomy can be scanned by the probe without loss of coupling. Since the probe is submerged, any scan length is possible. This approach was used to investigate osteoarthritis of the wrist, allowing accurate measurement of the volume of synovitis (Figure 2.5).

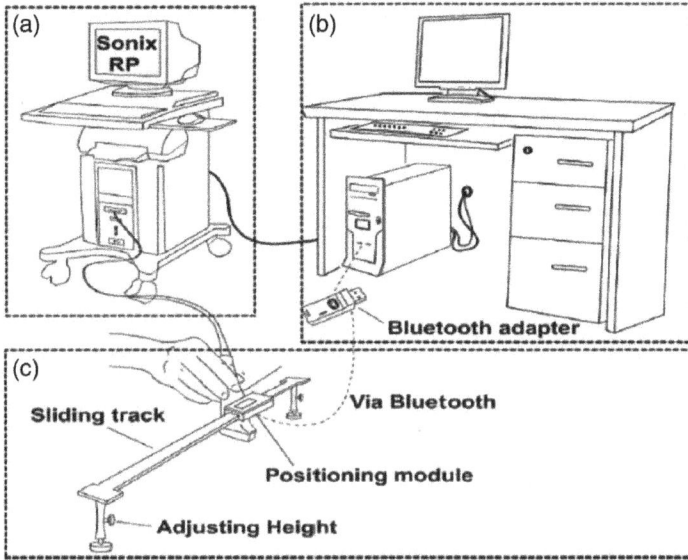

FIGURE 2.4 Schematic diagram of the mechanically tracked linear scanning mechanism used to generate a 3D ultrasound image. As illustrated in Part C, the ultrasound transducer is mounted on a position module, which can be translated linearly along a sliding track. The location along the track is communicated with the computer via Bluetooth. (Reproduced with permission by IEEE from Huang et al.[1])

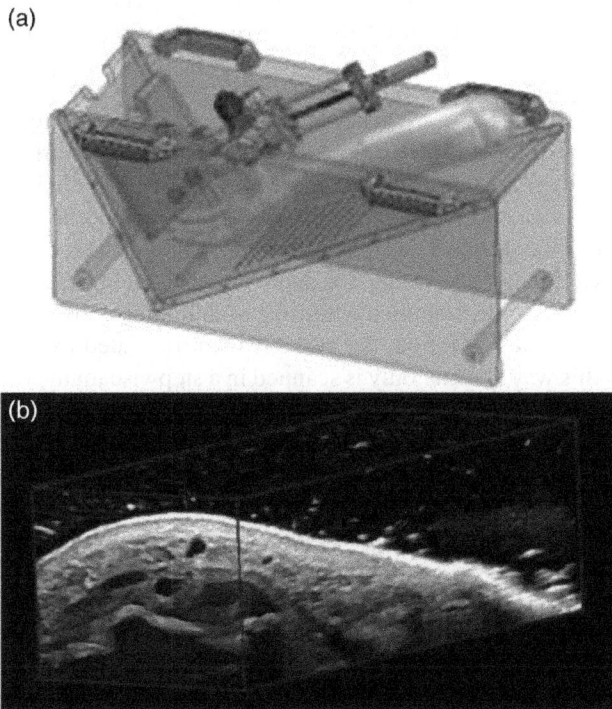

FIGURE 2.5 (a) Schematic diagram of the 3D ultrasound system for imaging the wrist. The 3D scanning assembly is contained in a water tank and the ultrasound transducer is submerged. (b) A 3D image of the carpometacarpal joint in an osteoarthritic patient's hand with effusion synovitis.

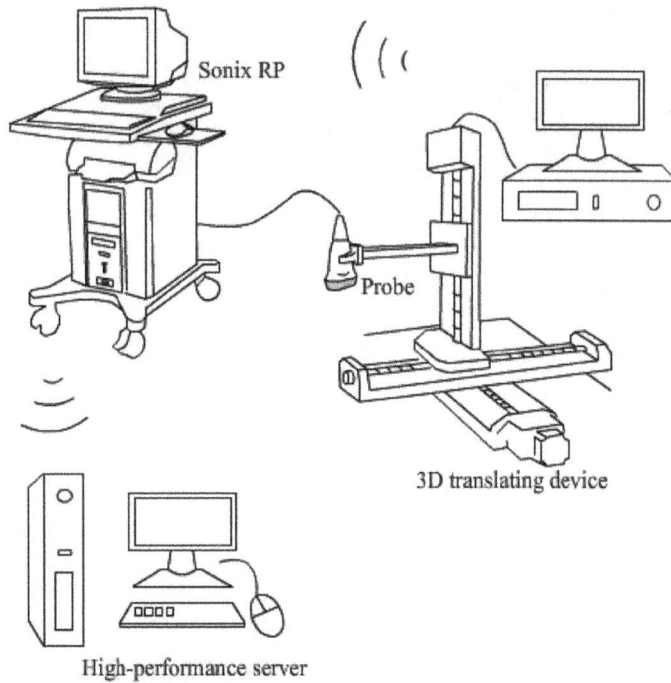

FIGURE 2.6 Schematic diagram of the system used for 3D ultrasound strain imaging. The imaging system uses a mechanical 3D translating device to control a conventional linear ultrasound transducer to collect precompression and post-compression radiofrequency echo signal frames, which are wirelessly transferred to the computer for analysis of the strain. (Reproduced with permission by IEEE, adapted from Zhaohong et al.[34])

Strain imaging using ultrasound approaches has been investigated for many years. A controlled mechanism that can record the applied compression (stress) as well as the location of the ultrasound probe is required to generate strain 3D US images. A linear scanning approach, which has been described by Chen et al.,[34] provides this capability. The system they developed consists of a mechanism with a translation device as well as additional orthogonal translation devices providing three-axis movement (see Figure 2.6). Using this approach, the system is used to image the anatomy at a location and then compression is applied while ultrasound images are recorded. After recording the stress-strain information at a location, the probe is then translated to the next location and the process is repeated. In this way, the anatomy is scanned in a stepwise manner, generating 3D images pre and post compression, from which a 3D strain map can be calculated.

3D ultrasound images have also been acquired for prostate,[35,36] gynecologic,[37] and breast[38,39] brachytherapy using the pull-back approach, in which the probe is translated in a linear manner. In this method, a biplane transrectal US transducer with a transverse array is mounted on an assembly that typically provides a mechanism to pull back the probe in a stepwise manner. By stepping back the ultrasound probe in equal steps (i.e., withdrawing the probe), the acquired 2D US images will be parallel to each other and can be reconstructed into a 3D US image using the 3D US linear scanning method.

The linear scanning approach using mechanical devices generates excellent 3D US images but involves bulky motorized devices. Another approach that does not need a mechanism to move the probe was described by Tank et al.[40] This approach makes use of a stationary conventional US probe and an acoustic reflector which is translated using a motorized mechanism. This allows the scanning mechanism that supports the reflector to be compact and small, but sufficient to cover the US beam. One approach requires positioning the US probe horizontally and the reflector oriented

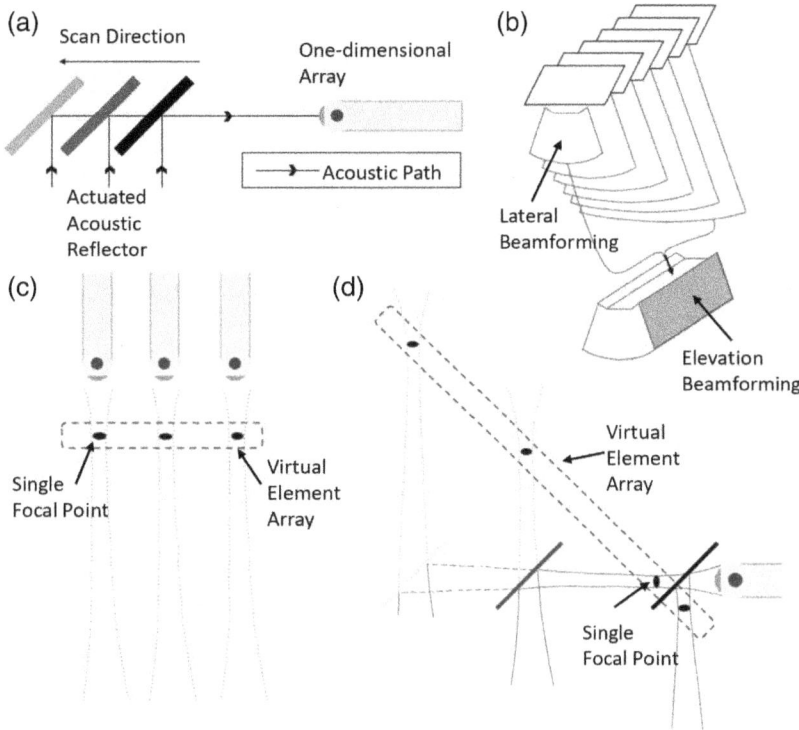

FIGURE 2.7 Schematic diagram of the reflection method to generate 3D ultrasound images. (a) The acoustic path with the actuated reflector. (b) Representation of the 3D beam forming. (c) Virtual element array in the linear scanning of the ultrasound transducer. (d) Diagonal view with respect to the scanning direction of the virtual element array in the proposed imaging mechanism. (Adapted from Tang et al.[40])

at 45° to the axis of the US beam (see Figure 2.7). Another approach is also described which makes use of a vertically positioned conventional US probe with two reflectors. One of these reflectors is stationary, reflecting the US beam horizontally and the other reflector is motorized and is used to scan the anatomy. This approach was used to image test phantoms, which showed the utility of this approach, but clinical use has not yet been demonstrated.

Since the 2D US images acquired to generate a 3D US image are parallel and have predefined regular spacing between the images, the 3D US image can be reconstructed while the 2D US images are being acquired, i.e., placement of the 2D US images into the correct position and orientation in the 3D image. However, the resolution of the 3D US image will not be isotropic and will depend on the in-plane (axial and lateral) and elevational resolution of the acquired 2D US images, and the inter-2D US image spacing. The resolution parallel to the acquired 2D US images will be the same as the resolution of the original 2D US images. But, the resolution in the direction of the 3D scanning will be equal to the elevational resolution of the 2D US image or worse if the spacing between acquired images is not optimized. Typically, the elevational resolution is poorer than the in-plane resolution of the acquired 2D US images, making the resolution of the 3D US image poorest in the 3D scanning direction. Thus, a transducer with good elevational resolution should be used for optimal results.

TILTING MECHANICAL SCANNERS: EXTERNAL FIXTURES

In this approach, the conventional US probe generating 2D US images is coupled to a motorized fixture. When the motor is activated, the probe is tilted about its face and conventional 2D US images

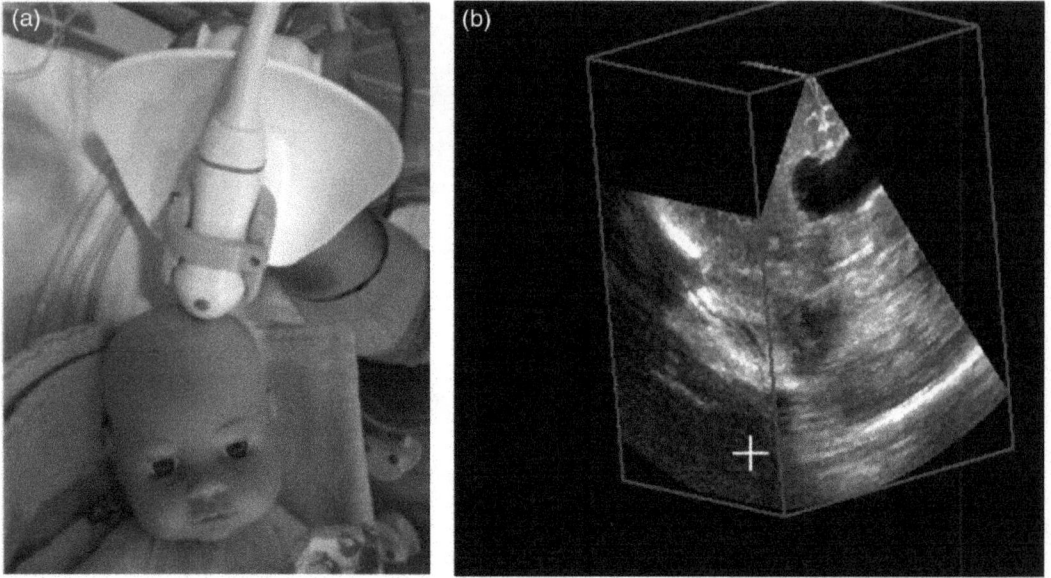

FIGURE 2.8 (a) A mechanical tilting method of the ultrasound transducer is used in a handheld fixture and used to image the brains of neonatal patients with post-hemorrhagic ventricular dilation. (b) 3D ultrasound image of a neonate brain using the external mechanical tilting fixture shown in panel (a).

are acquired into the computer. Thus, the acquired 2D US images are arranged as a fan (Figure 2.8) with an angular spacing that can be controlled by the user. To ensure that there are no artifacts in the 3D image, the probe tip must remain fixed on the skin of the patient while the US transducer is tilted.[41-43]

Since the user can control the rate of tilting of the US probe and the angular range of the tilt, the time required to acquire the 3D US images can be chosen by the user and optimized for the application. However, these parameters must take into account the 2D US image update rate, which depends on the US machine settings (i.e., the image depth setting and the number of focal zones). The angular separation between the acquired 2D US images is chosen to yield the desired image quality and avoid missed spatial information. With this information, the number of acquired 2D US images is determined by the total scan angle needed to cover the desired anatomy. For example, if the scan duration is 6 s causing the probe to tilt over an angle of 60° and the 2D US images are acquired at 30 Hz, the spacing between acquired 2D US images will be 0.33°. Since these parameters can be defined prior to the scan, the generation of the 3D US image can occur as the 2D US images are acquired, as their relative positions in 3D is known. This allows viewing of the 3D US image immediately after the completion of the scan, which is used to assess if a repeated scan is required.

Kishimoto et al. described a 3D US imaging application that made use of the tilting approach for imaging the neonatal brain ventricles of babies suffering from intraventricular hemorrhage and post-hemorrhagic ventricle dilatation.[43-45] The system consisted of a handheld motorized device (Figure 2.8b) that housed a conventional 2D US probe typically used for imaging the neonatal brain. The US probe was positioned against the neonate's fontanelle while the 2D US images were acquired into a personal computer while the probe was tilted. The acquired images were then reconstructed into a 3D US image simultaneously which was available for viewing immediately after the completion of the scan. In this approach, the total scan angle was 65° and the scan duration was 8.7 s. The acquisition of the 3D US image into an external computer allowed for post-processing of the images and segmentation of the enlarged ventricles for assessment of whether an intervention is required to reduce their volume.[46-48]

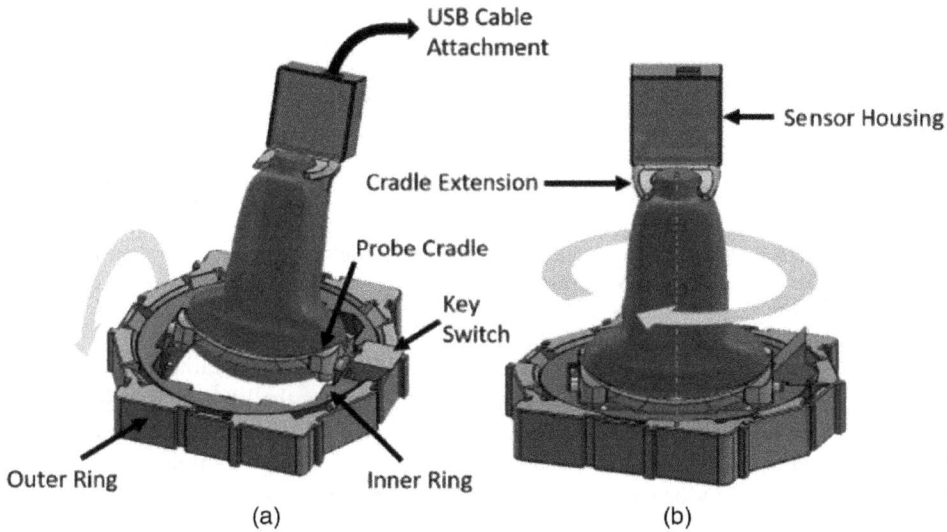

FIGURE 2.9 Schematic diagram of the fixture with an ultrasound transducer cradle, detachable sensor housing, and base with inner and outer ring plus a selector switch, which is used to restrict the transducer motion to a single axis of rotation. (a) Tilting of the transducer motion and (b) rotational motion of the transducer. (Reproduced with permission by IEEE, adapted from Morgan et al.[50])

Another approach for using the probe tilting to generate 3D US images was described by Herickhoff et al.[49,50] Their approach made use of a cradle that housed a detachable conventional US probe in a collar that provided the ability to rotate the probe around its axis as well as to tilt the probe by keeping the face of the probe at one location. This design made use of a three-axis gyroscope, three-axis accelerometer, and three-axis magnetometer to track the probe when it was tilted or rotated by the user (see Figure 2.9). Although the design did not include motors to move the probe, these could be easily added to generate automated movements. Nonetheless, this approach provided a low-cost method to generate high-quality 3D US images over a small region.

The tilting scanning approach will generate 3D images in which the resolution is also not isotropic. Since the US image resolution degrades in both the axial and elevational directions with imaging depth (i.e., the axial direction of the acquired 2D US images), the 3D US image resolution will degrade with imaging depth. The resolution will further degrade with imaging depth as the angle between the acquired US images increases.

ROTATIONAL SCANNERS

The endocavity 3D US approach used for transrectal and transvaginal imaging typically uses external fixture approaches integrated with endocavity US probes, which are rotated around their long axis over 180° or more to scan the anatomy (Figure 2.10). In this approach, the endocavity probe is mounted on a cradle that is integrated with a motorized fixture that is used to rotate or pull back the inserted probe.

Typical 3D US applications make use of end-firing transrectal probes for prostate biopsy. By rotating the transrectal US probe over about 180°, the acquired 2D US images will be arranged as shown in Figure 2.10a and intersect along the axis of rotation in the acquired 2D US images.[51–53] If the end-fired images are symmetric about the axis of the US probe, the intersection of the fan of 2D US images will be in the center of the 3D US image. However, if the 2D US images are not symmetric about the axis of the US probe, as for gynecologic US probes, the intersection of the fan of images will not be in the center of the acquired 3D US image, but at an angle with respect to that

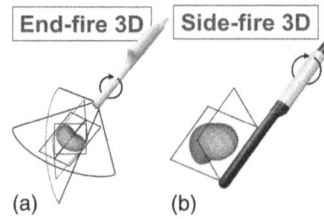

FIGURE 2.10   Schematic diagrams of two methods to generate a 3D ultrasound image. (a) The end-firing ultrasound transducer is rotated around its axis over 180°. (b) The side-firing transducer is rotated over about 180°.

central axis of the acquired image. Figure 2.11a shows a 3D US image obtained using an end-firing transrectal US probe during a prostate biopsy procedure showing the high-quality images that can be obtained to guide the biopsy needle to a suspicious target. As well, the 3D US image can also be used for registration to an MR image for 3D US-MRI fusion biopsy procedure.

3D US application of endocavity side-firing US probes that generate sagittal 2D US images (Figure 2.10b) is typically done in prostate brachytherapy. By rotating the US probe, the acquired images will be arranged as a fan and will intersect at the axis of rotation of the US probe, which will be a few mm from the edge of the acquired 2D US images. In the prostate brachytherapy application, a side-firing probe is typically rotated over 110°.[51,54,55] Figure 2.11b shows an example of a 3D US image of the prostate using the endocavity 3D US imaging approach with a side-firing probe used in prostate brachytherapy and cryosurgery.[51,54,56–62] This approach has also been used for the development of 3D US-guided gynecologic brachytherapy. In this application, transrectal and transvaginal conventional US probes can be integrated into an external motorized fixture to rotate and reconstruct the pelvic anatomy and allow visualization of the brachytherapy applicator and inserted brachytherapy needles.[63–66]

The resolution in both the 3D US images generated with the end-firing and side-firing endocavity US probes will not be isotropic. The use of both the US probes will result in the spatial sampling being highest near the axis of rotation of the probe and poorest away from the axis of rotation. As well, the resolution of the 3D US image will further degrade in the axial and elevational directions as the distance from the US probe is increased. The combination of these effects will cause the 3D

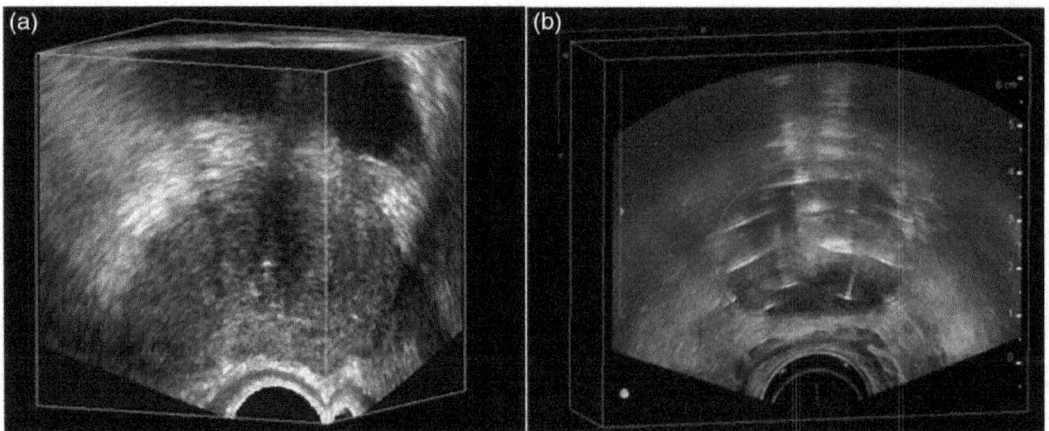

FIGURE 2.11   (a) End-fire 3D ultrasound image of the prostate used for biopsy. (b) Side-fire 3D ultrasound image of the prostate with brachytherapy needles used to verify the position of the needles for optimal dose delivery to the prostate and tumor.

US image resolution to vary spatially, with the best resolution near the US probe and the rotational axis, and the poorest being away from the US probe and rotational axis.

The acquisition of the 3D prostate US images allows for the integration of prostate segmentation, which is useful for planning the prostate biopsy and brachytherapy procedures as well as for fusion biopsy, in which the boundary of the prostate in the 3D US and MR images are registered and the target lesions identified in the MR image are superimposed on the 3D US image.[24,67–70]

To ensure that the 3D US image is geometrically correct and that the distance measurements in the 3D US image are correct for both the end-firing and side-firing US probe approaches, knowledge of the distance of the acquired US image to the axis of rotation is required. Thus, knowledge of the distance from the piezoelectric array inside the probe to the acquired 2D US image must be determined. This can be accomplished with information from the manufacturer or by calibration procedure using a test phantom.

The image quality of 3D US image obtained with the end-firing probe using the rotational scanning method is very sensitive to patient or probe motion during the scan as the axis of rotation is in the center of the 3D US image. Since the acquired 2D US images intersect along the probe's rotational axis, which will be at the center of the 3D US image, any motion during the scan will result in the acquired images not matching at the center of the image, which will result in a visible artifact and image distortion. Furthermore, artifacts in the center of the 3D US image will also be generated if the axis of rotation is not accurately known and if the piezoelectric array in generating a US image is at an angle to the axis of rotation. These geometric issues can be resolved through calibration using appropriate test phantoms and motion effects can be minimized by mounting the rotational scanning device in a stabilization fixture as typically done in prostate brachytherapy and in some prostate biopsy systems.

HYBRID MOTION SCANNERS

While the linear and tilting 3D US imaging methods generate excellent 3D US images, applications such as 3D US liver imaging require imaging between the ribs as well as below the ribs. Imaging between the ribs requires the use of the tilting scanning method, in which the probe is placed between the ribs and caused to tilt, sweeping out the volume, while imaging below the ribs can make use of the linear scanning method. Thus, a single motorized mechanism that can perform both scans as well as combine the linear and tilting method to sweep a large volume would be beneficial. This type of probe was reported by Neshat et al.[71] and described as a handheld motorized device with two motors to move the conventional 2D US probe through an angle of 60° and a translation of 3 cm linear extent simultaneously (Figure 2.12a). As well, the device could be used to only tilt the US probe and only translate the US probe. A modification of the hybrid scanning approach was also reported by Xing et al. who described a single motorized handheld device that could accommodate any conventional US probe and move it by translating and tilting the probe simultaneously. This device was used for guiding and assessing 3D US images of the liver in thermal ablation liver procedures (Figure 2.12b).[71–74]

ROBOTICALLY CONTROLLED 3D US ACQUISITION

Various specialty mechanical mechanisms have been developed and utilized for a variety of scanning methods, but these methods are constrained to the type of motion that is accommodated by the motorized device. However, robotic systems with a conventional 2D US probe mounted at the end effector of the robot provide scanning flexibility since they can control the motion of the US probe and also standardize the 3D US image acquisition. In addition, robotic systems can also provide the capability of controlling the generation of 3D US images from a remote site. Such a system was described by Janvier et al.[75] who used an industrial robot (CRS Robotics Corporation, Burlington, Ontario, Canada) with a force and torque sensor, which provided easy handling and

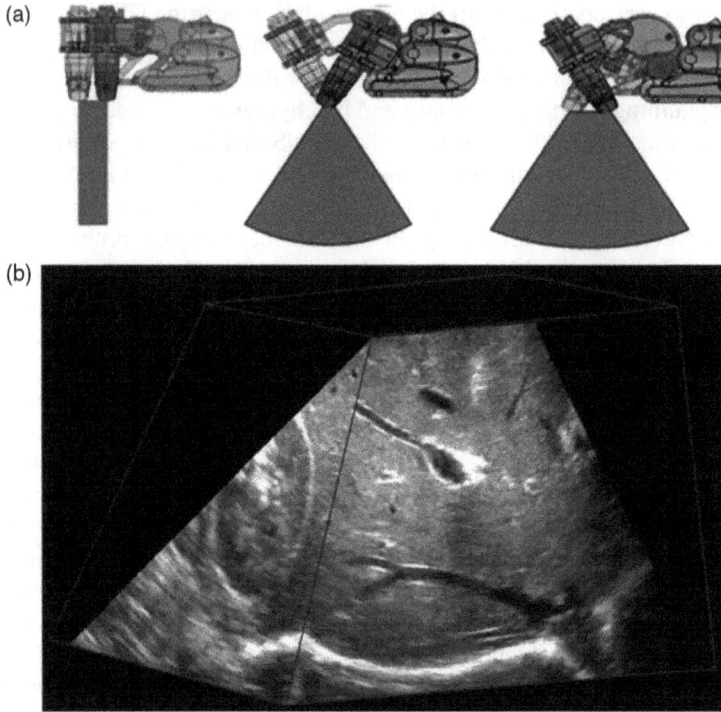

FIGURE 2.12 (a) Schematic diagram of the hybrid motions of the transducer, which include linear translation, tilting, and a combined motion of translation and tilting. (b) Example of 3D ultrasound image generated with the hybrid scanner. (Reproduced with permission by John Wiley and Sons, adapted from Neshat et al.[71])

precise positioning by the operator of the probe (see Figure 2.13). This capability also ensures patient security and safety, which are critical for remote control and manipulation. The use of a robotic approach to 3D US imaging may be applicable in imaging patients in remote sites where sonographers or radiologists are not available. Although this approach provided great imaging flexibility, the use of a robotic approach limits the use of the method as it adds to the cost of procedures and requires positioning of the patient to accommodate the robotic system.

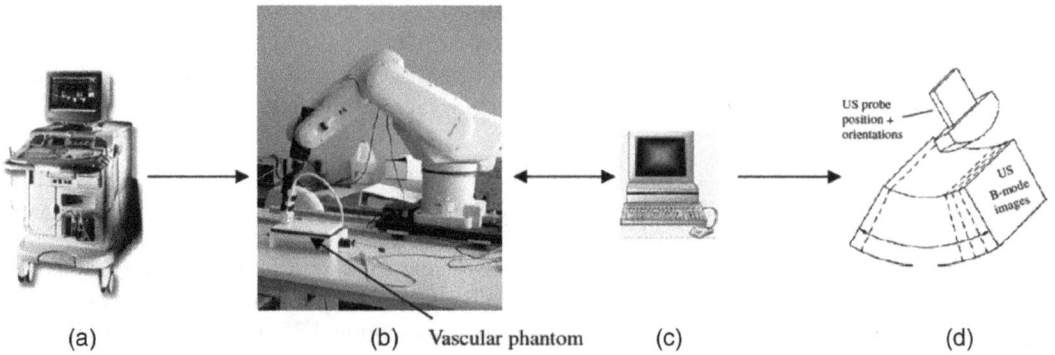

FIGURE 2.13 Schematic depiction of the robot-based 3D ultrasound imaging. (a) Ultrasound machine. (b) Robot with a mounted conventional ultrasound transducer. (c) Computer that is used to control the robot, acquire the ultrasound images, and reconstructed them into a 3D ultrasound image. (d) Depiction of the motion of the ultrasound transducer and the reconstructed 3D ultrasound image region. (Adapted from Janvier et al.[75])

Other applications of robotic acquisition of 3D US images were also described by Wei et al. for use in 3D US-guided prostate brachytherapy.[59] In this approach, a transrectal US probe was supported by the robot and rotated along its axis, sweeping the prostate as is done in the rotational imaging approach. Although this approach can generate 3D US images of the prostate for guiding a biopsy and a brachytherapy procedure, it was replaced with more compact mechanisms that could be mounted on the patient's bed for brachytherapy[62,76] or on the cart for prostate biopsy.[51,77]

INTERNAL FIXTURE-BASED MECHANICALLY SWEPT SYSTEMS

The most common commercial method for generating 3D US images is based on handheld probes with an internal motorized mechanism to sweep a one-dimensional (1D) piezoelectric array. These types of 3D US probes are now commonly available at many US device manufacturers. These 3D probes can have an internal mechanism to cause the motion of the 1D array to be tilted or translation. The internal 1D array is configured to sweep back and forth to generate two or three 3D US images/s providing 4D viewing of the anatomy with real-time display.

The housing of these 3D probes must contain the 1D array and motorized scanning mechanism that allows motion over a sufficient angle to cover a large volume of the anatomy. Thus, these probes are typically bulkier than conventional 2D US probes but are easier to use than 3D US mechanical scanning systems using external fixtures. However, these types of 3D US probes require a special US machine and software that can control the 3D scanning and reconstruction of the acquired 2D images into a 3D image. Although these 3D US probes are used extensively in obstetrics, vascular imaging, and urological imaging, their use is limited in case a larger volume needs to be imaged (e.g., carotid arterial imaging) or in case of image-guided intervention where seamless integration is required between the 3D US imaging device, tracking hardware, system software, and visualization software of the anatomy. In these cases, investigators have developed specialized external fixture-based and externally tracked (i.e., optical and electromagnetic) or motorized fixture systems for 3D US imaging.

REFERENCES

1. Prager RW, Ijaz UZ, Gee AH, Treece GM. Three-dimensional ultrasound imaging. *Proc Inst Mech Eng H.* 2010;224(2):193–223.
2. Treece GM, Gee AH, Prager RW, Cash CJ, Berman LH. High-definition freehand 3-D ultrasound. *Ultrasound Med Biol.* 2003;29(4):529–546.
3. Mozaffari MH, Lee WS. Freehand 3-D ultrasound imaging: A systematic review. *Ultrasound Med Biol.* 2017;43(10):2099–2124.
4. Sw S, Pavy HG Jr., von Ramm OT. High-speed ultrasound volumetric imaging system. Part I. Transducer design and beam steering. *IEEE Trans Ultrason Ferroelec Freq Control.* 1991;38:100–108.
5. Yen JT, Steinberg JP, Smith SW. Sparse 2-D array design for real time rectilinear volumetric imaging. *IEEE Trans Ultrason Ferroelec Freq Control.* 2000;47(1):93–110.
6. Fenster A, Tong S, Cardinal HN, Blake C, Downey DB. Three-dimensional ultrasound imaging system for prostate cancer diagnosis and treatment. *IEEE Trans Instrum Meas.* 1998;1(4):32–35.
7. Nelson TR, Pretorius DH. Three-dimensional ultrasound imaging. *Ultrasound Med Biol.* 1998;24(9):1243–1270.
8. Fenster A, Downey D. Three-dimensional ultrasound imaging. *Proc SPIE.* 2001;4549:1–10.
9. Rankin RN, Fenster A, Downey DB, Munk PL, Levin MF, Vellet AD. Three-dimensional sonographic reconstruction: Techniques and diagnostic applications. *Am J Roentgenol.* 1993;161(4):695–702.
10. Picot PA, Rickey DW, Mitchell R, Rankin RN, Fenster A. Three-dimensional colour Doppler imaging. *Ultrasound Med Biol.* 1993;19(2):95–104.
11. Downey DB, Fenster A. Three-dimensional power Doppler detection of prostate cancer [letter]. *Am J Roentgenol.* 1995;165(3):741.
12. Smith WL, Fenster A. Optimum scan spacing for three-dimensional ultrasound by speckle statistics. *Ultrasound Med Biol.* 2000;26(4):551–562.

13. Landry A, Spence JD, Fenster A. Measurement of carotid plaque volume by 3-dimensional ultrasound. *Stroke*. 2004;35(4):864–869.

14. Landry A, Fenster A. Theoretical and experimental quantification of carotid plaque volume measurements made by 3D ultrasound using test phantoms. *Med Phys*. 2002;29(10):2319–2327.

15. Landry A, Spence JD, Fenster A. Quantification of carotid plaque volume measurements using 3D ultrasound imaging. *Ultrasound Med Biol*. 2005;31(6):751–762.

16. Ainsworth CD, Blake CC, Tamayo A, Beletsky V, Fenster A, Spence JD. 3D ultrasound measurement of change in carotid plaque volume: A tool for rapid evaluation of new therapies. *Stroke*. 2005;35:1904–1909.

17. Krasinski A, Chiu B, Spence JD, Fenster A, Parraga G. Three-dimensional ultrasound quantification of intensive statin treatment of carotid atherosclerosis. *Ultrasound Med Biol*. 2009;35(11):1763–1772.

18. Graebe M, Entrekin R, Collet-Billon A, Harrison G, Sillesen H. Reproducibility of two 3-D ultrasound carotid plaque quantification methods. *Ultrasound Med Biol*. 2014;40(7):1641–1649.

19. Lusis AJ. Atherosclerosis. *Nature*. 2000;407(6801):233–241.

20. van Engelen A, Wannarong T, Parraga G, et al. Three-dimensional carotid ultrasound plaque texture predicts vascular events. *Stroke*. 2014;45(9):2695-+.

21. Zhou R, Fenster A, Xia Y, Spence JD, Ding M. Deep learning-based carotid media-adventitia and lumen-intima boundary segmentation from three-dimensional ultrasound images. *Med Phys*. 2019;46(7):3180–3193.

22. Gill J, Ladak H, Steinman D, Fenster A. Accuracy and variability assessment of semi-automatic technique for segmentation of the carotid arteries from 3D ultrasound images. *Med Phys*. 2001;27(6):1333–1342.

23. Mao F, Gill J, Downey D, Fenster A. Segmentation of carotid artery in ultrasound images: Method development and evaluation technique. *Med Phys*. 2000;27(8):1961–1970.

24. Ukwatta E, Yuan J, Buchanan D, et al. Three-dimensional segmentation of three-dimensional ultrasound carotid atherosclerosis using sparse field level sets. *Med Phys*. 2013;40(5):052903.

25. Chiu B, Krasinski A, Spence JD, Parraga G, Fenster A. Three-dimensional carotid ultrasound segmentation variability dependence on signal difference and boundary orientation. *Ultrasound Med Biol*. 2010;36(1):95–110.

26. Chiu B, Shamdasani V, Entrekin R, Yuan C, Kerwin WS. Characterization of carotid plaques on 3-dimensional ultrasound imaging by registration with multicontrast magnetic resonance imaging. *J Ultrasound Med*. 2012;31(10):1567–1580.

27. Ponikowska M, Swierkot J, Nowak B. The importance of ultrasound examination in early arthritis. *Reumatologia*. 2018;56(6):354–361.

28. Okano T, Mamoto K, Di Carlo M, Salaffi F. Clinical utility and potential of ultrasound in osteoarthritis. *Radiol Med*. 2019;124(11):1101–1111.

29. Papernick S, Dima R, Gillies DJ, Appleton T, Fenster A. Reliability and concurrent validity of three-dimensional ultrasound for quantifying knee cartilage volume. *Osteoarthr Cartilage*. 2021;29:S341–S341.

30. Bamber JC, Eckersley RJ, Hubregtse P, Bush NL, Bell DS, Crawford DC. Data processing for 3-D ultrasound visualization of tumour anatomy and blood flow. *SPIE*. 1992;1808:651–663.

31. Carson PL, Li X, Pallister J, Moskalik A, Rubin JM, Fowlkes JB. Approximate quantification of detected fractional blood volume and perfusion from 3-D color flow and Doppler power signal imaging. In: *1993 ultrasonics symposium proceedings*. Piscataway, NJ: IEEE; 1993:1023–1026.

32. King DL, King DLJ, Shao MY. Evaluation of in vitro measurement accuracy of a three-dimensional ultrasound scanner. *J Ultrasound Med*. 1991;10(2):77–82.

33. Huang QH, Yang Z, Hu W, Jin LW, Wei G, Li X. Linear tracking for 3-D medical ultrasound imaging. *IEEE Trans Cybern*. 2013;43(6):1747–1754.

34. Chen Z, Chen Y, Huang Q. Development of a wireless and near real-time 3D ultrasound strain imaging system. *IEEE Trans Biomed Circuits Syst*. 2016;10(2):394–403.

35. Schmid M, Crook JM, Batchelar D, et al. A phantom study to assess accuracy of needle identification in real-time planning of ultrasound-guided high-dose-rate prostate implants. *Brachytherapy*. 2013;12(1):56–64.

36. Batchelar D, Gaztanaga M, Schmid M, Araujo C, Bachand F, Crook J. Validation study of ultrasound-based high-dose-rate prostate brachytherapy planning compared with CT-based planning. *Brachytherapy*. 2014;13(1):75–79.

37. Schmid MP, Nesvacil N, Potter R, Kronreif G, Kirisits C. Transrectal ultrasound for image-guided adaptive brachytherapy in cervix cancer: An alternative to MRI for target definition? *Radiother Oncol*. 2016;120(3):467–472.

38. Poulin E, Gardi L, Barker K, Montreuil J, Fenster A, Beaulieu L. Validation of a novel robot-assisted 3DUS system for real-time planning and guidance of breast interstitial HDR brachytherapy. *Med Phys.* 2015;42(12):6830–6839.
39. Poulin E, Gardi L, Fenster A, Pouliot J, Beaulieu L. Towards real-time 3D ultrasound planning and personalized 3D printing for breast HDR brachytherapy treatment. *Radiother Oncol.* 2015;114(3):335–338.
40. Tang Y, Tsumura R, Kaminski JT, Zhang HK. Actuated reflector-based 3-D ultrasound imaging with synthetic aperture focusing. *IEEE Trans Ultrason Ferroelec Freq Control.* 2022;69(8):2437–2446.
41. Delabays A, Pandian NG, Cao QL, et al. Transthoracic real-time three-dimensional echocardiography using a fan-like scanning approach for data acquisition: Methods, strengths, problems, and initial clinical experience. *Echocardiography.* 1995;12(1):49–59.
42. Gilja OH, Thune N, Matre K, Hausken T, Odegaard S, Berstad A. In vitro evaluation of three-dimensional ultrasonography in volume estimation of abdominal organs. *Ultrasound Med Biol.* 1994;20(2):157–165.
43. Kishimoto J, Fenster A, Lee DS, de Ribaupierre S. In vivo validation of a 3-d ultrasound system for imaging the lateral ventricles of neonates. *Ultrasound Med Biol.* 2016;42(4):971–979.
44. Kishimoto J, de Ribaupierre S, Salehi F, Romano W, Lee DS, Fenster A. Preterm neonatal lateral ventricle volume from three-dimensional ultrasound is not strongly correlated to two-dimensional ultrasound measurements. *J Med Imaging.* 2016;3(4):046003.
45. Kishimoto J, de Ribaupierre S, Lee DS, Mehta R, St Lawrence K, Fenster A. 3D ultrasound system to investigate intraventricular hemorrhage in preterm neonates. *Phys Med Biol.* 2013;58(21):7513–7526.
46. Qiu W, Yuan J, Kishimoto J, et al. User-guided segmentation of preterm neonate ventricular system from 3-D ultrasound images using convex optimization. *Ultrasound Med Biol.* 2015;41(2):542–556.
47. Qiu W, Chen Y, Kishimoto J, et al. Automatic segmentation approach to extracting neonatal cerebral ventricles from 3D ultrasound images. *Med Image Anal.* 2017;35:181–191.
48. Qiu W, Chen Y, Kishimoto J, et al. Longitudinal analysis of pre-term neonatal cerebral ventricles from 3D ultrasound images using spatial-temporal deformable registration. *IEEE Trans Med Imaging.* 2017;36(4):1016–1026.
49. Herickhoff CD, Morgan MR, Broder JS, Dahl JJ. Low-cost volumetric ultrasound by augmentation of 2D systems: Design and prototype. *Ultrason Imaging.* 2018;40(1):35–48.
50. Morgan MR, Broder JS, Dahl JJ, Herickhoff CD. Versatile low-cost volumetric 3-D ultrasound platform for existing clinical 2-D systems. *IEEE Trans Med Imaging.* 2018;37(10):2248–2256.
51. Bax J, Cool D, Gardi L, et al. Mechanically assisted 3D ultrasound guided prostate biopsy system. *Med Phys.* 2008;35(12):5397–5410.
52. Natarajan S, Marks LS, Margolis DJ, et al. Clinical application of a 3D ultrasound-guided prostate biopsy system. *Urol Oncol.* 2011;29(3):334–342.
53. Cool D, Sherebrin S, Izawa J, Chin J, Fenster A. Design and validation of a 3D transrectal ultrasound prostate biopsy system. *Med Phys.* 2008;35(10):4695–4707.
54. Tong S, Downey DB, Cardinal HN, Fenster A. A three-dimensional ultrasound prostate imaging system. *Ultrasound Med Biol.* 1996;22(6):735–746.
55. Tong S, Cardinal HN, McLoughlin RF, Downey DB, Fenster A. Intra- and inter-observer variability and reliability of prostate volume measurement via two-dimensional and three-dimensional ultrasound imaging. *Ultrasound Med Biol.* 1998;24(5):673–681.
56. Downey DB, Chin JL, Fenster A. Three-dimensional US-guided cryosurgery. *Radiology.* 1995;197(P):539.
57. Chin JL, Downey DB, Elliot TL, et al. Three dimensional transrectal ultrasound imaging of the prostate: Clinical validation. *Can J Urol.* 1999;6(2):720–726.
58. Chin JL, Downey DB, Mulligan M, Fenster A. Three-dimensional transrectal ultrasound guided cryoablation for localized prostate cancer in nonsurgical candidates: A feasibility study and report of early results. *J Urol.* 1998;159(3):910–914.
59. Wei Z, Wan G, Gardi L, Mills G, Downey D, Fenster A. Robot-assisted 3D-TRUS guided prostate brachytherapy: System integration and validation. *Med Phys.* 2004;31(3):539–548.
60. Wei Z, Ding M, Downey DB, Fenster A. 3D TRUS Guided Robot Assisted Prostate Brachytherapy. In: Duncan J, Gerig G, eds. *MICCAI 2005.* Berlin, Heidelberg: Springer-Verlag; 2005:17–24.
61. Hrinivich WT, Hoover DA, Surry K, et al. Three-dimensional transrectal ultrasound guided high-dose-rate prostate brachytherapy: A comparison of needle segmentation accuracy with two-dimensional image guidance. *Brachytherapy.* 2016;15(2):231–239.
62. Hrinivich WT, Hoover DA, Surry K, et al. Accuracy and variability of high-dose-rate prostate brachytherapy needle tip localization using live two-dimensional and sagittally reconstructed three-dimensional ultrasound. *Brachytherapy.* 2017;16(5):1035–1043.

63. Rodgers JR, Surry K, Leung E, D'Souza D, Fenster A. Toward a 3D transrectal ultrasound system for verification of needle placement during high-dose-rate interstitial gynecologic brachytherapy. *Med Phys.* 2017;44(5):1899–1911.

64. Rodgers JR, Bax J, Surry K, et al. Intraoperative 360-deg three-dimensional transvaginal ultrasound during needle insertions for high-dose-rate transperineal interstitial gynecologic brachytherapy of vaginal tumors. *J Med Imaging.* 2019;6(2):025001.

65. Van Elburg DJ, Roumeliotis M, Morrison H, Rodgers JR, Fenster A, Meyer T. Dosimetry of a sonolucent material for an ultrasound-compatible gynecologic high-dose-rate brachytherapy cylinder using Monte Carlo simulation and radiochromic film. *Brachytherapy.* 2021;20(1):265–271.

66. Rodgers JR, Mendez LC, Hoover DA, Bax J, D'Souza D, Fenster A. Feasibility of fusing three-dimensional transabdominal and transrectal ultrasound images for comprehensive intraoperative visualization of gynecologic brachytherapy applicators. *Med Phys.* 2021;48(10):5611–5623.

67. Sun Y, Yuan J, Qiu W, Rajchl M, Romagnoli C, Fenster A. Three-dimensional nonrigid MR-TRUS registration using dual optimization. *IEEE T Med Imaging.* 2015;34(5):1085–1095.

68. Sun Y, Qiu W, Romagnoli C, Fenster A. Three-dimensional non-rigid landmark-based magnetic resonance to transrectal ultrasound registration for image-guided prostate biopsy. *J Med Imaging.* 2015;2(2):925002.

69. Qiu W, Yuan J, Ukwatta E, Fenster A. Rotationally resliced 3D prostate TRUS segmentation using convex optimization with shape priors. *Med Phys.* 2015;42(2):877–891.

70. Qiu W, Yuan J, Ukwatta E, Sun Y, Rajchl M, Fenster A. Dual optimization based prostate zonal segmentation in 3D MR images. *Med Image Anal.* 2014;18(4):660–673.

71. Neshat H, Cool DW, Barker K, Gardi L, Kakani N, Fenster A. A 3D ultrasound scanning system for image guided liver interventions. *Med Phys.* 2013;40(11):112903.

72. Gillies D, Bax J, Barker K, et al. Development of a 3D ultrasound guidance and verification system for focal liver tumor ablation therapy. *Med Phys.* 2019;46(11):5387–5387.

73. Gillies D, Bax J, Barker K, et al. Towards a 3D ultrasound guidance system for focal liver tumor ablation therapy. *Med Phys.* 2019;46(6):E224–E225.

74. Gillies DJ, Bax J, Barker K, Gardi L, Kakani N, Fenster A. Geometrically variable three-dimensional ultrasound for mechanically assisted image-guided therapy of focal liver cancer tumors. *Med Phys.* 2020;47(10):5135–5146.

75. Janvier MA, Durand LG, Cardinal MH, et al. Performance evaluation of a medical robotic 3D-ultrasound imaging system. *Med Image Anal.* 2008;12(3):275–290.

76. Bax J, Smith D, Bartha L, et al. A compact mechatronic system for 3D ultrasound guided prostate interventions. *Med Phys.* 2011;38(2):1055–1069.

77. Smith WL, Surry KJ, Mills GR, Downey DB, Fenster A. Three-dimensional ultrasound-guided core needle breast biopsy. *Ultrasound Med Biol.* 2001;27(8):1025–1034.

3 Freehand 3D Ultrasound
Principle and Clinical Applications

Elvis C. S. Chen, Leah A. Groves, Hareem Nisar,
Patrick K. Carnahan, Joeana Cambranis-Romero, and
Terry M. Peters

INTRODUCTION: THREE-DIMENSIONAL FREEHAND ULTRASOUND

Enabled by the advances in surgical metrology and the miniaturization of electronics and sensors, three-dimensional (3D) ultrasound (US) is becoming a preferred volumetric imaging modality due to its nonionizing, portable, and interactive nature, leading to a cost-effective imaging alternative to computed tomography (CT) and magnetic resonance imaging (MRI). While the conventional two-dimensional (2D) US is routinely used clinically, the main advantages of 3D US [1–3] include the ability to: (i) display an arbitrary 2D image slice through the 3D US volume, allowing views that are not possible with 2D US imaging alone; (ii) provide 3D visualization including surface or volume rendering from an arbitrary vantage point, thereby assisting the clinician in the interpretation and perception of the anatomy; (iii) provide a means for quantitative 3D volumetric measurements; and (iv) provide a means for image fusion with other volumetric imaging modalities. 3D US has been used clinically for neurosurgery [4], breast [5, 6], liver [7–9], cardiovascular [10], pediatric [11], prostate [12], and musculoskeletal [13] applications among others.

Categorically, there are five mechanisms to acquire 3D US volume: (i) 2D array transducer, (ii) mechanical 3D probe, (iii) 2D US transducer actuated using mechanical localizers, (iv) freehand, and (v) the sensorless approach. Each of these is briefly summarized in the subsequent text.

2D array transducers: Real-time 3D US can be acquired directly using transducers based on a 2D array of piezoelectric materials arranged in a planer or ring pattern [14, 15]. 3D US transducers that allow explicit imaging in 3D in real time are relatively large and expensive, and their image resolution is not as good as their 2D counterparts [1].

Mechanical 3D probe: Mechanical 3D probe embeds a 1D array of piezoelectric materials within a probe casing [1] that captures 3D US volume in near real time. The 1D array of transducers is computer controlled to rotate, tile, or translate to acquire a series of 2D images over the examined area. During the acquisition, the mechanical 3D probe must be held stationary, leading to potential latent errors during the volume reconstruction if subtle movements have occurred. A specialized form of mechanical 3D US is a technique where single piezoelectric transducer is spun mechanically at an oblique angle, thereby generating a "2.5D" conic image [16]. 3D volume is obtained by moving the single piezoelectric transducer along a known linear path.

Mechanically actuated 3D US: A conventional 2D US transducer can be mechanically actuated (translation, rotation, or a combination of both) in a predefined motion, where the poses (orientation and translation) of the successively acquired 2D US images are known through the kinematics and the encoder of the mechanical linkage system. Among all the techniques, mechanically actuated 3D US produces volumes with high fidelity, due to its fine control of the probe movement (refer to Chapter 1 for detail). Clinical considerations, such as ergonomics and OR planning, must be optimized as the external fixtures (i.e., the mechanical linkage system) are typically large and heavy.

Freehand 3D US: In the freehand approach, a spatial tracking sensor is attached to the conventional 2D/3D US transducer such that a set of 2D US images (or 3D volumes) are acquired along the locations known within a global reference frame. This allows the set of B-mode images to be

DOI: 10.1201/9781003299462-4

reconstructed into a regular 3D voxel array for visualization and analysis [17]. Two technical components of the freehand approach are spatial calibration and the reconstruction of a 3D voxel array from the irregularly acquired 2D US images. This approach allows the acquisition of a large 3D US volume without restricting probe movement, enables the clinicians to choose an optimal view to image an anatomical surface, and avoids imaging artifacts such as signal dropout (i.e., a phenomenon where the anatomical surface is parallel to the direction of the US wave, rendering the surface under US invisible). A specialized acquisition protocol needs to be developed and adhered to by the sonographer to acquire a densely sampled dataset.

Sensorless approach: Due to the finite thickness of the US beam profile and the presence of speckles, it is possible to infer the relative pose of successively acquired 2D US images without an external tracking system. Known as sensorless 3D US, this technique has regained interest within the research community due to recent advances in learning-based techniques [18, 19]. However, earlier theoretical works have shown that it may not be possible to fully reconstruct a 3D US volume using only speckle correlation analysis [20, 21]. While the sensorless approach is the most convenient, requiring no extrinsic apparatus, it still suffers from spatial drift [22] even with the latest learning-based approaches.

Among these 3D US approaches, freehand US is perhaps the most flexible one as it can be applied to both rigid and flexible (e.g., endoscopic) US with wide clinical applications. The technological basis for freehand 3D US is spatial tracking and calibration, US volume reconstruction, and visualization.

TECHNICAL OVERVIEW OF FREEHAND 3D ULTRASOUND

In the first step of freehand US imaging, a tracking pose sensor is rigidly attached to the US transducer. Since the pose sensor can only directly determine its own pose (orientation and position), its geometrical relationship with the US image coordinate system must be determined through a process known as probe calibration. Accurate probe calibration and real-time tracking allow the pose of each 2D US image to be known in a common coordinate system, through which a 3D volume reconstruction algorithm is used to format the set of 2D images into a 3D rectangular grid.

SPATIAL TRACKING SYSTEM

Several types of tracking systems exist, with optical and magnetic systems being the most widely available. A comprehensive review and brief history of tracking systems used for surgical intervention have been presented by Birkfellner et al. [23]. Each type of tracking system differs concerning the physics and mathematical principles used to derive the tracking information. In general, an optical tracking system based on cameras and infrared lights operates on the principle of triangulation and fiducial-based registration [24], whereas the video metric tracking system operates on the principle of pattern recognition and homography. Optical tracking systems offer submillimeter accuracy[1] and high acquisition rate (>60 Hz) (Figure 3.1). Its main drawback is the requirement for the line of sight between the cameras and the optical pose sensor, which may not be not feasible in some clinical scenarios. The geometry of the optical pose sensor must follow vendor-specific criteria to maintain tracking accuracy and distinguish between multiple optical pose sensors [25] (Figure 3.2). Optical pose sensors can be either passive or active: active optical pose sensor generates its own infrared signals, thus it is typically tethered (i.e., wired) to other components of the optical tracking system, whereas passive optical pose sensor is wireless.

Magnetic tracking employs the principle of sensing the strength and orientation of an artificially generated magnetic field [26] (Figure 3.3). For minimally invasive surgeries where surgical instruments, such as flexible endoscope or needle tip, must be tracked inside the human body, magnetic tracking has emerged as the method of choice because small magnetic pose sensors can be

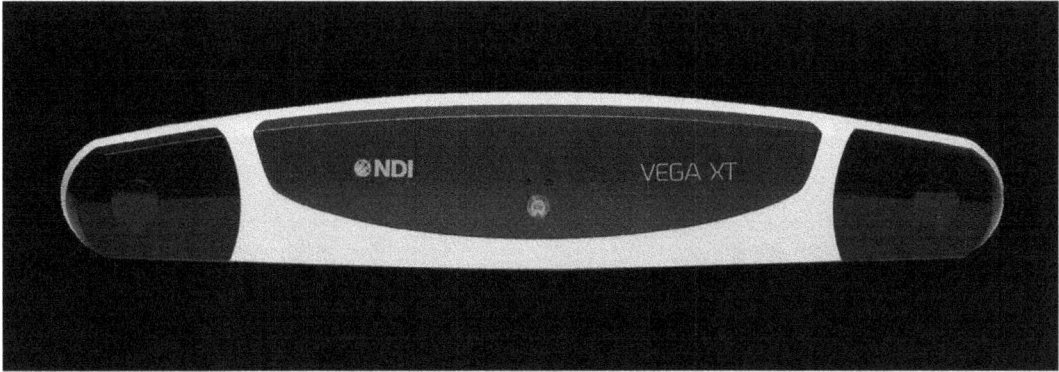

FIGURE 3.1   An example of optical tracking system (Vega, NDI, Canada) with a volumetric accuracy up to 0.1 mm RMS and acquisition rate up to 400 Hz. (Image courtesy of Northern Digital Inc.)

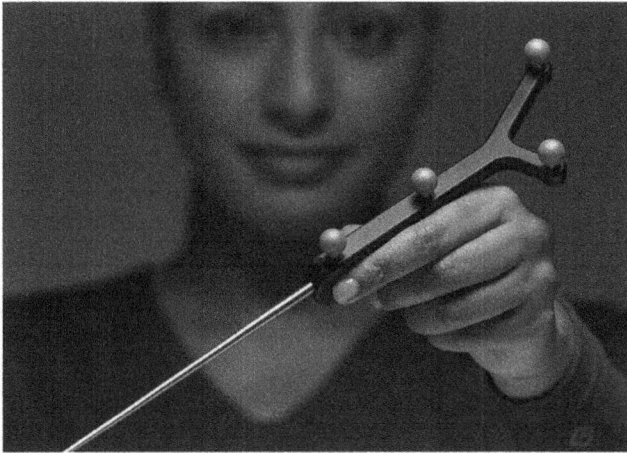

FIGURE 3.2   The silver spheres reflect infrared lights that can be sensed by the cameras of an optical tracking system. (Image courtesy of Northern Digital Inc.)

FIGURE 3.3   An example of a magnetic tracking system (Tabletop Field Generator, NDI, Canada). The form factor allows it to be placed between the patient and the table. It includes a thin barrier to minimize distortions. (Image courtesy of Northern Digital Inc.)

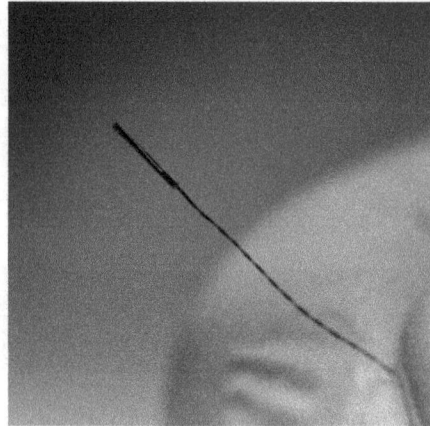

FIGURE 3.4 An example of a 5-DoF magnetic sensor with a dimension of 0.3 × 2.5 mm, small enough to be integrated into a 3F catheter.

embedded into these instruments without the line of sight requirement (Figure 3.4). Magnetic pose sensors are typically tethered to other components of the magnetic tracking system. It also offers submillimeter accuracy, but the accuracy may be compromised by the presence of ferromagnetic materials inside the surgical field. Major drawbacks regarding clinical applications include the need to place components of magnetic tracker, such as the magnetic field generator, close to the patient, and the need to integrate a fragile magnetic pose sensor into the surgical instruments, which in most cases are tethered by a wire. Careful consideration for clinical utilization of magnetic tracking system in a specific clinical scenario is required [26].

ULTRASOUND PROBE CALIBRATION

Once a pose sensor is rigidly attached to the US transducer (Figure 3.5), the geometrical relationship between the US image coordinate system and the pose sensor must be determined via a spatial calibration procedure. US probe calibration remains an area of active research [1, 27]. An array of calibration techniques and methods for accuracy assessment exist in the current literature. US probe calibration establishes the geometrical relationship between the location of an object depicted by the US image and the coordinate system of the tracking pose sensor via a homogeneous transform:

$$^{sensor}\begin{pmatrix} X \\ Y \\ Z \\ 1 \end{pmatrix} = {}^{sensor}T_{img} \cdot {}^{img}\begin{pmatrix} u \\ v \\ 0 \\ 1 \end{pmatrix}$$

where $^{img}\begin{pmatrix} u \\ v \\ 0 \\ 1 \end{pmatrix}$ denotes the location of the object in the US image indexed by the pixel loca-

tion $\begin{pmatrix} u \\ v \end{pmatrix}$, the $^{sensor}\begin{pmatrix} X \\ Y \\ Z \\ 1 \end{pmatrix}$ denotes the location of the same object but specified in the local

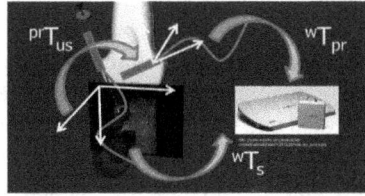

FIGURE 3.5 Ultrasound probe calibration. The geometrical relation between the coordinate system of the ultrasound image to that of the pose sensor must be accurately determined: $^{pr}T_{US}$ denotes the 4×4 homogeneous transformation of the probe calibration, $^{w}T_{pr}$ denotes the real-time pose information of the sensor attached to the ultrasound transducer, and $^{w}T_{s}$ denotes the real-time pose information of another surgical instrucment, such as the needle.

coordinate system of the pose sensor, and T is the homogeneous transformation denoting the probe calibration (Figure 3.6):

$$^{sensor}T_{img} = \begin{pmatrix} 1 & 0 & 0 & t_x \\ 0 & 1 & 0 & t_y \\ 0 & 0 & 1 & t_z \\ 0 & 0 & 0 & 1 \end{pmatrix} \begin{pmatrix} & & & 0 \\ & R_{3\times3} & & 0 \\ & & & 0 \\ 0 & 0 & 0 & 1 \end{pmatrix} \begin{pmatrix} s_x & 0 & 0 & 0 \\ 0 & s_y & 0 & 0 \\ 0 & 0 & s_z & 0 \\ 0 & 0 & 0 & 1 \end{pmatrix}$$

which is composed of anisotropic scaling factors (s_x, s_y, s_z) followed by orthonormal rotation $R_{3\times3}$ and translation (t_x, t_y, t_z). The US fiducial is image-based (2D), thus a 0 is appended for the z-component and the scaling factor s_z is arbitrary. The standard notation $^{J}T_{I}$ denotes the transformation that maps the coordinate system I to coordinate system J. The objective of US probe calibration is to determine the transformation $^{sensor}T_{img}$ efficiently and accurately. Considering the variance of the clinical environment and sensor attachment, ideally, the pose sensor should be calibrated every time before clinical deployment [17]. Consideration for the clinical deployment of a probe calibration technique includes the time required to derive a calibration, accuracy, and costs of the calibration phantom. Open-source implementations of US probe calibration techniques are available at refs [28, 29].

Probe calibration and real-time tracking are the basis for US volume reconstruction. During US volume reconstruction, every pixel in every B-mode image has to be located with respect to the reconstruction volume. This is achieved by transforming each pixel (^{img}p) first to the coordinate system of the pose sensor, then to the spatial tracker and, lastly, to the reconstruction volume:

$$^{vol}x = {}^{vol}T_{tracker} \cdot {}^{tracker}T_{pose} \cdot {}^{pose}T_{img} \cdot {}^{img}p$$

where ^{vol}x is the location of the pixel ^{img}p represented in the reconstructed volume *vol*. Note that the transform $^{vol}T_{tracker}$ is included mostly as a matter of mathematical convenience: if omitted, the reconstructed volume is aligned within the spatial tracker's coordinate system.

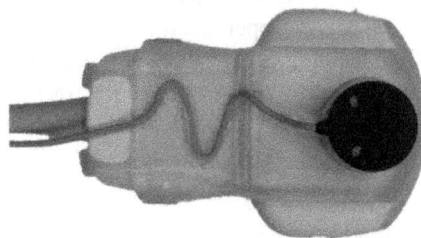

FIGURE 3.6 A clinical ultrasound transducer augmented with a magnetic pose sensor.

VOLUME RECONSTRUCTION

In 3D freehand US, the 2D B-mode image can be at any relative position and orientation, thus the US pixels lie at irregular locations in the rectangular grid of voxels. US reconstruction techniques interpolate these unstructured and scattered data and reconstruct a 3D rectangular grid of voxels: if an output voxel coincides with more than one 2D US image pixel, the voxel intensity must be "averaged" based on the intensities of these intersecting pixels. If an output voxel does not coincide with any 2D B-mode pixel, its output intensity must be interpolated based on neighboring voxel intensities. Considerations for any volume reconstruction include the quality and accurate representation of the underlying anatomy, the computational requirement (speed and memory usage) of the reconstruction algorithm, and the real-time requirements (e.g., used for intraoperative visualization or preoperative planning).

Freehand US volume reconstruction algorithms can be categorized into three types [2, 17]: (i) pixel-based methods, (ii) voxel-based methods, or (iii) function-based methods.

Pixel-Based Methods

Pixel-based methods assign the intensity of the output voxel in two steps. In the "distribution step," each pixel in the input 2D image is traversed and the pixel intensity is assigned to the coincident voxel or spatially nearest voxel. Overlapping intensities can be computed as a weighted average or as the maximum [30]. As an output voxel may not coincide with an input 2D pixel, thus considered as a "hole," the empty voxel is filled in the "hold-filling" step based on its proximity to a neighboring voxel that has an assigned intensity. The computational requirement for pixel-based methods is relatively low, but blank regions may still exist even after the hold-filling step. Other drawbacks of pixel-based methods include blurring due to the intensity of the output voxel computed as a mean over many surrounding pixels and sharp transition in intensity value across large holes [30].

Voxel-Based Methods

In the voxel-based method, the voxel in the output volume is traversed and the voxel intensities are estimated as a function of nearby images. This type of approach assumes that some information for all input 2D images is available, thus the memory requirement may be demanding. The intensity of a voxel can be estimated based on the nearest neighbor or interpolation between several input pixel values [2]. Although voxel-based methods are fast to compute and avoid blank regions in the output volume, reconstruction artifacts can often be observed in slices through the output volume due to misalignment of the 2D US images.

Function-Based Methods

These methods choose a 3D mathematical function, such as a polynomial, and determine the coefficients of the function such that it passes through the pixel intensity values. The output 3D volume is generated by evaluating this function at regular intervals to produce a voxel array. Examples include the use of radial basis function [17], Rayleigh reconstruction and interpolation [31], and Benzier interpolation [32]. The computational complexity for function-based methods is often high.

Trade-offs must be made between the real-time requirement and the quality of the volume reconstruction. With the advancement in computation hardware and GPU acceleration, it is now possible to achieve real-time volume reconstruction [33–35]. An open-source implementation of a 3D US reconstruction algorithm with advanced hole-filling techniques is available in ref [28]. Other methods of improving the quality of reconstructed US volume include regularization of the tracking pose signal [36].

VISUALIZATION

Reconstructed US volume can be rendered in several manners including multiplanar slicing, surface rendering, and volume rendering [37]. In multiplanar slicing, 2D US images are extracted from the 3D US volume that are either parallel to the bounding box of the 3D volume or at an oblique angle. 2D US images extracted from 3D US volume allow 2D views of the anatomy that is otherwise unavailable with the standard 2D US. While multiplanar slicing is computationally efficient, it still requires physicians to mentally reconstruct the 2D slices in 3D space [38]. It provides representative 2D views of the anatomy provided that the underlying anatomy is sufficiently sampled using 2D US and subject to reconstruction artifacts such as blurring that is inherent to the volume reconstruction process.

Surface rendering of the anatomy provides direct and interactive visualization of regions of interest. After the segmentation or classification, the polygonal representation of the isosurface can be generated and rendered using the standard computer graphics surface-rendering technique. In this approach, the fidelity of the visualization largely depends on accurate segmentation of the anatomy in US. To enhance the acuity of the visualization, it may be advantageous to apply a 2D segmentation algorithm to the original B-mode images followed by a surface reconstruction to extract the anatomical isosurface. Alternatively, 3D segmentation on the reconstructed volume or 2D segmentation on multi-view slices of the 3D volume may be performed to extract the anatomy, but these methods may be affected by blurring inherent in the volume reconstruction process, leading to inaccuracies in segmentation.

Without segmentation, 3D US volume can be directly visualized using the volume-rendering technique; however, advanced transfer functions need to be hand-crafted to disambiguate the various materials and structures contained within the US volume for effective visualization and perception [37]. The standard 1D transfer function may be suitable for visualizing noisy 3D US volume [39] when used in conjunction with ray-depth information to improve depth cues [40].

CLINICAL APPLICATIONS: VASCULAR IMAGING

Freehand 3D US is particularly suited for superficial vascular imaging, such as the detection and quantification of carotid atherosclerosis [41–43] due to the superficial location of the vessels and the ability of US to depict them. Quantitative 3D metrics such as total plaque area, total plaque volume, and vessel wall volume are used to assess carotid plaque burden [43]. Freehand 3D US, coupled with deep-learning-based image processing techniques, has the potential to derive these 3D quantities in near real time.

Groves et al. proposed a 3D freehand US system for vascular imaging with automatic segmentation of the carotid artery and internal jugular vein [44]. A Mask R-CNN was developed to delineate the vessel lumens from the 2D US in real time with the ability to differentiate between the internal jugular vein and the carotid artery. After the length of the vessels is scanned, these 2D segmentations of vessels are placed in a 3D coordinate system based on the tracking information from which the vessel surface model can be reconstructed (Figure 3.7).

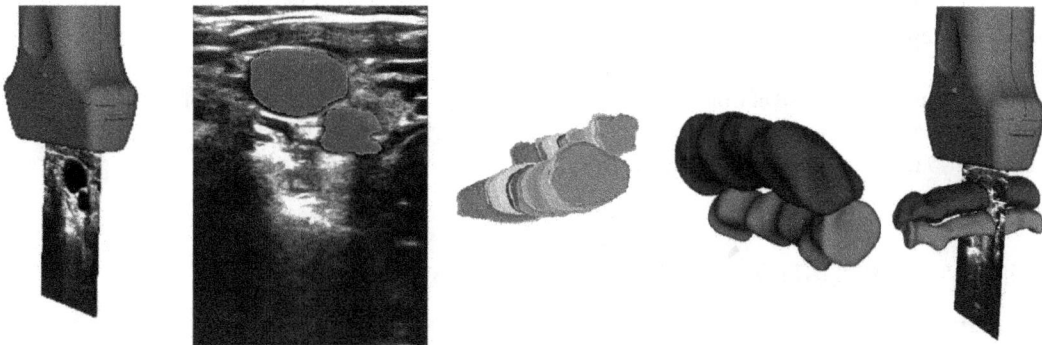

FIGURE 3.7 Vessel reconstruction from tracked ultrasound. From left: Tracked ultrasound, real-time vessel segmentation from 2D B-mode ultrasound, vessel segmentation placed in 3D using pose information, vessel reconstruction based on segmentation, and vessel visualization with tracked ultrasound.

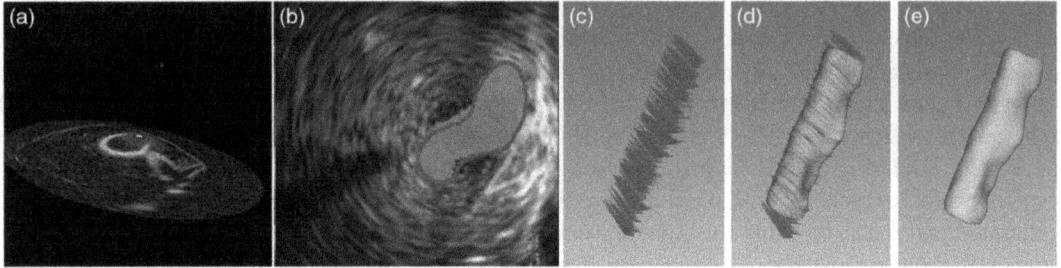

FIGURE 3.8 Inferior vena cava (IVC) reconstruction using tracked intracardiac echo (ICE). (a) The ICE ultrasound has a conic shape. (b) A U-Net-based vessel lumen segmentation. (c) Vessel segmentation reconstructed in 3D based on the conic shape and magnetic tracking. (d) Vessel surface reconstruction. (e) Vascular path visualization.

The same principle of vessel reconstruction for vascular mapping can be applied to intravascular ultrasound (IVUS) and intracardiac echocardiography (ICE) for transcatheter cardiovascular interventions. Vascular navigation is fundamental to transcatheter cardiac interventions such as transcatheter aortic valve implantation (TAVI), caval valve implantation, mitral and tricuspid valve annuloplasty, and repair and replacement surgeries [45]. While transcatheter cardiac interventions are typically performed under fluoroscopy guidance, additional radiation exposure nonetheless can be harmful to the patient, clinical staff, and medical trainees, even when used in conjunction with various shielding techniques [46, 47]. Toward minimizing radiation and providing a 3D vascular roadmap, Nisar et al. instrumented a 3D US system based on a magnetically tracked ICE and used it to generate a 3D vascular path of an inferior vena cava phantom. The 3D vascular path can be visualized and used in conjunction with a magnetically tracked guidewire [48] for vessel navigation (Figure 3.8).

CLINICAL APPLICATION: MULTIMODAL IMAGE REGISTRATION AND FUSION

Image registration is crucial for image-guided intervention and multimodal data fusion. Conspicuity of liver metastasis in 2D US is often low due to its size, location, and echogenicity of the metastasis being similar to the surrounding tissue [49]. Image fusion between preprocedural CT or MRI improves the conspicuity of liver tumors and the feasibility of percutaneous radiofrequency ablation of tumors not identifiable on B-mode US [50]. Hepatic vasculature, which is readily visible in US and CT, can be used as the basis for image fusion. Figure 3.9 depicts the workflow to achieve fusion between preprocedural CT and intraprocedural US using hepatic vasculatures. In each modality, the hepatic vasculatures were automatically segmented using learning-based methods, followed by vessel surface reconstructions in 3D. Registration between these two sets of vasculatures can be achieved using vessel bifurcation points, vessel centerline, or vessel surfaces. Once registered, visualization of the tumor can be depicted in a virtual reality environment or by the superimposition of images.

CLINICAL APPLICATION: 3D FUSION ECHOCARDIOGRAPHY

The current standard of care diagnostic imaging for mitral valve procedures primarily consists of transesophageal echocardiography (TEE) as it provides a clear view of the mitral valve leaflets and surrounding tissue [51, 52]. However, TEE has limitations in signal dropout and artifacts, particularly for structures lying below the valve such as chordae tendineae. Even with TEE with a matrix-array transducer, the field of view is limited, and structures further away from the US transducer

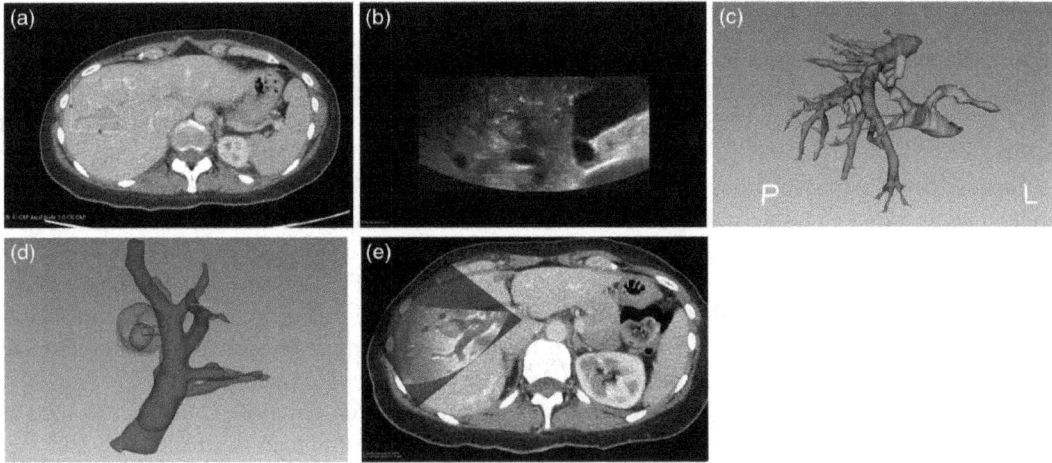

FIGURE 3.9 Image fusion between preprocedural CT and intraprocedural 3D US. (a) Hepatic vasculatures were segmented using learning-based methods from CT. (b) From 2D US. (c) Reconstructed vasculature surfaces. (d) CT-US registration using vessel surfaces. (e) Fusion of US superimposed onto CT.

suffer from poor spatial resolution. A complete description of the heart can be obtained only by combining the information provided by different acoustic windows [53]. 3D fusion echocardiography (3DFE) merges multiple real-time 3D US volumes from different transducer positions to improve image quality and completeness of the reconstructed image [54]. 3DFE can be achieved via tracking or volume-to-volume registration if adjacent volumes are sufficiently overlapped.

Carnahan et al. presented a 3DFE technique that involved three key steps: volume acquisition, groupwise volume registration, and blending [55]. In volume acquisition, standard diagnostic TEE volumes were acquired to image the mitral valve, with a minimum requirement of a single mid-esophageal view and five transgastric views targeting 80% overlap between adjacent volumes. Groupwise registration with a cost function using orientation and phase information [56] was used to compound all TEE images into a single volume. Volume blending was needed to provide voxel intensity coherence for visualization purposes (Figure 3.10).

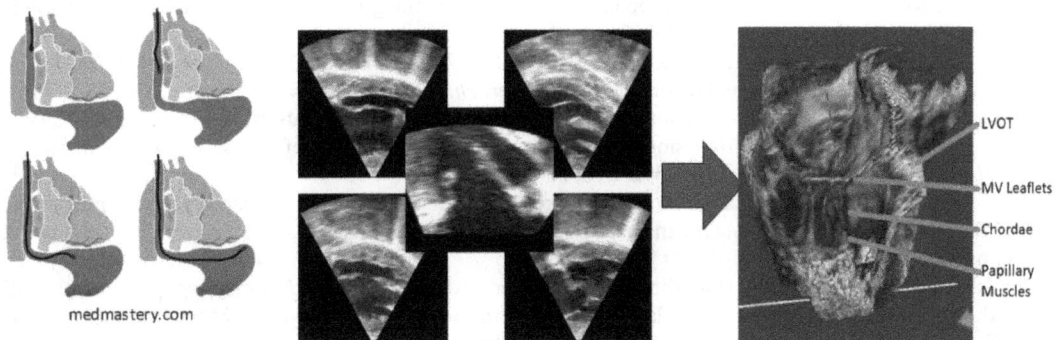

FIGURE 3.10 3D fusion echocardiogram. Standard diagnostic real-time 3D transesophageal echocardiography (TEE) volumes were acquired with a single mid-esophageal view and five transgastric views. With sufficient overlap between adjacent volumes (~80%), TEE volumes can be compounded into a single volume with large FOV providing a complete view of the heart.

SUMMARY AND FUTURE PERSPECTIVES

In this chapter, we have presented a condensed view on the technical component of 3D freehand US including spatial tracking, probe calibration, volume reconstruction, and visualization. Specific clinical focus of 3D freehand US was given on vascular navigation, multimodal image fusion, and 3DFE.

Despite being introduced decades ago [57], 3D freehand US has still not achieved widespread clinical adaptability in modern surgical theatres, despite its known benefits. The seamless integration of this complex system into the surgical workflow remains a challenge, and a robust and accurate intraprocedural probe calibration technique is still elusive. Furthermore, the tracking system required by 3D freehand US takes up valuable real estate in the operating room, and constraints such as maintaining line of sight for the optical tracking system must be met. To date, the clinical cost-benefits of 3D freehand US have not been demonstrated at a wide scale.

As computational resources continue to increase, traditional computational bottlenecks such as deploying advanced volume reconstruction and visualization techniques are becoming less significant. Capitalizing on these resources, the unmet technical needs for 3D freehand US include real-time image processing of tracked US video, accounting for motion that naturally occurs during US scanning (breathing and cardiac motion), and deformable registration between 3D US and preprocedural modalities such as CT and MRI. Efforts to address these challenges include the use of learning-based methods for sensorless 3D US [19, 58] and automatic probe placement [59, 60], which are gaining attention. In addition, there is a growing interest in combining augmented reality and US imaging [61–63].

NOTE

1 We adhere to ISO 5725-1 in which **accuracy** is described by a combination of "trueness" and "precision". Trueness refers to the closeness of agreement between the arithmetic mean of a large number of measurements and the true or accepted reference value. Precision refers to the closeness of agreement between these measurements.

REFERENCES

1. L. Mercier, T. Langø, F. Lindseth, and D. L. Collins, "A review of calibration techniques for freehand 3-D ultrasound systems," *Ultrasound Med Biol*, vol. 31, no. 4, pp. 449–471, 2005.
2. O. V. Solberg, F. Lindseth, H. Torp, R. E. Blake, and T. A. Nagelhus Hernes, "Freehand 3D ultrasound reconstruction algorithms—A review," *Ultrasound Med Biol*, vol. 33, no. 7, pp. 991–1009, 2007.
3. M. H. Mozaffari, and W.-S. Lee, "Freehand 3-D ultrasound imaging: A systematic review," *Ultrasound Med Biol*, vol. 43, no. 10, pp. 2099–2124, 2017.
4. G. Unsgaard, O. M. Rygh, T. Selbekk, T. B. Müller, F. Kolstad, F. Lindseth, and T. A. N. Hernes, "Intraoperative 3D ultrasound in neurosurgery," *Acta Neurochir (Wien)*, vol. 148, no. 3, pp. 235–253, 2006.
5. D. O. Watermann, M. Földi, A. Hanjalic-Beck, A. Hasenburg, A. Lüghausen, H. Prömpeler, G. Gitsch, and E. Stickeler, "Three-dimensional ultrasound for the assessment of breast lesions," *Ultrasound Obstet Gynecol*, vol. 25, no. 6, pp. 592–598, 2005.
6. V. Giuliano, and C. Giuliano, "Improved breast cancer detection in asymptomatic women using 3D-automated breast ultrasound in mammographically dense breasts," *Clin Imaging*, vol. 37, no. 3, pp. 480–486, 2013.
7. T. Lange, S. Eulenstein, M. Hünerbein, H. Lamecker, and P.-M. Schlag, "Augmenting intraoperative 3D ultrasound with preoperative models for navigation in liver surgery," in Medical Image Computing and Computer-Assisted Intervention – MICCAI 2004, 2004, pp. 534–541.
8. G. P. Penney, J. M. Blackall, M. S. Hamady, T. Sabharwal, A. Adam, and D. J. Hawkes, "Registration of freehand 3D ultrasound and magnetic resonance liver images," *Med Image Anal*, vol. 8, no. 1, pp. 81–91, 2004.
9. H. Neshat, D. W. Cool, K. Barker, L. Gardi, N. Kakani, and A. Fenster, "A 3D ultrasound scanning system for image guided liver interventions," *Med Phys*, vol. 40, no. 11, p. 112903, 2013.

10. B. Chiu, M. Egger, J. D. Spence, G. Parraga, and A. Fenster, "Development of 3D ultrasound techniques for carotid artery disease assessment and monitoring," *Int J Comput Assist Radiol Surg*, vol. 3, no. 1–2, pp. 1–10, 2008.

11. M. Riccabona, "Editorial review: Pediatric 3D ultrasound," *J Ultrason*, vol. 14, no. 56, pp. 5–20, 2014.

12. S. Natarajan, L. S. Marks, D. J. A. Margolis, J. Huang, M. L. Macairan, P. Lieu, and A. Fenster, "Clinical application of a 3D ultrasound-guided prostate biopsy system," *Urol Oncol*, vol. 29, no. 3, pp. 334–342, 2011.

13. S. Papernick, R. Dima, D. J. Gillies, T. Appleton, and A. Fenster, "Reliability and concurrent validity of three-dimensional ultrasound for quantifying knee cartilage volume," *Osteoarthr Cartil Open*, vol. 2, no. 4, p. 100127, 2020.

14. A. C. Dhanantwari, S. Stergiopoulos, L. Song, C. Parodi, F. Bertora, P. Pellegretti, and A. Questa, "An efficient 3D beamformer implementation for real-time 4D ultrasound systems deploying planar array probes," in *IEEE Ultrasonics Symposium*, 2004, vol. 2, pp. 1421–1424.

15. D. T. Yeh, O. Oralkan, I. O. Wygant, M. O'Donnell, and B. T. Khuri-Yakub, "3-D ultrasound imaging using a forward-looking CMUT ring array for intravascular/intracardiac applications," *IEEE Trans Ultrason Ferroelectr Freq Control*, vol. 53, no. 6, pp. 1202–1211, 2006.

16. H. Nisar, T. M. Peters, and E. C. S. Chen, "Characterization and calibration of foresight ICE," 2001.

17. R. N. Rohling, "3D Freehand Ultrasound Reconstruction and Spatial Compounding," Ph.D., University of Cambridge, 1998.

18. M. Luo, X. Yang, H. Wang, L. Du, and D. Ni, "Deep motion network for freehand 3D ultrasound reconstruction," in Medical Image Computing and Computer Assisted Intervention – MICCAI 2022, 2022, pp. 290–299.

19. H. Guo, S. Xu, B. Wood, and P. Yan, "Sensorless Freehand 3D Ultrasound Reconstruction via Deep Contextual Learning," in *Medical Image Computing and Computer Assisted Intervention – MICCAI 2020. MICCAI 2020*, A. L. Martel, P. Abolmaesumi, D. Stoyanov, D. Mateus, M. A. Zuluaga, S. K. Zhou, D. Racoceanu, and L. Joskowicz, Eds. 2020, pp. 463–472.

20. P.-C. Li, C.-Y. Li, and W.-C. Yeh, "Tissue motion and elevational speckle decorrelation in freehand 3D ultrasound," *Ultrason Imaging*, vol. 24, no. 1, pp. 1–12, 2002.

21. W. Smith, and A. Fenster, "Statistical analysis of decorrelation-based transducer tracking for three-dimensional ultrasound," *Med Phys*, vol. 30, no. 7, pp. 1580–1591, 2003.

22. A. Lang, "Improvement of speckle-tracked freehand 3-D ultrasound through the use of sensor fusion," Master of Applied Science, Queen's University, Canada, 2009.

23. W. W. Birkfellner, J. B. Hummel, E. Wilson, and K. Cleary, "Tracking Devices," in *Image-Guided Interventions*, T. M. Peters and K. Cleary, Eds. 2008, pp. 23–44.

24. B. K. P. Horn, "Closed-form solution of absolute orientation using unit quaternions," *J Opt Soc Am A*, vol. 4, no. 4, pp. 629–642, 1987.

25. A. Brown, A. Uneri, T. De Silva, A. Manbachi, and J. H. Siewerdsen, "Design and validation of an open-source library of dynamic reference frames for research and education in optical tracking," *J Med Imaging*, vol. 5, no. 2, 2018.

26. A. M. Franz, T. Haidegger, W. W. Birkfellner, K. Cleary, T. M. Peters, and L. Maier-Hein, "Electromagnetic tracking in medicine—A review of technology, validation, and applications," *IEEE Trans Med Imaging*, vol. 33, no. 8, pp. 1702–1725, 2014.

27. P.-W. Hsu, R. W. Prager, A. H. Gee, and G. M. Treece, "Freehand 3D Ultrasound Calibration: A Review," in *Advanced Imaging in Biology and Medicine*, Berlin, Heidelberg: Springer Berlin Heidelberg, 2009, pp. 47–84.

28. A. Lasso, T. Heffter, A. Rankin, C. Pinter, T. Ungi, and G. Fichtinger, "PLUS: Open-source toolkit for ultrasound-guided intervention systems," *IEEE Trans Biomed Eng*, vol. 61, no. 10, pp. 2527–2537, 2014.

29. K. Gary, L. Ibanez, S. Aylward, D. G. Gobbi, M. B. Blake, and K. Cleary, "IGSTK: An open source software toolkit for image-guided surgery," *Computer (Long Beach Calif*, vol. 39, no. 4, pp. 46–53, 2006.

30. T. Vaughan, A. Lasso, T. Ungi, and G. Fichtinger, "Hole filling with oriented sticks in ultrasound volume reconstruction," *J Med Imaging*, vol. 2, no. 3, p. 034002, 2015.

31. Q. Huang, Y. Huang, W. Hu, and X. Li, "Bezier interpolation for 3-d freehand ultrasound," *IEEE Trans Hum Mach Syst*, vol. 45, no. 3, pp. 385–392, 2015.

32. J. M. Sanches, and J. S. Marques, "A Rayleigh reconstruction/interpolation algorithm for 3D ultrasound," *Pattern Recognit Lett*, vol. 21, no. 10, pp. 917–926, 2000.

33. Z. Chen, and Q. Huang, "Real-time freehand 3D ultrasound imaging," *Comput Methods Biomech Biomed Eng Imaging Vis*, vol. 6, no. 1, pp. 74–83, 2018.

34. T. Wang, J. Wu, and Q. Huang, "Enhanced extended-field-of-view ultrasound for musculoskeletal tissues using parallel computing," *Curr Med Imaging Rev*, vol. 10, no. 4, pp. 237–245, 2015.

35. Y. Dai, J. Tian, di Dong, G. Yan, and H. Zheng, "Real-time visualized freehand 3D ultrasound reconstruction based on GPU," *IEEE Trans Information Technol Biomed*, vol. 14, no. 6, pp. 1338–1345, 2010.

36. M. Esposito, C. Hennersperger, R. Gobl, L. Demaret, M. Storath, N. Navab, M. Baust, and A. Weinmann, "Total variation regularization of pose signals with an application to 3D freehand ultrasound," *IEEE Trans Med Imaging*, vol. 38, no. 10, pp. 2245–2258, 2019.

37. D. Mann, J. J. Caban, P. J. Stolka, E. M. Boctor, and T. S. Yoo, "Multi-dimensional transfer functions for effective visualization of streaming ultrasound and elasticity images," 2011, p. 796439.

38. J. N. Welch, J. A. Johnson, M. R. Bax, R. Badr, S. So, T. Krummel, and R. Shahidi, "Real-time freehand 3D ultrasound system for clinical applications," 2001, pp. 724–730.

39. P. Ljung, J. Krüger, E. Groller, M. Hadwiger, C. D. Hansen, and A. Ynnerman, "State of the art in transfer functions for direct volume rendering," *Computer Graphics Forum*, vol. 35, no. 3, pp. 669–691, 2016.

40. "US8425422B2: Adaptive volume rendering for ultrasound color flow diagnostic imaging – Google Patents." [Online]. Available: https://patents.google.com/patent/US8425422. [Accessed: 15-Mar-2023].

41. G. Heiss, A. R. Sharrett, R. Barnes, L. E. Chambless, M. Szklo, and C. Alzola, "Carotid atherosclerosis measured by B-mode ultrasound in populations: Associations with cardiovascular risk factors in the ARIC study," *Am J Epidemiol*, vol. 134, no. 3, pp. 250–6, 1991.

42. J. Persson, J. Formgren, B. Israelsson, and G. Berglund, "Ultrasound-determined intima-media thickness and atherosclerosis. Direct and indirect validation," *Arterioscler Thromb*, vol. 14, no. 2, pp. 261–264, 1994.

43. J. D. Spence, "Measurement of carotid plaque burden," *Curr Opin Lipidol*, vol. 31, no. 5, pp. 291–298, 2020.

44. L. A. Groves, B. VanBerlo, N. Veinberg, A. Alboog, T. M. Peters, and E. C. S. Chen, "Automatic segmentation of the carotid artery and internal jugular vein from 2D ultrasound images for 3D vascular reconstruction," *Int J Comput Assist Radiol Surg*, vol. 15, no. 11, pp. 1835–1846, 2020.

45. B. D. Prendergast, H. Baumgartner, V. Delgado, O. Gérard, M. Haude, A. Himmelmann, B. Iung, M. Leafstedt, J. Lennartz, F. Maisano, E. A. Marinelli, T. Modine, M. Mueller, S. R. Redwood, O. Rörick, C. Sahyoun, E. Saillant, L. Søndergaard, M. Thoenes, K. Thomitzek, M. Tschernich, A. Vahanian, O. Wendler, E. J. Zemke, and J. J. Bax, "Transcatheter heart valve interventions: Where are we? Where are we going?," *Eur Heart J*, vol. 40, no. 5, pp. 422–440, 2019.

46. N. Theocharopoulos, J. Damilakis, K. Perisinakis, E. Manios, P. Vardas, and N. Gourtsoyiannis, "Occupational exposure in the electrophysiology laboratory: Quantifying and minimizing radiation burden," *Br J Radiol*, vol. 79, no. 944, pp. 644–651, 2006.

47. G. Christopoulos, L. Makke, G. Christakopoulos, A. Kotsia, B. V. Rangan, M. Roesle, D. Haagen, D. J. Kumbhani, C. E. Chambers, S. Kapadia, E. Mahmud, S. Banerjee, and E. S. Brilakis, "Optimizing radiation safety in the cardiac catheterization laboratory," *Catheter Cardiovasc Interv*, vol. 87, no. 2, pp. 291–301, 2016.

48. R. Piazza, H. Nisar, J. T. Moore, S. Condino, M. Ferrari, V. Ferrari, T. M. Peters, and E. C. S. Chen, "Towards electromagnetic tracking of j-tip guidewire: Precision assessment of sensors during bending tests," in Medical Imaging 2020: Image-Guided Procedures, Robotic Interventions, and Modeling, 2020, p. 5.

49. C. J. Harvey, and T. Albrecht, "Ultrasound of focal liver lesions," *Eur Radiol*, vol. 11, no. 9, pp. 1578–1593, 2001.

50. Y. Minami, H. Chung, M. Kudo, S. Kitai, S. Takahashi, T. Inoue, K. Ueshima, and H. Shiozaki, "Radiofrequency ablation of hepatocellular carcinoma: Value of virtual CT sonography with magnetic navigation," *Am J Roentgenol*, vol. 190, no. 6, pp. W335–W341, 2008.

51. P. M. Shah, "Current concepts in mitral valve prolapse: Diagnosis and management," *J Cardiol*, vol. 56, no. 2, pp. 125–133, 2010.

52. A. Linden, J. Seeburger, T. Noack, V. Falk, and T. Walther, "Imaging in cardiac surgery: Visualizing the heart," *Thorac Cardiovasc Surg*, vol. 65, no. 03, pp. S213–S216, 2017.

53. V. Grau, and J. A. Noble, "Adaptive Multiscale Ultrasound Compounding Using Phase Information," 2005, pp. 589–596.

54. C. Szmigielski, K. Rajpoot, V. Grau, S. G. Myerson, C. Holloway, J. A. Noble, R. Kerber, and H. Becher, "Real-time 3D fusion echocardiography," *JACC Cardiovasc Imaging*, vol. 3, no. 7, pp. 682–690, 2010.

55. P. Carnahan, J. Moore, D. Bainbridge, C. S. Chen, and T. M. Peters, "Multi-view 3D transesophageal echocardiography registration and volume compounding for mitral valve procedure planning," *Applied Sciences*, vol. 12, no. 9, p. 4562, 2022.

56. V. Grau, H. Becher, and J. A. Noble, "Phase-Based Registration of Multi-view Real-Time Three-Dimensional Echocardiographic Sequences," 2006, pp. 612–619.

57. S. Sherebrin, A. Fenster, R. N. Rankin, and D. Spence, "Freehand three-dimensional ultrasound: Implementation and applications," 1996, pp. 296–303.

58. M. Luo, X. Yang, X. Huang, Y. Huang, Y. Zou, X. Hu, N. Ravikumar, A. F. Frangi, and D. Ni, "Self context and shape prior for sensorless freehand 3D ultrasound reconstruction," 2021, pp. 201–210.

59. R. Droste, L. Drukker, A. T. Papageorghiou, and J. A. Noble, "Automatic probe movement guidance for freehand obstetric ultrasound," 2020, pp. 583–592.

60. C. Zhao, R. Droste, L. Drukker, A. T. Papageorghiou, and J. A. Noble, "Visual-assisted probe movement guidance for obstetric ultrasound scanning using landmark retrieval," in Medical Image Computing and Computer Assisted Intervention – MICCAI 2021, 2021, pp. 670–679.

61. N. Cattari, S. Condino, F. Cutolo, M. Ferrari, and V. Ferrari, "In situ visualization for 3D ultrasound-guided interventions with augmented reality headset," *Bioengineering*, vol. 8, no. 10, p. 131, 2021.

62. G. Samei, K. Tsang, C. Kesch, J. Lobo, S. Hor, O. Mohareri, S. Chang, S. L. Goldenberg, P. C. Black, and S. Salcudean, "A partial augmented reality system with live ultrasound and registered preoperative MRI for guiding robot-assisted radical prostatectomy," *Med Image Anal*, vol. 60, p. 101588, 2020.

63. F. Giannone, E. Felli, Z. Cherkaoui, P. Mascagni, and P. Pessaux, "Augmented reality and image-guided robotic liver surgery," *Cancers (Basel)*, vol. 13, no. 24, p. 6268, 2021.

Section II

3D Ultrasound Applications

4 3D Ultrasound Algorithms and Applications for Liver Tumor Ablation

Derek J. Gillies and Shuwei Xing

INTRODUCTION

LIVER CANCER

Liver is the largest solid organ in the body located inferiorly to the diaphragm and normally resting against the lateral and anterior abdominal walls. It performs more than 500 vital functions, although its primary functions are to produce bile for digestion, process nutrients and drugs, and filter blood from the stomach and intestines. The liver is largely divided into two lobes (i.e., right and left) by the falciform ligament; however, it can be further subdivided into smaller segments in relation to the hepatic arterial, portal vein, and biliary drainage. The French surgeon and anatomist Claude Couinaud first proposed the division into eight functionally independent segments, each of which has its own vascular inflow, outflow, and biliary drainage, which can allow for independent removal and resection.[1,2]

Primary liver cancer is the sixth most diagnosed cancer and the third leading cause of cancer death on a global scale, with approximately 906,000 new cases and 830,000 deaths as of 2020.[3] In primary liver cancer, hepatocellular carcinoma (HCC) is the most common subtype comprising 75–85% of cases, followed by intrahepatic cholangiocarcinoma (10–15%) and other rare types. Both incidence and mortality rates of HCC are two to three times higher among men than in women in most regions, and it is highly prevalent in low-income and developing countries (e.g., Mongolia, China, the Republic of Korea, and sub-Saharan Africa).[4] This is predominantly due to the incidence of hepatitis B and/or C viruses[5] and is often preceded by cirrhosis (scarring).[6]

Liver is also a frequent site for metastatic cancer originating from other parts of the body, such as the lung, breast, pancreas, gastrointestinal tract, and lymphatic system. In colorectal cancer, which ranked third in terms of incidence and second in terms of mortality in 2020, at least 25–50% of patients develop colorectal liver metastases during the course of their illness.[3] This is in part due to the direct connection between the liver and nearby intestines via the abundant blood supply provided by the portal vein.[7,8]

Although early detection of liver cancer can be difficult, increased screening through the use of ultrasound (US) imaging is one approach that can be beneficial since the liver rests on the abdominal wall. The organ is easily accessible from this location and can be treated using open surgery or percutaneous (through the skin) approaches. Early detection of cancer can typically lead to improved survival outcomes and allow for the increased use of curative intent techniques such as resection, transplantation, and image-guided interventions.

TREATMENT OF LIVER CANCER

Liver Surgery: Liver Resection and Transplantation

Surgery in the forms of partial resection and total organ transplantation have traditionally been used for providing curative intent options for patients with liver cancer. Those diagnosed with early

DOI: 10.1201/9781003299462-6

stage HCC with a single lesion less than 3 cm,[9] sometimes with up to three lesions, are considered for resection as sufficient remaining liver tissue will remain to preserve function necessary for survival.[10] The presence of comorbidities, such as cirrhosis, can increase the risk of HCC recurrence after liver resection and this typically motivates the need for other treatment options, like transplantation. Since transplantation has the potential to remove localized tumors and underlying cirrhosis, it is often a desirable curative method. Although preferred, poor liver function, multifocal HCC, low patient eligibility, frequent and severe complications, and long waiting time for liver donors make it harder to benefit widespread clinical needs.

Both resection and transplantation are limited to approximately 10–20% of patients.[10] Even if a patient is eligible for these types of procedures, complication rates have been observed up to 26% in resection and 33% in transplantation,[11] leading to high costs clinically and financially. As a result, these existing drawbacks often lead to a preferential selection of other treatment alternatives, such as image-guided interventions, for focal tumor extent.

Chemotherapy and Radiation Therapy

Advanced-stage liver cancer that has spread beyond the organ to local nodes and distant sites or has invaded the portal or hepatic veins typically has very few options for curative therapy. The prognosis is often poor and conventional therapies, such as resection and transplantation, often can't handle the burden of disease. Systemic therapies like chemotherapy with a molecular target agent, sorafenib, have not been shown to be overly effective in treating advanced-stage HCC. Unfortunately, the slowing of tumor proliferation and angiogenesis often only extends overall survival by 2–7 months.[12,13]

External beam radiation has increasingly been used in the treatment of liver cancer due to advances in planning and delivery that have improved sparing of the radiosensitive healthy tissue. Radiation therapy is most often ideal for early to intermediate-stage liver cancer and is recommended when other techniques are not possible, or it can be used as a bridge to transplantation.[14] Breathing motion and liver position changes can make the widespread use of radiation therapy difficult, but advances in imaging and liver motion tracking have led to more investigation into conformal radiation delivery, dose escalation, and fewer fractions of therapy.[15]

Localized Treatments: Transarterial Chemoembolization

Intermediate-stage liver cancer typically refers to a multinodular disease that has either more than 3 lesions or 2 to 3 lesions with at least one greater than 5 cm.[16] Patients showing signs of this stage are often treated with embolization,[17] with the most common form being transarterial chemoembolization (TACE) which focuses on inducing tumor necrosis by acute arterial occlusion with the addition of chemotherapeutic drugs.[10] These are typically salvage or bridge therapies for patients with preserved liver function and do not usually have curative intent as necrosis is often unachievable for larger tumors due to incomplete embolization and tumor angiogenesis.[9] Approaches that combine TACE with other procedures have been investigated for tumors larger than 3 cm to make patients eligible for other curative approaches,[18] with the use of TACE post-resection showing benefits compared to other anti-recurrence therapies.[19] Embolization can also be performed using radiation sources, such as Yttrium-90, with evidence suggesting potential benefits for patients with advanced tumor stages and few other treatment options.[20]

Localized Treatments: Liver Tumor Ablation

Ablation techniques are considered the best treatment alternatives for HCC patients who are not eligible for surgical techniques.[10] Tumor ablation is defined as the direct application of therapies to eradicate or substantially destroy focal tumors, either using energy-based or chemical approaches.[21] These ablation approaches can include the use of radiofrequency, microwave, freezing (cryo), laser, high-frequency US, irreversible electroporation, and ethanol.[22] Therapy is applied through the use of applicators, such as electrodes in radiofrequency ablation (RFA), antennas in microwave ablation (MWA), or fibers in laser ablation, to provide a focal ablation region. These procedures can be used in a palliative setting for pain management, but also have the potential for curative intent on early stage or small tumors.

Most common treatment approaches for tumor ablation are either RFA or MWA. These thermal methods focus on the production of heat for tissue ablation but have different mechanisms of heat production that require different equipment and application techniques. RFA techniques use applicators that are needle-like and typically use a single monopolar active applicator, occasionally separating at the tip into multiple tines for a larger ablation volume, with a 375–500 kHz alternating current dissipating at one or more grounding pads to produce resistive heat.[20,23] MWA uses needle-like applicators without the need for grounding pads and generates microwaves with a frequency between 915 MHz and 2.45 GHz to produce frictional heat from oscillating water molecules.[23] Multiple applicators can be placed in both techniques to allow for large volumes to be treated, depending on the size and geometry of the targeted lesion; however, MWA can also allow for simultaneous activation of applicators to exploit electromagnetic (EM) field overlap. Aside from the current size of the MWA applicators, this method has the potential to offer improved performance over RFA.[23] Although both methods have numerous benefits, one limitation is the proximity of lesions to large vessels as sufficient heating cannot be achieved due to the heat sink effect (local cooling due to nearby blood flow), which requires other forms of therapy.

The placement accuracy of therapeutic applicators is critical for procedure success due to the percutaneous nature of these procedures and the focal volume of therapy. Intended treatment ablation volumes are often expanded by 5–10 mm beyond the tumor extent to allow for some forgiveness in applicator placement and account for microscopic spread of disease not visualized by the observed gross volume. Although the minimally invasive nature of these procedures leads to faster patient recovery times, less complications, and an overall less traumatic experience for the patient, current approaches have resulted in local cancer recurrence with an observed range of 6–39% for patients treated for HCC or colorectal liver metastases.[14,24–26] This undesirable outcome and variability in performance has been associated with the main method of treatment delivery, namely, the guidance and placement of therapeutic applicators.

CLINICAL NEED FOR IMPROVED APPLICATOR GUIDANCE

Medical imaging is an invaluable tool that has been shown to support physicians in achieving desired diagnostic or therapeutic outcomes across the entire body with no exception for image-guided interventions in the management of liver cancer. Procedures that incorporate 3D information, such as computed tomography (CT) imaging,[27] magnetic resonance imaging (MRI),[28] and EM tool tracking,[29] have been shown to improve targeting accuracy, which leads to higher clinical success rates on first ablation attempts overall when compared to conventional techniques.[29,30] As the most commonly used imaging modality, CT provides a versatile and widely accessible utility that can span many aspects of ablation procedures. This includes preoperative identification of cancer (typically using an intravenous contrast agent),[31] intraoperative guidance for applicator insertions and verification of placement,[32] and postoperative follow-up for ongoing assessments.[33] MRI has also been used as it offers advantages such as improved soft-tissue contrast, nonionizing radiation, and the potential for real-time treatment monitoring.[28]

Although a variety of CT and MR-guided interventional systems have been developed, intraoperative access to the patient is often limited, the need for real-time imaging typically reduces the achievable resolution and contrast,[34] procedure times are long, and the costs of advanced imaging equipment reduce its availability and widespread adoption. The use of 2D US has the potential to respond to this demand due to the affordability, real-time imaging, and nonionizing nature of the modality, but conventional 2D US-guided procedures can be highly subjective and require a high level of training and experience to achieve consistent results. Simultaneous manipulation of

2D US probes and applicators while physicians mentally relate 2D US images on a limited field of view to locations inside patients may introduce image guidance variability and inevitably local cancer recurrence. Therefore, a clinical need exists for a cost-effective intraoperative approach that can help physicians improve image guidance to lesions for sufficient cancer therapy without local recurrence.

3D US FOR IMAGE-GUIDED LIVER TUMOR ABLATIONS

3D US is an alternative method for image guidance that can provide (nearly) real-time intraoperative imaging that is portable and it is able to provide multiplanar images for visualizing complex anatomy and focal therapy ablation applicators. Since the first 3D US image was demonstrated in the 1970s and the first commercial product (Kretz Combison 330) was launched in 1989, the quality of 3D US imaging has extensively improved. In US-guided liver tumor interventions, 3D US imaging has advantages over 2D US imaging for the following reasons:

- *Reduced operator-dependence*: Conventional 2D US imaging requires the operator to sweep the probe back and forth across the patient while mentally reconstructing the anatomy from multiple slices. 3D US imaging can acquire volumetric information and display the complex 3D anatomy directly.
- *Visualize arbitrary planes*: Interventional procedures may require an arbitrary image plane for guiding procedures more effectively. In 3D US imaging, unconventional planes can be visualized, such as pseudocoronal views and slices in-plane with angulated applicators.
- *Quantitative measurement*: Quantitative measurements in 2D US are based on heuristic formula with strong prior assumptions which are prone to error.[35] 3D US imaging provides a more accurate and reproducible way to measure the complex anatomy, thus improving physician confidence. During interventional procedures, this is helpful for operators to diagnose or compare with preoperative CT or MRI.

CLINICAL WORKFLOW OF 3D US-ASSISTED LIVER TUMOR ABLATION SYSTEMS

Similar to image-guided interventions as described by Peters' et al.,[36] the proposed intraoperative clinical workflow using 3D US-assisted liver tumor ablation systems can be divided into following points:

- 3D US acquisition
- Registration of 2D/3D US with preoperative CT/MRI
- Applicator placement planning
- Applicator insertion, monitoring, and tracking
- Assessment of overlap between ablation coverage and tumor extent
- Applicator adjustment, if necessary
- Treatment

In the following sections, this workflow and recent developments will be described, including systems for 3D US acquisition of liver, image-based applicator tracking/segmentation, registration of 2D/3D US with preoperative CT/MRI, and liver tumor coverage assessment.

3D US ACQUISITION OF THE LIVER

Development of reconstruction approaches and hardware designs has facilitated the growth of different 3D US acquisition systems. Fenster et al.[37,38] proposed a taxonomy of 3D US systems in terms

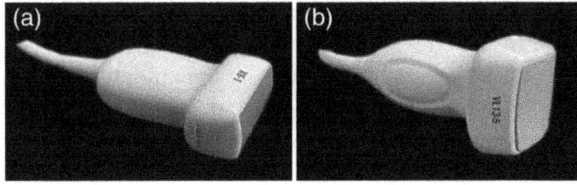

FIGURE 4.1 (a) Philips X6-1 2D array 3D US transducer. (b) Philips VL13-5 internal mechanical 3D US transducer.

of scanning methods and similarly, the acquisition of liver 3D US can be grouped into three categories: 3D US transducer, mechanical 3D US, and freehand 3D US.

1. When using a 3D US transducer, a 2D phased array of transducer elements is designed to enable stationary acquisition of 3D US. Although many commercially available 3D US transducers, such as the Philips X6-1 (shown in Figure 4.1a), have been applied to the liver, the additional high cost can limit the widespread use in focal liver tumor ablations and requires a hand to support the probe during the procedure.
2. Mechanical 3D US usually motorize an ordinary 1D array transducer along a predefined scanning trajectory to acquire a sequence of 2D US images. Knowledge of the precise position and orientation of the image planes can be monitored using mechanical encoders, reconstructing high-quality 3D US images. One commercial mechanical 3D US solution is Philips VL13-5 (Figure 4.1b) which uses motion within the transducer to acquire images. Some mechanical 3D US systems have been developed which demonstrate that external mounting and motion (Figure 4.2) can be used to create 3D US images.[39,40] This has the added benefit of using transducers already present clinically when using customizable 3D printed holders, reducing the potential cost of adding 3D US to the workflow and future-proofing the system for advancements in transducer technology. In contrast with the phased array, some mechanical 3D US systems can manipulate the geometry of scanning to accommodate variable clinical situations. Tilting the probe about a single point can reduce the physical footprint of the 3D scan to acquire intercostal images of the liver, avoiding reflection artifacts from ribs.[39] Conversely, translating the probe during tilting can create elongated hybrid geometries for larger fields of view, which is often beneficial for supporting better clinical decisions and facilitating software aids like image registration. These scanners can be paired with counterbalanced support systems, allowing physicians

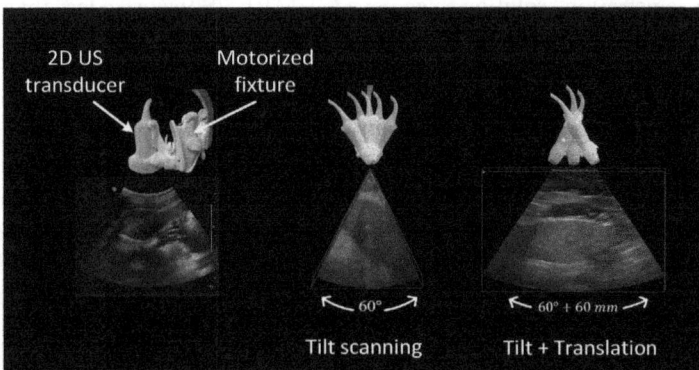

FIGURE 4.2 Mechatronic 3D US scanner. External motors are used to rotate and translate the transducer in variable motions to obtain geometrically variable 3D US images.

to acquire 3D US images for procedure planning and evaluation followed directly by guidance of the applicator insertion without switching the transducer.[41]

3. Due to the bulkiness of the mechanical scanning apparatus, freehand scanning approach has been investigated to build 3D US images from 2D US images. This approach can be beneficial as it has the potential to record arbitrary distances and orientations, which can be useful in some clinical cases such as long blood vessel or bone reconstructions.[42,43] Freehand scanning can be classified into two approaches: external tracking and image correlation.

  • External tracking devices, such as optical, EM, or visual tracking systems, can be rigidly mounted to conventional US transducers to continuously measure the position and orientation of each US image. External tracking devices are often associated with a high cost and additional environmental limitations that can hinder performance, such as line of sight constraints for optical or visual tracking systems and magnetic field distortion of EM systems when close to ferromagnetic materials. However, due to the appeal of freehand tracking, 3D US acquisition with EM tracking is becoming more popular than optical or visual systems as EM field generators are becoming more robust to ferromagnetic materials. One such approach is shown by Hennersperger et al.,[42] demonstrating the clinical feasibility of EM tracking-based systems when used in conjunction with associated software containing 3D US reconstruction and registration.

  • Image correlation relies on calculating the relative position and orientation between subsequent US image frames prior to 3D US reconstruction.[44] This image-based estimation of US transducer trajectories has been a challenging problem over many decades as out-of-plane motion is difficult to estimate robustly and accurately. Even though promising results have been observed using deep-learning methods,[45,46] robust clinical feasibility remains an ongoing problem.

LIVER US SYSTEMS

Commercially available systems have been developed to combine hardware and software innovations for interventional liver procedures. The most common approach, as found in the Virtual Navigator (Esaote, Genoa, Italy), Real-time Virtual Sonography (Hitachi Medical, Tokyo, Japan), and Emprint™ SX Ablation Platform with Aim™ navigation (Medtronic, Dublin, Ireland), uses external EM tracking systems to monitor the US probe and applicator during insertion. Image fusion with US is also possible with the Virtual Navigator (real-time PET/CT-US and contrast-enhanced US) to improve localization of actively uptaking lesions, especially when the target lesion is poorly visualized in B-mode US imaging. The PercuNav system (Philips Healthcare, the Netherlands) has provided an "auto-registration" solution to achieve CT/MRI-US fusion in less than 1 min, and their miniature EM tracking sensors attached to the tips of ablation applicators further improve localization intraoperatively. The PercuNav also takes advantage of EM tracking to acquire 3D US images to facilitate registration and treatment evaluation.

Mechanically assisted 3D US systems have been developed[39,40] that consist of a motor-driven 3D US scanning subsystem supported by a counterbalanced mechatronic arm on a portable cart (Figure 4.3). When compared to EM tracking systems, mechanical tracking systems offer a useful alternative as use of internal mechanical encoders is less susceptible to environmental conditions. Tracked mechatronic arms with EM brakes offer the potential for reproducible US probe placement, repeat 3D US imaging within the same coordinate space to monitor changes, and targeted image-guided procedures when paired with applicator guides (Figure 4.4). Known applicator guide trajectories can be superimposed onto digital real-time US images, facilitating workflows analogous to 3D US image-guided prostate biopsies for tumor targeting. Using software for image acquisition, visualization, and guidance, an approach to intraoperatively evaluate estimated ablation zone tumor coverage and reposition applicators has been developed. Through the use of phantom and liver

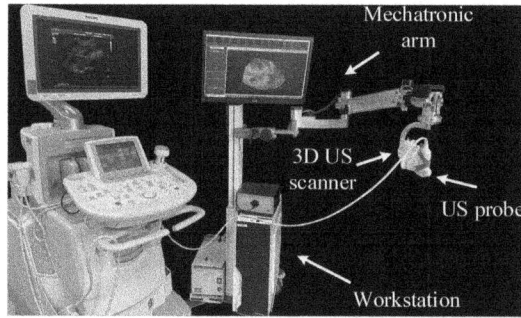

FIGURE 4.3 Mechanically assisted 3D US system capable of using commercially available 2D US imaging.

tumor ablation clinical patient trials, the feasibility of 3D US has been demonstrated by using this system and it supports the use of guidance systems to improve interventional procedures. Advanced image processing techniques, such as image segmentation and registration of preprocedural CT/MRI, have also been integrated into the mechatronic 3D US liver system, offering unique challenges and opportunities during interventional liver procedures, as will be described in subsequent sections.

APPLICATOR SEGMENTATIONS

In focal liver tumor ablations, complete tumor coverage by the ablation zone is required to achieve sufficient tumor eradication. Needle applicators for these procedures typically have a diameter of 1.5–2.8 mm (12–17 gauge) and range in length from 12 to 30 cm to accommodate for the variety of patient thicknesses and depth of lesions. Considering the ablation zone is generated around the applicator tip by physical or chemical effects, accurately identifying the applicator (including the shaft and tip) is crucial for applicator guidance, intraprocedural evaluation of tumor coverage, and overall clinical outcomes. Real-time 2D US is required during applicator insertion to assist with guidance and the unrestricted nature of the clinical procedure often results in a complex, changing

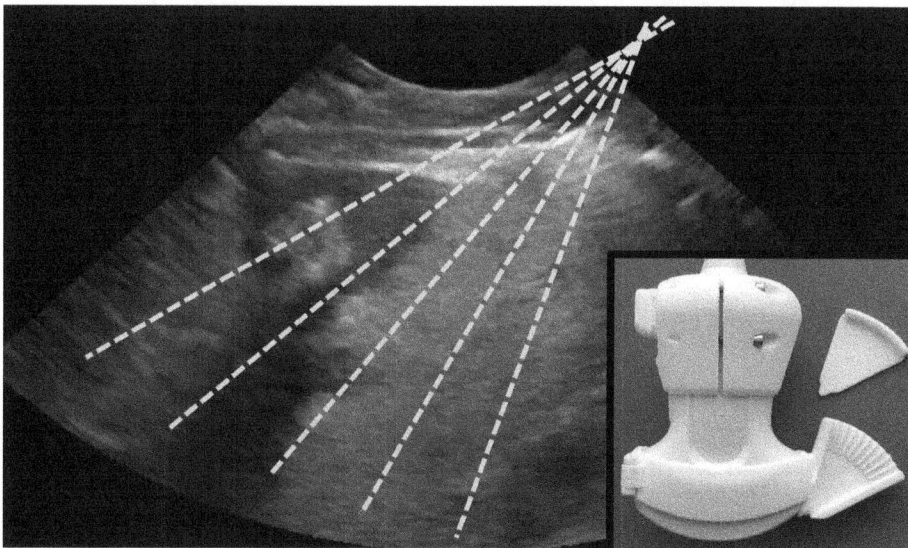

FIGURE 4.4 Example of using an applicator guide mounted to a 2D US transducer, with known insertion trajectories, to guide interventional liver procedures. Possible paths can be superimposed onto real-time images intraoperatively.

landscape as the applicators drift in and out of the image plane. Steep insertion angles of applicators cause variable acoustic reflections and visibility, increasing the probability of readjustment post insertion. Once guided to a location within the body, 3D US can be used to identify the applicator shaft and tip to provide the required information for a volumetric assessment of tumor coverage; however, quick and reproducible identification remains a challenging unmet clinical need. One solution for improving guidance in the 2D US image space, as well as improving verification in the 3D US image space, is advanced image processing in the form of applicator image segmentation.

CLINICAL CHALLENGES OF APPLICATOR IDENTIFICATION

- *Out-of-plane 2D imaging*: Monitoring real-time, in-plane 2D US images is essential for physicians to track applicator insertions. Manipulation of the US transducer and applicator occurs simultaneously during advancement to the tumor centroid and is highly dependent on the physician's experience. As a result, this dynamic process can lead to misidentified tumor centroid locations and undesirable applicator orientations using 2D US images. In most settings, out-of-plane US images can be easily identified when partial visibility of the applicator occurs on many image planes, which can be corrected by the physician. However, due to the elevational resolution of US and thickness of the applicator (Figure 4.5), US images displaying the "complete" applicator may falsely identify the central axis of the applicator. To further complicate the problem, many applicator tips are the "tri-bevel" shape (Figure 4.6) and can confuse the exact applicator tip location. Fortunately, additional spatial information and appreciation of the local vicinity can reduce applicator localization error caused by out-of-plane imaging when using 3D US images.
- *Patient tissue properties and steepness of the applicator*: Since US imaging depends on differences in acoustic impedance to visualize boundaries, structures with similar properties will blend together. Figure 4.7(a) shows one clinical case of an RFA applicator inserted into a tumor, which may have similar properties as the applicator and results in a partially invisible portion of the applicator inside the tumor. Another source of visibility degradation occurs when applicators are inserted parallel to the ultrasonic waves. These steep insertion angles reduce the amount of reflected information back to the US transducer, decreasing the signal observable over the background (Figure 4.7b).
- *Others*: Common US artifacts, such as side lobes, reverberation, and "Bayonet" artifact, can also affect the applicator's visibility, as described in Reusz et al.[47] From a software perspective of automatic applicator segmentation, the reverberation artifact may be the most common one to confuse modern algorithms due to the duplication of visualized instruments. Clinically, the physician often interprets these artifacts correctly and rarely misidentifies applicator localization.

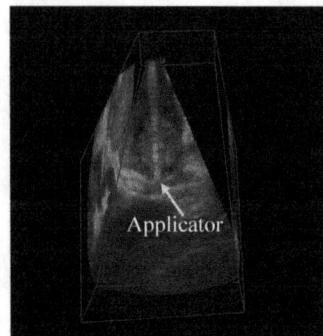

FIGURE 4.5 The thickness appearance problem of applicators in US images due to the elevational resolution.

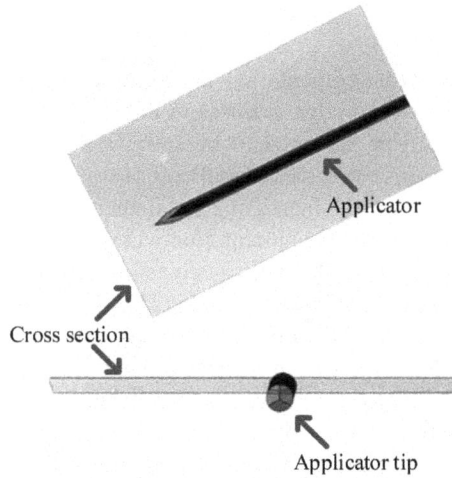

FIGURE 4.6 The pseudo-applicator tip ("tri-bevel tip") can be visualized in-plane during 2D US imaging (top) even if the central axis is out of plane (bottom).

RELATED WORK IN APPLICATOR SEGMENTATION

Applicator segmentation approaches have been investigated for many other US interventional procedures, such as in the prostate, breast, heart, and anesthetic administration. Unfortunately, many of these previously investigated techniques are not readily transferable since applicator insertions within the liver pose a unique problem due to the insertion requirements for depths up to 30 cm and large angles. Conventional algorithms need to be modified and with the rapid development of deep learning and its promising performance achieved in other segmentation tasks, deep-learning-based approaches that predominantly use convolutional neural networks (CNN) have been gradually investigated for applicator segmentation in US images. Thus, state-of-the-art work in applicator segmentation can be divided in two categories: classical and deep-learning-based methods.

In terms of the time relevance of input images, these methods can be further sub-classified into time-independent and time-dependent methods. In time-independent methods, the input is usually a static 2D or 3D US image and the timestamps of acquired images do not need to have any relationships for the method's performance. In contrast, the time-dependent method highly relies on the correlation between adjacent US images to localize the applicator. Due to the real-time demand for data acquisition, time-dependent methods have mostly focused on 2D US images.

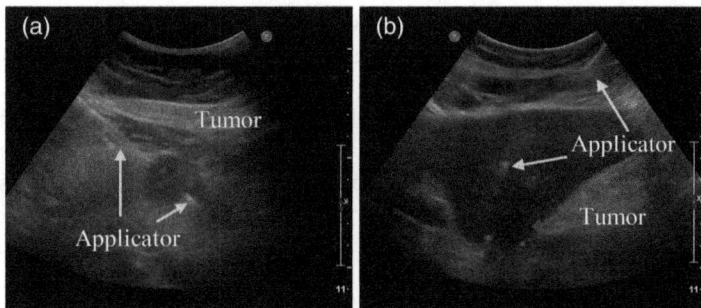

FIGURE 4.7 Clinical images of inserted radiofrequency ablation applicators demonstrating the invisibility challenges in the (a) tumor and (b) liver parenchyma.

Classical Methods

Time-independent Methods

The general hypothesis for applicator segmentation in a static US image is that the applicator has a relatively higher intensity with strong edge features in contrast to other neighboring structures. In most cases, the applicators can be assumed to be straight,[48] but longer insertions or thinner applicators have led to investigations into curved applicator segmentations.[49,50] Doppler US can be used when applicators are difficult to see, increasing the detectable signal when using an external vibration source. Similar to other common segmentation workflows, a pre-processing step is often necessary to extract or enhance applicator features, such as image thresholding,[51] Gabor filtering,[52] variance filters, and image normalization. Following this initial step, one or multiple basic models are applied to localize the applicator shaft and tip, such as Hough and Radon transform-based models developed for 2D and 3D US images.[52–54] These models can efficiently find the line-shaped object in a parameterized space, and for some cases with partially occluded applicators, this model can still robustly localize single or multiple applicators.

These methods are useful for determining an applicator's trajectory, but the applicator tip needs other techniques to increase identification accuracy. The random sample consensus (RANSAC) model has been shown as an alternative method to segment a single applicator.[55] This model is based on a voting procedure within a defined search radius, so it can not only localize the applicator shaft and tip accurately, but it also is less sensitive to speckle or noise in US images. The Otsu method can be used to determine the applicator tip location by maximizing the interclass variance between the assumed hyperintense applicator against the hypointense background. This method was evaluated on a clinical liver dataset of 3D US images and was able to provide verification of applicators in less than 0.3 s to a median accuracy of 5° trajectory error and 4 mm tip localization error.[56]

Time-dependent Methods

Unlike time-independent methods, time-dependent approaches use more advanced features extracted from adjacent US frames to facilitate applicator localization, such as spatiotemporal, spectral, and statistical features. Ayvali et al.[57] developed a pyramidal Lucas–Kanade optical flow-based tracking method to localize the applicator tip using a circular Hough transform. Considering a key assumption of an optical flow algorithm is based on a brightness constancy assumption, local image structure (i.e., edges and corners) can be used as features to be tracked. In situations when anatomical structures appear as straight lines (e.g., gallbladder or liver diaphragm boundaries) or unavoidable US artifacts are present (e.g., reverberation), applicator segmentation based on intensity information alone turns into a more challenging problem. This has motivated investigations into natural tremor motions to detect more robust feature descriptors for applicator segmentation based on minute displacements between adjacent US frames. Subtle displacements arising from tremor motion have extractable periodic patterns to indicate the possibility of applicator position. A block-based approach for needle trajectory detection was demonstrated by Beigi et al.[58] using spectral properties. After testing in-vivo, this approach was found too sensitive to US noise and led to further developments into their CASPER approach.[59] The strength of optical flow and differential flow methods was exploited again to derive spatiotemporal and spectral features from phase images using a sequence of US frames, resulting in applicator tracking once a support vector machine (SVM) model was trained. Since prior knowledge was still not incorporated in this approach, perturbation from minute tissue movement surrounding the applicator can degrade the quality of the segmentation. To counteract this, time-domain statistical features can be used to train an SVM with posterior probabilities[60] and the resulting probability map can be used with a Hough transform for applicator localization.

Deep-Learning-Based Methods

Time-independent Method

Supervised deep-learning techniques using pre-identified images to generate training, validation, and testing datasets are the most common approach for segmenting applicators in most medical images, including liver US images. One of the earliest approaches was demonstrated by Hacihaliloglu et al.[61] using a faster region-based CNN approach to detect the applicator for in-plane 2D US images. Since the learning model can only provide a bounding box to localize the applicator, trajectory and tip determination still requires the use of conventional features, which is computationally expensive. Gillies and Rodgers et al.[62] proposed a modified UNet-based model to segment needle-shaped interventional instruments across many common interventional procedures, such as liver tumor ablation, kidney biopsy and ablation, and prostate and gynecologic brachytherapy, in real-time 2D US images. Largest island with linear fit and RANSAC methods were applied to demonstrate the capacity for real-time segmentation performance. Again, these methods highly relied on spatial features (i.e., image intensity and shape) to localize the applicator, so these time-independent methods also struggle with imperceptible or low-intensity applicator appearances.

In 3D US, extension of common deep-learning architectures for 2D US can be applied. CNNs have been shown to be effective with high accuracy when using locally extracted raw data from three orthogonal planes, even when the applicator appears small relative to the background.[59] When using a CNN to output the cross-sectional, sometimes oblique, US image plane containing the full-length applicator, RANSAC models can be applied to determine the applicator localization. To address the computational limitation of fully convolutional NNs (FCNs), Yang et al.[63] proposed a patch-of-interest (POI)-FuseNet framework, which consists of a POI selector for coarse segmentation and a FuseNet for finer classification. To make use of 2D and 3D FCN features to hierarchically extract contextual information, two-channel modes in the FuseNet can be used. One is a slice-based 2D UNet framework while the other is a multi-scale 3D UNet framework. Their results, including applicator segmentation accuracy and computation time, showed great potential in clinical use.

Time-dependent Methods

Information incorporated from consecutive image frames can be used to enhance computationally efficient deep-learning algorithms. Needle tip identification can be improved by augmenting images with a spatial total variation regularization denoising image processing step and passing to a real-time object identification architecture such as the you-only-look-once (YOLO) network.[64] Learning and identifying motion artifacts from breathing and arterial pulsation can further improve robustness by using a long short-term memory (LSTM)-based framework to localize the applicator tip.[65] When used in a coarse-to-fine search strategy, the YOLO framework can also be designed to detect invisible applicators after a coarse detection step using more robust spatiotemporal features, such as those generated using speckle dynamics from consecutive 2D US frames.[66] Some groups have investigated the use of robot-assisted applicator insertions, which provide another set of unique conditions that can be exploited since the applicator slowly advances into the target along an approximately straight line compared to freehand approaches. Chen et al.[67] proposed a two-channel encoder and single-channel decoder network architecture for this application using two adjacent US frames, demonstrating the potential for more automated clinical workflows. Analogous to the UNet in 2D applications, the VNet offers potential for automatic applicator detection in 3D US images. When used in a time-dependent workflow, the preceding time step can inform selection of the most probable applicator during segmentation.[68] Inspired by conventional 2D US visualization, two perpendicular cross-sectional planes (i.e., parallel and perpendicular to the transducer) of the 3D US image containing full-length applicators can then be displayed, visualizing a complete applicator for improving placement verification.

APPLICATOR SEGMENTATION SUMMARY

Tracking needle-shaped applicators intraprocedurally is a common clinical need in US-guided liver tumor ablation procedures to improve guidance for providing optimal therapy, while minimizing risks to healthy tissue and avoiding adverse events like local cancer recurrence. In commonly used 2D US imaging, time-dependent and deep-learning-based approaches have significantly improved applicator segmentation. Taking advantage of prior US frames can enable memory functions to localize the applicator, which can improve the robustness of the algorithm when used clinically. State-of-the-art work can not only identify human imperceptible applicator tip locations, but also can address challenging artifacts in routine ablation procedures, such as abrupt changes of image appearance due to patient movement, breathing, and pulsation. Correctly identifying the applicator shaft is also critical for guidance, which can be helpful to overlay therapeutic ablation zone information for providing real-time tumor coverage evaluation. Although most applicator segmentation work in the liver has been focused on 2D US, 3D US-based applicator segmentations are being investigated for improved insertion verification using static 3D US images. Robot-assisted procedures do not have the same challenge as real-time 3D display for physicians, so real-time 3D US guidance could be investigated further in the future to exploit the increase amount of information without the complication of 3D display. Unfortunately, access to large 3D US liver datasets still hinders development and validation of the clinical robustness of 3D US-based methods. Overall, recent efforts focusing on automatic image-based applicator segmentation in 2D and 3D US images have achieved promising success and we believe this low-cost solution has the potential for widespread accessibility for liver tumor ablation procedures.

IMAGE REGISTRATION OF 2D/3D US WITH PREPROCEDURAL CT/MRI

Image registration has been a fundamental and challenging task in image-guided interventions with the developments of modern imaging and the demand for incorporating more information into clinical workflows. The process of image registration refers to bring two or more images into the same spatial coordinate system and in liver tumor ablation procedures, registration predominantly means to align diagnostic images (CT or MRI) of the patient with intraprocedural US images (Figure 4.8). Since tumors have different visibility and contrast across multiple image modalities, image registration offers a utility to address clinical scenarios when the liver tumor may present with very poor to zero visibility under current US-guided focal liver tumor ablation procedures.

IMAGE REGISTRATION CLINICAL CHALLENGES

1. *Multimodality registration*: At the beginning of an image-guided liver procedure, an initial US screening is usually performed to identify the tumor and its location. Unfortunately, when tumor visibility is poor, it is difficult for the physician to measure the tumor size and shape accurately, not to mention subsequent guiding of the applicator insertion. Since the liver tumor can usually present better in CT or MRI than in US imaging, the physician will conventionally replace US with CT or MRI to perform the procedure. This often requires access to an image-guidance surgical suite and is often not possible, especially in developing countries, considering the limited availability of expensive imaging equipment and suites. Contrast-enhance US has an increasing potential to improving visibility,[69] but image registration techniques between the US and diagnostic CT/MRI images could be a cost-effective method to provide complementary information from multiple modalities for improving intraoperative guidance.

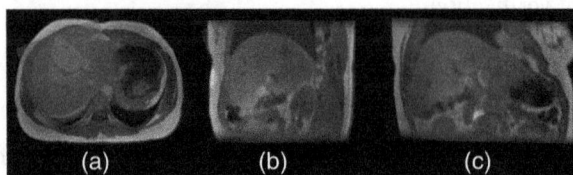

FIGURE 4.8 US-MRI registration. (a) Axial view (b) Coronal view. (c) Sagittal view.

2. *Real-time targeting*: Although improving visibility can aid with identification and local-ization, the physician still needs to be adaptive and monitor for changing positions of the liver tumor and applicator in real time. Real-time monitoring is not usually a challeng-ing problem for cases with good US visibility, but when multimodal images are used for cases with ultrasonically invisible or poorly visible tumors, repeat static images can often interrupt clinical workflows and extend procedure times proportional to the number of readjustments. Real-time registration of 2D US with CT or MRI could offer improved vis-ibility while advancing the applicator; however, multimodality (US to CT/MRI) and multi-dimensionality registration (2D to 3D) is currently one of the most challenging problems for medical image registration.

RELATED WORK IN IMAGE REGISTRATION

Similar to applicator segmentation, image registration can be broadly categorized into classical and deep-learning methods. Commonly, the taxonomy of rigid (up to six degrees-of-freedom) and nonrigid registration has been used to specify the level of deformation an algorithm is allowed to manipulate. Classical methods can be subcategorized based on their feature space (i.e., shape feature-based or intensity-based) and differentiate using their transformation model, optimization strategy, and similarity metric.

Classical Methods

1. *Feature-based registration*: In the context of liver tumor ablation, the most significant structures in US imaging are the liver surface and major internal vessels, such as the hepatic and portal veins. Anatomically, these internal liver structures are often assumed to be uniformly distributed and allow for image registration deformation estimations for the liver as a whole between 3D US and MRI or CT.[70] After identifying the vessel features in the image, a common approach is to develop a similarity metric to optimize the vessel cen-terline match between two images for image registration. To align the extracted centerline features from both modalities, iterative closet point (ICP),[71,72] coherent point drift (CPD),[73] and their variants[74,75] are widely applied. The simplicity of an ICP approach can offer advantages due to its low computational complexity,[74,76] but considering the CPD principle regards the alignment of two centerline point sets as a probability density estimation prob-lem, CPD-based optimization has gained much attention as the approach is quite robust to the variance of the segmented vessel.[77,78] Image registration using centerline features can also be refined using a live surface rendering to enhance the image alignment.[74,79] Nazem et al.[79] developed a two-stage incremental-based method to register CT and US images by using an ICP algorithm to achieve point-based alignment, followed by a Kalman filter approach to overcome the noise and outliers. Liver bifurcations are another commonly used feature to provide acceptable registration accuracy on its own or as an initial align-ment before a complicated fine-tuned optimization.[80,81]

2. *Intensity-based registration*: Feature-based registrations are often highly reliant on time-consuming image pre-processing steps, which hinder further clinical application in liver tumor ablation procedures to some extent. Therefore, intensity-based approaches based on the brightness of visualized structures have been developed extensively to automate the image registration workflow.

   • Initially, developers directly used image intensity as the feature to perform multimodal image registration. Similarity metrics turned out to be the key factor in achieving a reasonable registration result, typically using a correlation ratio (CR)[82] or mutual infor-mation (MI).[83,84]

   • Several investigators developed a "remapping" approach to enhance registration by converting both image modalities to a third, artificial modality. This is advantageous as commonly used similarity metrics in mono-modality registration problems can be repurposed to the intermediate modality. Penney et al.[85] mapped US and MR images to an intermediate vessel probability representation, followed by a registration carried out using normalized cross-correlation. Wachinger et al.[86] proposed a structural representation called Laplacian images which had superior theoretical properties to identify similar internal structures from multiple modalities.

- Instead of creating an artificial modality, changing a multimodal registration to a monomodal registration problem can also be performed by choosing a desired dominant image domain and converting the matching image to a pseudo-version first. Pseudo-US images can be generated from CT or MR images to reproduce major US imaging properties through intensity mapping[87,88] or other similarity metrics such as the linear correlation of linear combination (LC^2) approach.[89]

- Local structure descriptors were investigated with the idea that multimodal images can be transformed into a representation that is independent of the original imaging modality. Heinrich et al.[90] first proposed a modality-independent neighborhood descriptor (MIND) based on the concept of local self-similarities (LSS)[91] and later introduced a "self-similarity context" (SSC) image descriptor.[92] Yang et al.[93] proposed a local structure orientation descriptor (LSOD) based on intra-image similarity to address considerable modality differences.

Deep-Learning-Based Methods

Recently, with the rapid advancement of deep-learning techniques, several learning-based methods have been developed. Learning-based methods are less computationally expensive for registration due to the pre-trained transformation model, which can significantly benefit the real-time demand of liver tumor ablation procedures. Even though some conventional multimodal registration approaches have achieved clinically acceptable alignments, expensive computations have been the main factor that hinder clinical application.

The generality of the transformation model highly relies on the training data. For example, the trained model based on a local database may be challenging to achieve great results to other unseen data at other facilities. Therefore, compared to classical registration methods, the database is often a critical component when comparing learning-based approaches and many methods can be differentiated by their (1) feature space, (2) similarity metric, (3) transformation model and regularization, and (4) database.

1. *Establishing a feature space*: Extracting representative and invariant features from different modalities is critical to enable accurate and robust image registrations. In addition to directly using common features inspired by classical registration approaches (vessel centerlines, image intensity, etc.), high-level implicit features can be extracted to design more robust registration methods when using networks like the fully connected network, UNet, and residual network. Point correspondence between different image modalities, motivated by visual odometry in computer vision, is one approach that can be developed based on 2D and 3D UNet architectures. Markova et al.[94] demonstrated that a dense keypoint descriptor can be developed to facilitate a fast, generic, and fully automatic registration approach between 2D US and 3D CT in liver tumor ablation procedures.

2. *Similarity metricv*: The similarity metric aims to measure the matching accuracy between image pairs and is typically used to construct a cost function for optimizing the transformation model. As noted previously, many different similarity metrics have been proposed, such as MI, CR, sum of squared distance (SSD), and cross-correlation (CC) with variants. However, these similarity metrics may not be robust enough to achieve satisfying results in many scenarios as spatial and temporal variability in the intraprocedural imaging, partly due to user dependency, is complex to summarize with simple statistical properties or information theory-based measures.

Different physical acquisitions have the potential to generate statistical correlation between imaging structures that simply do not correspond to the same anatomical structures,[95] so instead of using hand-crafted similarity metrics, neural networks can be used to extract high-dimensional representative features to generate custom similarity metrics. Simonovsky et al.[96] proposed learning a general similarity metric through a classification-based CNN by discriminating between aligned and misaligned patches from different modalities, which was shown to outperform an MI similarity metric. Belghazi et al.[97] presented a robust mutual information neural estimator (MINE) that is trainable through back propagation and linearly scalable in dimensionality and sample size, showing great potential to improve multimodal registration. Interestingly, even though the similarity metric and optimizer are a significant part of training deep-learning networks, some groups have shown it can potentially be avoided altogether. Sun et al.[98] and Walsum et al.[99] proposed a displacement vector network without explicitly selecting a similarity metric and were able to demonstrate success on mono-modal and simulated multimodal image datasets, but multimodal image registration on clinical data still remains a challenge.

3. *Transformation model and regularization*: Over the past few decades, transformation models in classical registration methods have been developed extensively, but these mature models cannot directly transfer into deep-learning frameworks. In the beginning, CNN-based methods that did not consider any characteristics of physical models learned their simple transformation models through supervision of ground truth data. Reasonable results could sometimes be achieved, but robustness was often not guaranteed. This led to further development of more advanced transformation models, such as the general adversarial network (GAN)-based model and the reinforcement learning (RL)-based model. In general, GANs train two networks simultaneously by producing inferences with a generator network in an adversarial setting where a discriminator network outputs likelihood probabilities. For an image registration problem, many investigators[100] proposed using the generator network as the transform model and the discriminator network as the similarity metric to recover a more complex range of deformations. In RL-based methods, instead of performing a single regression that maps the raw image data to registration parameters, the registration task can be decomposed into a sequence of classification problems. Although these above approaches aim to learn a general model that can follow the imaging characteristics of more traditional physical models, such as typology-preserving or diffeomorphism, there is no existing method to confirm their functions. In order to make transformation models more explainable and follow certain characteristics, Dalca et al.[101] and Krebs et al.,[102] almost at the same time, proposed a probabilistic generative model for diffeomorphic registration based on variational inference. Their work successfully built a connection between classical methods and learning-based methods to achieve state-of-the-art results.

4. *Database*: As goes the saying of "*there is no such thing as perfect*", how to collect enough data for training can be quite challenging with our current understanding. To make matters worse, training data must often be manually aligned by experts to provide a known solution, which is difficult in reality. Images of phantoms can be useful for exploration and development in an ideal setting, but often times these images do not capture the complexity of clinical scenarios and nearly all methods decline in performance when used clinically. Although artificially augmenting clinical data could guarantee obtaining sufficient amounts of data, whether those data could reflect the generality of real clinical data has been controversial. Thus, the highest chance for clinical success is currently dependent on a trusted clinical image dataset with accurate registrations for training. Currently, registration models based on weakly labeled data[95] or no labeled data[101] have demonstrated feasibility and accuracy that is comparable or better than classical methods, but these solutions are not yet mature.

Summary

In classical registration approaches, feature-based multimodal registration methods are still one of the most widely used solutions in clinical and research fields, especially when 3D US is available. In contrast with some intensity-based registration methods, extracted liver features are more robust and the optimization strategy is computationally inexpensive. However, since current feature-based registration methods are highly reliant on the success of feature extraction, which is quite time-consuming on its own, it is essential to automate liver vessel segmentation, center line and bifurcation identification, or liver surface segmentation steps for clinical practicality. In addition, most feature-based methods assume structures are present in both modalities, but due to different fields of view and different imaging properties, this requirement may be difficult to meet between US and CT/MRI. Intensity-based multimodal registration approaches have been developed significantly and have shown promising results; however, considering robustness and computation time, this solution has not been widely employed clinically. Deep-learning-based approaches are young and active area of investigation, so clinical applications are actively being investigated. Fortunately, deep-learning-based approaches have achieved quite promising results and more importantly, the limitation of expensive computation in classical registration field can be addressed. With much more image data being collected and the advancement of each registration component, we believe deep-learning-based registration will have a big impact on liver tumor ablation procedures in the future.

Liver Tumor Coverage Assessment

In focal liver tumor ablation procedures, one or multiple applicators are percutaneously inserted into the tumor and the applicator tip is heated or frozen to induce irreversible cellular death.[103–105] Clinically, to ensure complete tumor eradication, not only does the targeted focal tumor need to be entirely covered by the ablation zone, but a safety margin of 5–10 mm is also required to avoid the occurrence of the residual tumor.[106,107] During the procedure, the physician usually estimates the ablation zone using thermal properties of therapy applicators and settings on the ablation system, which are provided by the manufacturers of the ablation devices.[108] Therefore, the estimated tumor coverage by the ablation zone depends on knowledge of the position of the applicator with respect to the tumor margins.

2D US-based Tumor Coverage Evaluation

As mentioned previously, 2D US is a commonly used imaging modality to guide the applicator insertion as it provides real-time imaging. It is widely available, has a small footprint, and is relatively inexpensive.[109–111] In conventional focal liver tumor ablation procedures, a single in-plane US image will usually be captured once applicators have been inserted into a satisfactory position to perform tumor coverage evaluation. Clinically, to balance the processing time spent on evaluation and procedure efficacy, the ablation margin is measured by calculating the distance between a few feature points (two or three pairs) manually picked on the tumor and virtual ablation zone boundaries. However, the evaluation based on these chosen points may not be sufficient in 2D US, requiring the physician to stop the ongoing procedure to obtain the ablation margin. To reduce the physicians' burden and create a smoother workflow, augmenting the 2D in-plane US image with the virtual ablation zone can provide a more intuitive feeling of tumor coverage (Figure 4.9a). The real-time mapped virtual ablation zone can better assist applicator guidance and readjustment, especially for inexperienced physicians, further improving clinical patient throughput.

To achieve real-time tumor coverage estimation, the applicator first needs to be tracked in real time before the associated ablation zone can be virtually overlaid. Therefore, accurate applicator segmentation in 2D US images is essential for integrating real-time tumor coverage evaluation.

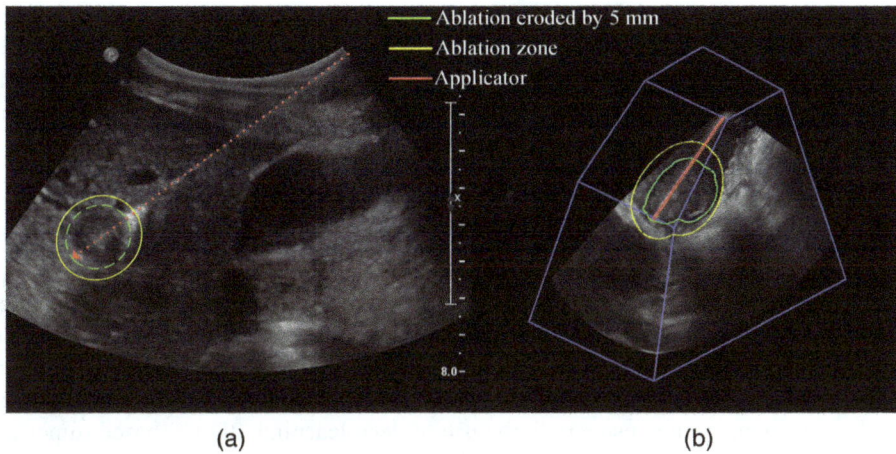

FIGURE 4.9   2D (a) and 3D (b) US-based tumor coverage evaluation.

As previously discussed, pure image-based applicator segmentation is still at an early stage, hence we have not seen much related work based on 2D US imaging alone. Fusion of 2D US images with preprocedural CT or MRI has been employed by several groups to improve intraprocedural tumor coverage evaluation.[93,112]Makino et al.[113] demonstrated the feasibility of US-CT/MRI fusion imaging for quantitative intraprocedural RFA evaluation. In their study, the target tumor volume with a virtual safety margin is overlaid on real-time US images to assess tumor coverage in combination with contrast-enhanced US. External devices or sensors (EM tracking, optical tracking, and inertial measurement unit) have also been used for assisting applicator localization with the intention to benefit tumor coverage evaluation,[44] but this 3D pose information is better used when 3D US is employed.

3D US-Based Tumor Coverage Evaluation

Using acquired 2D US images alone may not be sufficient to provide volumetric coverage information of the entire tumor, particularly for cases with multiple applicators, as the estimated ablation zone is determined by the applicators' 3D spatial configuration (Figure 4.9b). For cases with irregular tumor shapes, incomplete tumor coverage may occur, which may only be detected well after the treatment. Park et al.[114] showed the feasibility of fusing 2D US with 3D US images for RFA guidance in liver cancer. In contrast to the use of a fusion approach, Stippel et al.[115] evaluated the ablation volume and diameter using an EM-tracked 3D guidance tool. Results showed that compared to the use of conventional 2D US, their approach could facilitate the exact placement of the applicator in the tumor. However, all these approaches depend on the accurate registration of intraprocedural 2D US with another modality with 3D information. In addition, diagnostic CT/MRI images are usually obtained up to a month before the procedure, which may result in inaccuracy due to changes in the size and shape of the tumor during that month.

To address these issues, the use of intraprocedural 3D US[37,116,117] could be a promising solution, which can directly help to identify the applicator location in the tumor, especially when facilitated with an applicator segmentation algorithm. Rose et al.[118] and Xu et al.[119] showed that 3D US, particularly with an interactive multiplanar reformatted image display to show the quasi-coronal view, provides useful positional and configurational information in 3D space. Sjølie et al.[120] used an optically tracked US probe to acquire 2D US images for 3D US reconstruction and subsequently guide the applicator placement. By using multiplanar reformatted views, the physical off-center distance was evaluated intraprocedurally to assist the adjustment of the applicator. Xing et al.[41]

further expanded on a mechatronic arm-supported 3D US liver tumor ablation system to employe surface- and volume-based metrics for evaluating tumor coverage and to identify any tumor region undertreated. They also proposed an intuitive visualization technique called margin uniformity that shows the quantitative ablation margin information and visualizes the spatial relative information between the applicator and the liver tumor using a 2D plot alone.

Summary

Assessing the expected treatment volume coverage intraoperatively is a unique opportunity in liver ablation procedures that can help a physician decide if an applicator position is placed appropriately prior to initiating an ablation treatment. 2D US-based tumor coverage evaluation could be an approach that would be useful during real-time operation, but whether the use of external tracking devices can be clinically acceptable is still under investigation. However, advancements in real-time 2D US applicator segmentation, especially the use of deep learning, 2D US-based tumor coverage evaluation is showing growing signs of clinical feasibility. Tumor coverage evaluation purely relying on 2D US images can be easily integrated with clinically available US machines, as long as US image data can be accessed. The feasibility and clinical applicability of 3D US has also been shown to be helpful when assessing volumetric tumor coverage evaluations intraoperatively. Since there is no conflict between 2D and 3D US-based tumor coverage evaluation, their combination could provide required complementary information to enhance the efficacy of liver tumor ablation procedures.

CONCLUSION

In liver tumor ablation, conventional procedures are often highly user-dependent and can lead to local cancer recurrence when sufficient guidance is not achieved. 3D US can act as an intermediate, even sole, image guidance modality to improve the overall clinical workflow during planning, guidance, and verification of applicator placements within the body. Hardware and advanced image-processing software advancements for 3D US acquisition, applicator segmentation, image registration, and tumor coverage assessment are potential avenues to facilitate widespread adoption and increase the scope of modern ablation procedures. With the rapid development of deep-learning techniques, applicator segmentation and image registration are significantly improving in performance to facilitate new paradigms in the way ablation procedures are envisioned. Although most 3D US-assisted liver systems are still in the prototype stage, combinations of technologies are driving innovative techniques for reducing local cancer recurrence, increasing the potential complexity of procedures, and providing greater access to quality healthcare around the globe.

REFERENCES

1. Botero AC, Strasberg SM. Division of the left hemiliver in man: Segments, sectors, or sections. *Liver Transplant Surg.* 1998;4(3):226–231. doi:10.1002/LT.500040307
2. Couinaud C. Liver anatomy: Portal (and suprahepatic) or biliary segmentation. *Dig Surg.* 1999;16(6):459–467. doi:10.1159/000018770
3. Sung H, Ferlay J, Siegel RL, et al. Global cancer statistics 2020: GLOBOCAN estimates of incidence and mortality worldwide for 36 cancers in 185 countries. *CA Cancer J Clin.* 2021;71(3):209–249. doi:10.3322/CAAC.21660
4. Chuang SC, Vecchia CL, Boffetta P. Liver cancer: Descriptive epidemiology and risk factors other than HBV and HCV infection. *Cancer Lett.* 2009;286(1):9–14. doi:10.1016/J.CANLET.2008.10.040
5. El-Serag HB, Rudolph KL. Hepatocellular carcinoma: Epidemiology and molecular carcinogenesis. *Gastroenterology.* 2007;132(7):2557–2576. doi:10.1053/J.GASTRO.2007.04.061
6. Hesseltine CW, Shotwell OL, Ellis JJ, Stubblefield RD. Aflatoxin formation by *Aspergillus flavus*. *Bacteriol Rev.* 1966;30(4):795–805. doi:10.1128/BR.30.4.795-805.1966

7. Zhou H, Liu Z, Wang Y, et al. Colorectal liver metastasis: Molecular mechanism and interventional therapy. *Signal Transduct Target Ther.* 2022;7(1):1–25. doi:10.1038/s41392-022-00922-2

8. Martin J, Petrillo A, Smyth EC, et al. Colorectal liver metastases: Current management and future perspectives. *World J Clin Oncol.* 2020;11(10):761. doi:10.5306/WJCO.V11.I10.761

9. Fitzmorris P, Shoreibah M, Anand BS, Singal AK. Management of hepatocellular carcinoma. *J Cancer Res Clin Oncol.* 2015;141(5):861–876. doi:10.1007/S00432-014-1806-0

10. El-Serag HB, Marrero JA, Rudolph L, Reddy KR. Diagnosis and treatment of hepatocellular carcinoma. *Gastroenterology.* 2008;134(6):1752–1763. doi:10.1053/J.GASTRO.2008.02.090

11. Schoenberg MB, Bucher JN, Vater A, et al. Resection or transplant in early hepatocellular carcinoma: A systematic review and meta-analysis. *Dtsch Arztebl Int.* 2017;114(31-32):519. doi:10.3238/ARZTEBL.2017.0519

12. Llovet JM, Ricci S, Mazzaferro V, et al. Sorafenib in advanced hepatocellular carcinoma. *N Engl J Med.* 2008;359(4):378–390. doi:10.1056/NEJMoa0708857

13. Cheng AL, Guan Z, Chen Z, et al. Efficacy and safety of sorafenib in patients with advanced hepatocellular carcinoma according to baseline status: Subset analyses of the phase III sorafenib Asia-Pacific trial. *Eur J Cancer.* 2012;48(10):1452–1465. doi:10.1016/J.EJCA.2011.12.006

14. Dawson LA. Overview: Where does radiation therapy fit in the spectrum of liver cancer local-regional therapies? *Semin Radiat Oncol.* 2011;21(4):241–246. doi:10.1016/J.SEMRADONC.2011.05.009

15. Brock KK. Imaging and image-guided radiation therapy in liver cancer. *Semin Radiat Oncol.* 2011;21(4):247–255. doi:10.1016/J.SEMRADONC.2011.05.001

16. El-Serag HB. Hepatocellular carcinoma. *N Engl J Med.* 2011;365(12):1118–1127. doi:10.1056/NEJMra1001683

17. Llovet JM, Real MI, Montaña X, et al. Arterial embolisation or chemoembolisation versus symptomatic treatment in patients with unresectable hepatocellular carcinoma: A randomised controlled trial. *Lancet.* 2002;359(9319):1734–1739. doi:10.1016/S0140-6736(02)08649-X

18. Wang X, Hu Y, Ren M, Lu X, Lu G, He S. Efficacy and safety of radiofrequency ablation combined with transcatheter arterial chemoembolization for hepatocellular carcinomas compared with radiofrequency ablation alone: A time-to-event meta-analysis. *Korean J Radiol.* 2016;17(1):93–102. doi:10.3348/KJR.2016.17.1.93

19. Zhong C, Guo R, Li J, et al. A randomized controlled trial of hepatectomy with adjuvant transcatheter arterial chemoembolization versus hepatectomy alone for stage III: A hepatocellular carcinoma. *J Cancer Res Clin Oncol.* 2009;135(10):1437–1445. doi:10.1007/s00432-009-0588-2

20. Sangro B, Carpanese L, Cianni R, et al. Survival after yttrium-90 resin microsphere radioembolization of hepatocellular carcinoma across Barcelona clinic liver cancer stages: A European evaluation. *Hepatology.* 2011;54(3):868–878. doi:10.1002/HEP.24451

21. Ahmed M, Solbiati L, Brace CL, et al. Image-guided tumor ablation: Standardization of terminology and reporting criteria—A 10-year update. *J Vasc Interv Radiol.* 2014;25(11):1691–1705.e4. doi:10.1016/J.JVIR.2014.08.027

22. Wells SA, Hinshaw JL, Lubner MG, Ziemlewicz TJ, Brace CL, Lee FT. Liver ablation: Best practice. *Radiol Clin.* 2015;53(5):933–971. doi:10.1016/J.RCL.2015.05.012

23. Brace CL. Radiofrequency and microwave ablation of the liver, lung, kidney, and bone: What are the differences? *Curr Probl Diagn Radiol.* 2009;38(3):135–143. doi:10.1067/J.CPRADIOL.2007.10.001

24. Tanis E, Nordlinger B, Mauer M, et al. Local recurrence rates after radiofrequency ablation or resection of colorectal liver metastases. Analysis of the European Organisation for Research and Treatment of Cancer #40004 and #40983. *Eur J Cancer.* 2014;50(5):912–919. doi:10.1016/J.EJCA.2013.12.008

25. Sung NH, Lee SY, Moon SC, et al. Comparing the outcomes of radiofrequency ablation and surgery in patients with a single small hepatocellular carcinoma and well-preserved hepatic function. *J Clin Gastroenterol.* 2005;39(3):247–252. doi:10.1097/01.MCG.0000152746.72149.31

26. Siperstein AE, Berber E, Ballem N, Parikh RT. Survival after radiofrequency ablation of colorectal liver metastases: 10-year experience. *Ann Surg.* 2007;246(4):559–565. doi:10.1097/SLA.0B013E318155A7B6

27. Passera K, Selvaggi S, Scaramuzza D, Garbagnati F, Vergnaghi D, Mainardi L. Radiofrequency ablation of liver tumors: Quantitative assessment of tumor coverage through CT image processing. *BMC Med Imaging.* 2013;13(1):3. doi:10.1186/1471-2342-13-3

28. Clasen S, Pereira PL. Magnetic resonance guidance for radiofrequency ablation of liver tumors. *J Magn Reson Imaging.* 2008;27(2):421–433. doi:10.1002/JMRI.21264

29. Krücker J, Xu S, Glossop N, et al. Electromagnetic tracking for thermal ablation and biopsy guidance: Clinical evaluation of spatial accuracy. *J Vasc Interv Radiol.* 2007;18(9):1141–1150. doi:10.1016/J.JVIR.2007.06.014

30. Liu F, Liang P, Yu X, et al. A three-dimensional visualisation preoperative treatment planning system in microwave ablation for liver cancer: A preliminary clinical application. *Int J Hyperth*. 2013;29(7):671–677. doi:10.3109/02656736.2013.834383

31. Solomon SB, Silverman SG. Imaging in interventional oncology. *Radiology*. 2010;257(3):624–640. doi:10.1148/radiol.10081490

32. Leng S, Christner JA, Carlson SK, et al. Radiation dose levels for interventional CT procedures. *Am J Roentgenol*. 2011;197(1):W97–W103. doi:10.2214/AJR.10.5057

33. Sainani NI, Gervais DA, Mueller PR, Arellano RS. Imaging after percutaneous radiofrequency ablation of hepatic tumors: Part 1, normal findings. *Am J Roentgenol*. 2013;200(1):184–193. doi:10.2214/AJR.12.8478

34. Yutzy SR, Duerk JL. Pulse sequences and system interfaces for interventional and real-time MRI. *J Magn Reson Imaging*. 2008;27(2):267–275. doi:10.1002/JMRI.21268

35. Prager RW, Ijaz UZ, Gee AH, Treece GM. Three-dimensional ultrasound imaging. *Proc Inst Mech Eng Part H J Eng Med*. 2010;224(2):193–223. doi:10.1243/09544119JEIM586

36. Cleary K, Peters TM. Image-guided interventions: Technology review and clinical applications. *Annu Rev Biomed Eng*. 2010;12(1):119–142. doi:10.1146/annurev-bioeng-070909-105249

37. Fenster A, Downey DB. 3-D ultrasound imaging: A review. *IEEE Eng Med Biol Mag*. 1996;15(6):41–51. doi:10.1109/51.544511

38. Fenster A, Downey DB, Cardinal HN. Three-dimensional ultrasound imaging. *Phys Med Biol*. 2001;46(5):R67. doi:10.1088/0031-9155/46/5/201

39. Gillies DJ, Bax J, Barker K, Gardi L, Kakani N, Fenster A. Geometrically variable three-dimensional ultrasound for mechanically assisted image-guided therapy of focal liver cancer tumors. *Med Phys*. 2020;47(10):5135–5146. doi:10.1002/MP.14405

40. Neshat H, Cool DW, Barker K, Gardi L, Kakani N, Fenster A. A 3D ultrasound scanning system for image guided liver interventions. *Med Phys*. 2013;40(11):112903. doi:10.1118/1.4824326

41. Xing S, Romero JC, Cool DW, et al. 3D US-based evaluation and optimization of tumor coverage for US-guided percutaneous liver thermal ablation. *IEEE Trans Med Imaging*. 2022; 41(11):3344–3356. doi:10.1109/TMI.2022.3184334

42. Hennersperger C, Karamalis A, Navab N. Vascular 3D+T Freehand Ultrasound Using Correlation of Doppler and Pulse-Oximetry Data. In: *Information Processing in Computer-Assisted Interventions. IPCAI 2014. Lecture Notes in Computer Science*. Vol 8498 LNCS. Springer Verlag; 2014:68–77. doi:10.1007/978-3-319-07521-1_8

43. Salehi M, Prevost R, Moctezuma J-L, Navab N, Wein W. Precise Ultrasound Bone Registration With Learning-Based Segmentation and Speed of Sound Calibration. In: *Lecture Notes in Computer Science (Including Subseries Lecture Notes in Artificial Intelligence and Lecture Notes in Bioinformatics)*. Vol 10434 LNCS. Springer Verlag; 2017:682–690. doi:10.1007/978-3-319-66185-8_77

44. Prevost R, Salehi M, Jagoda S, et al. 3D freehand ultrasound without external tracking using deep learning. *Med Image Anal*. 2018;48:187–202. doi:10.1016/J.MEDIA.2018.06.003

45. Prevost R, Salehi M, Sprung J, Ladikos A, Bauer R, Wein W. Deep Learning for Sensorless 3D Freehand Ultrasound Imaging. In: *Lecture Notes in Computer Science (Including Subseries Lecture Notes in Artificial Intelligence and Lecture Notes in Bioinformatics)*. Vol 10434 LNCS. Springer Verlag; 2017:628–636. doi:10.1007/978-3-319-66185-8_71

46. Guo H, Chao H, Xu S, Wood BJ, Wang J, Yan P. Ultrasound volume reconstruction from freehand scans without tracking. *IEEE Trans Biomed Eng*. 2022;70(3):1–11. doi:10.1109/TBME.2022.3206596

47. Reusz G, Sarkany P, Gal J, Csomos A. Needle-related ultrasound artifacts and their importance in anaesthetic practice. *Br J Anaesth*. 2014;112(5):794–802. doi:10.1093/BJA/AET585

48. de Jong TL, Klink SJC, Moelker A, Dankelman J, van den Dobbelsteen JJ. Needle Deflection in Thermal Ablation Procedures of Liver Tumors: A CT Image Analysis. In: Webster RJ, Fei B, eds. *Medical Imaging 2018: Image-Guided Procedures, Robotic Interventions, and Modeling*. SPIE; 2018:48. doi:10.1117/12.2292884

49. Greer JD, Adebar TK, Hwang GL, Okamura AM. Real-time 3D curved needle segmentation using combined B-mode and power Doppler ultrasound. In: *Lecture Notes in Computer Science (Including Subseries Lecture Notes in Artificial Intelligence and Lecture Notes in Bioinformatics)*. Vol 8674 LNCS; 2014:381–388. doi:10.1007/978-3-319-10470-6_48

50. Adebar TK, Okamura AM. 3D segmentation of curved needles using Doppler ultrasound and vibration. In: *Lecture Notes in Computer Science (Including Subseries Lecture Notes in Artificial Intelligence and Lecture Notes in Bioinformatics)*. Vol 7915 LNCS.; 2013:61–70. doi:10.1007/978-3-642-38568-1_7

51. Draper KJ, Blake CC, Gowman L, Downey DB, Fenster A. An algorithm for automatic needle localization in ultrasound-guided breast biopsies. *Med Phys*. 2000;27(8):1971–1979. doi:10.1118/1.1287437

52. Okazawa SH, Ebrahimi R, Chuang J, Rohling RN, Salcudean SE. Methods for segmenting curved needles in ultrasound images. *Med Image Anal*. 2006;10(3):330–342. doi:10.1016/J.MEDIA.2006.01.002

53. Novotny PM, Stoll JA, Vasilyev NV, et al. GPU based real-time instrument tracking with three-dimensional ultrasound. *Med Image Anal*. 2007;11(5):458–464. doi:10.1016/J.MEDIA.2007.06.009

54. Barva M, Uherčík M, Mari JM, et al. Parallel integral projection transform for straight electrode localization in 3-D ultrasound images. *IEEE Trans Ultrason Ferroelectr Freq Control*. 2008;55(7):1559–1569. doi:10.1109/TUFFC.2008.833

55. Kaya M, Bebek O. Needle Localization Using Gabor Filtering in 2D Ultrasound Images. In: *2014 IEEE International Conference on Robotics and Automation (ICRA)*. IEEE; 2014:4881–4886. doi:10.1109/ICRA.2014.6907574

56. Gillies DJ, Awad J, Rodgers JR, et al. Three-dimensional therapy needle applicator segmentation for ultrasound-guided focal liver ablation. *Med Phys*. 2019;46(6):13548. doi:10.1002/mp.13548

57. Ayvali E, Desai JP. Optical flow-based tracking of needles and needle-tip localization using circular Hough transform in ultrasound images. *Ann Biomed Eng*. 2015;43(8):1828–1840. doi:10.1007/s10439-014-1208-0

58. Beigi P, Rohling R, Salcudean SE, Ng GC. Spectral analysis of the tremor motion for needle detection in curvilinear ultrasound via spatiotemporal linear sampling. *Int J Comput Assist Radiol Surg*. 2016;11(6):1183–1192. doi:10.1007/s11548-016-1402-7

59. Pourtaherian A, Ghazvinian Zanjani F, Zinger S, et al. Improving Needle Detection in 3D Ultrasound Using Orthogonal-Plane Convolutional Networks. In: *Lecture Notes in Computer Science (Including Subseries Lecture Notes in Artificial Intelligence and Lecture Notes in Bioinformatics)*. Vol 10434 LNCS. Springer Verlag; 2017:610–618. doi:10.1007/978-3-319-66185-8_69

60. Beigi P, Rohling R, Salcudean T, Lessoway VA, Ng GC. Detection of an invisible needle in ultrasound using a probabilistic SVM and time-domain features. *Ultrasonics*. 2017;78:18–22. doi:10.1016/J.ULTRAS.2017.02.010

61. Mwikirize C, Nosher JL, Hacihaliloglu I. Convolution neural networks for real-time needle detection and localization in 2D ultrasound. *Int J Comput Assist Radiol Surg*. 2018;13(5):647–657. doi:10.1007/s11548-018-1721-y

62. Gillies DJ, Rodgers JR, Gyacskov I, et al. Deep learning segmentation of general interventional tools in two-dimensional ultrasound images. *Med Phys*. 2020;47(10):4956–4970. doi:10.1002/MP.14427

63. Yang H, Shan C, Bouwman A, Kolen AF, de With PHN. Efficient and robust instrument segmentation in 3D ultrasound using patch-of-interest-FuseNet with hybrid loss. *Med Image Anal*. 2021;67:101842. doi:10.1016/J.MEDIA.2020.101842

64. Mwikirize C, Nosher JL, Hacihaliloglu I. Learning needle tip localization from digital subtraction in 2D ultrasound. *Int J Comput Assist Radiol Surg*. 2019;14(6):1017–1026. doi:10.1007/s11548-019-01951-z

65. Mwikirize C, Kimbowa AB, Imanirakiza S, Katumba A, Nosher JL, Hacihaliloglu I. Time-aware deep neural networks for needle tip localization in 2D ultrasound. *Int J Comput Assist Radiol Surg*. 2021;16(5):819–827. doi:10.1007/s11548-021-02361-w

66. Zade AAT, Aziz MJ, Majedi H, Mirbagheri A, Ahmadian A. Spatiotemporal analysis of speckle dynamics to track invisible needle in ultrasound sequences using convolutional neural networks. *bioRxiv*. August 2022: 502–579. doi:10.1101/2022.08.02.502579

67. Chen S, Lin Y, Li Z, Wang F, Cao Q. Automatic and accurate needle detection in 2D ultrasound during robot-assisted needle insertion process. *Int J Comput Assist Radiol Surg*. 2022;17(2):295–303. doi:10.1007/s11548-021-02519-6

68. Arif M, Moelker A, van Walsum T. Automatic needle detection and real-time bi-planar needle visualization during 3D ultrasound scanning of the liver. *Med Image Anal*. 2019;53:104–110. doi:10.1016/J.MEDIA.2019.02.002

69. Meloni MF, Livraghi T, Filice C, Lazzaroni S, Calliada F, Perretti L. Radiofrequency ablation of liver tumors: The role of microbubble ultrasound contrast agents. *Ultrasound Q*. 2006;22(1):41–47. http://www.ncbi.nlm.nih.gov/pubmed/16641792.

70. Porter BC, Rubens DJ, Strang JG, Smith J, Totterman S, Parker KJ. Three-dimensional registration and fusion of ultrasound and MRI using major vessels as fiducial markers. *IEEE Trans Med Imaging*. 2001;20(4):354–359. doi:10.1109/42.921484

71. Besl PJ, McKay ND. Method for Registration of 3-D Shapes. In: Schenker PS, ed. *Proc. SPIE 1611, Sensor Fusion IV: Control Paradigms and Data Structures*. Vol 1611. SPIE; 1992:586–606. doi:10.1117/12.57955

72. Zhang Z. Iterative point matching for registration of free-form curves and surfaces. *Int J Comput Vis.* 1994;13(2):119–152. doi:10.1007/BF01427149

73. Myronenko A, Song X. Point set registration: Coherent point drifts. *IEEE Trans Pattern Anal Mach Intell.* 2010;32(12):2262–2275. doi:10.1109/TPAMI.2010.46

74. Penney GP, Blackall JM, Hayashi D, Sabharwal T, Adam A, Hawkes DJ. Overview of an ultrasound to CT or MR registration system for use in thermal ablation of liver metastases. In: *Proceedings of the Annual Conference on Medical Image Understanding and Analysis.* 2001:65–68.

75. Aylward SR, Jomier J, Guyon J-P, Weeks S. Intra-operative 3D ultrasound augmentation. In: *Proceedings IEEE International Symposium on Biomedical Imaging.* IEEE; 2002:421–424. doi:10.1109/ISBI.2002.1029284

76. Lange T, Eulenstein S, Hünerbein M, Lamecker H, Schlag P-M. Augmenting Intraoperative 3D Ultrasound With Preoperative Models for Navigation in Liver Surgery. In: *Lecture Notes in Computer Science.* Vol 3217. Springer Verlag; 2004:534–541. doi:10.1007/978-3-540-30136-3_66

77. Thomson BR, Smit JN, Ivashchenko OV, et al. MR-to-US Registration Using Multiclass Segmentation of Hepatic Vasculature With a Reduced 3D U-Net. In: *Lecture Notes in Computer Science (Including Subseries Lecture Notes in Artificial Intelligence and Lecture Notes in Bioinformatics).* Vol 12263 LNCS. Springer Science and Business Media Deutschland GmbH; 2020:275–284. doi:10.1007/978-3-030-59716-0_27

78. Smit JN, Kuhlmann KFD, Thomson BR, Kok NFM, Fusaglia M, Ruers TJM. Technical note: Validation of 3D ultrasound for image registration during oncological liver surgery. *Med Phys.* 2021;48(10):5694–5701. doi:10.1002/MP.15080

79. Nazem F, Ahmadian A, Seraj ND, Giti M. Two-stage point-based registration method between ultrasound and CT imaging of the liver based on ICP and unscented Kalman filter: A phantom study. *Int J Comput Assist Radiol Surg.* 2014;9(1):39–48. doi: 10.1007/s11548-013-0907-6

80. Keil M, Oyarzun Laura C, Wesarg S. Ultrasound B-Mode Segmentation for Registration with CT in Percutaneous Hepatic Interventions. In: *Lecture Notes in Computer Science (Including Subseries Lecture Notes in Artificial Intelligence and Lecture Notes in Bioinformatics).* Vol 7761 LNCS. Springer Verlag; 2013:91–97. doi:10.1007/978-3-642-38079-2_12

81. Oyarzun Laura C, Drechsler K, Erdt M, et al. Intraoperative Registration for Liver Tumor Ablation. In: *Lecture Notes in Computer Science (Including Subseries Lecture Notes in Artificial Intelligence and Lecture Notes in Bioinformatics).* Vol 7029 LNCS. Springer, Berlin, Heidelberg; 2012:133–140. doi:10.1007/978-3-642-28557-8_17

82. Roche A, Pennec A, Malandain G, Ayache N. Rigid registration of 3-D ultrasound with MR images: A new approach combining intensity and gradient information. *IEEE Trans Med Imaging.* 2001;20(10):1038–1049. doi:10.1109/42.959301

83. Viola P, Wells WM. Alignment by maximization of mutual information. *Int J Comput Vis.* 1997;24(2):137–154. doi:10.1023/A:1007958904918

84. Maes F, Collignon A, Vandermeulen D, Marchal G, Suetens P. Multimodality image registration by maximization of mutual information. *IEEE Trans Med Imaging.* 1997;16(2):187–198. doi:10.1109/42.563664

85. Penney GP, Blackall JM, Hamady MS, Sabharwal T, Adam A, Hawkes DJ. Registration of freehand 3D ultrasound and magnetic resonance liver images. *Med Image Anal.* 2004;8(1):81–91. doi:10.1016/J.MEDIA.2003.07.003

86. Wachinger C, Navab N. Entropy and Laplacian images: Structural representations for multi-modal registration. *Med Image Anal.* 2012;16(1):1–17. doi:10.1016/J.MEDIA.2011.03.001

87. Wein W, Khamene A, Clevert D-A, Kutter O, Navab N. Simulation and Fully Automatic Multimodal Registration of Medical Ultrasound. In: *Lecture Notes in Computer Science (Including Subseries Lecture Notes in Artificial Intelligence and Lecture Notes in Bioinformatics).* Vol 4791 LNCS. Springer Verlag; 2007:136–143. doi:10.1007/978-3-540-75757-3_17

88. Wein W, Brunke S, Khamene A, Callstrom MR, Navab N. Automatic CT-ultrasound registration for diagnostic imaging and image-guided intervention. *Med Image Anal.* 2008;12(5):577–585. doi:10.1016/J.MEDIA.2008.06.006

89. Fuerst B, Wein W, Müller M, Navab N. Automatic ultrasound–MRI registration for neurosurgery using the 2D and 3D LC2 metric. *Med Image Anal.* 2014;18(8):1312–1319. doi:10.1016/J.MEDIA.2014.04.008

90. Heinrich MP, Jenkinson M, Bhushan M, et al. MIND: Modality independent neighbourhood descriptor for multi-modal deformable registration. *Med Image Anal.* 2012;16(7):1423–1435. doi:10.1016/j.media.2012.05.008

91. Shechtman E, Irani M. Matching local self-similarities across images and videos. Proceedings of IEEE Computer Society Conference on Computer Vision and Pattern Recognition. Minneapolis, Minnesota, June 18–23, 2007. doi:10.1109/CVPR.2007.383198

92. Heinrich MP, Jenkinson M, Papież BW, Brady SM, Schnabel JA. Towards Realtime Multimodal Fusion for Image-Guided Interventions Using Self-Similarities. In: *Lecture Notes in Computer Science (Including Subseries Lecture Notes in Artificial Intelligence and Lecture Notes in Bioinformatics)*. Vol 8149 LNCS. Springer, Berlin, Heidelberg; 2013:187–194. doi:10.1007/978-3-642-40811-3_24

93. Yang M, Ding H, Kang J, Cong L, Zhu L, Wang G. Local structure orientation descriptor based on intra-image similarity for multimodal registration of liver ultrasound and MR images. *Comput Biol Med.* 2016;76:69–79. doi:10.1016/j.compbiomed.2016.06.025

94. Markova V, Ronchetti M, Wein W, Zettinig O, Prevost R. Global Multi-modal 2D/3D Registration via Local Descriptors Learning. May 2022. doi:10.48550/arxiv.2205.03439

95. Hu Y, Modat M, Gibson E, et al. Weakly-supervised convolutional neural networks for multimodal image registration. *Med Image Anal.* 2018;49:1–13. doi:10.1016/J.MEDIA.2018.07.002

96. Simonovsky M, Gutiérrez-Becker B, Mateus D, Navab N, Komodakis N. A deep metric for multimodal registration. *Miccai 2016.* 2016;1(October):191–200. doi:10.1007/978-3-319-46726-9_2

97. Belghazi MI, Baratin A, Rajeswar S, et al. Mutual information neural estimation. *35th Int Conf Mach Learn ICML 2018.* 2018;2:864–873.

98. LS B, Zhang S *Deformable MRI-Ultrasound Registration Using 3D Convolutional Neural Network.* Springer International Publishing; 2018. doi:10.1007/978-3-030-01045-4

99. YS B, Moelker A, Niessen WJ. *Towards Robust CT-Ultrasound Registration Using Deep Learning.* Springer International Publishing; 2018. doi:10.1007/978-3-030-02628-8

100. Mahapatra D, Antony B, Sedai S, Garnavi R. Deformable medical image registration using generative adversarial networks. In: *2018 IEEE 15th International Symposium on Biomedical Imaging (ISBI 2018).* IEEE; 2018:1449–1453. doi:10.1109/ISBI.2018.8363845

101. Dalca AV, Balakrishnan G, Guttag J, Sabuncu MR. Unsupervised learning of probabilistic diffeomorphic registration for images and surfaces. *Med Image Anal.* 2019;57:226–236. doi:10.1016/j.media.2019.07.006

102. Krebs J, Delingette H, Mailhe B, Ayache N, Mansi T. Learning a probabilistic model for diffeomorphic registration. *IEEE Trans Med Imaging.* 2019;38(9):2165–2176. doi:10.1109/TMI.2019.2897112

103. Goldberg SN. Radiofrequency tumor ablation: Principles and techniques. *Eur J Ultrasound.* 2001;13(2):129–147. doi:10.1016/S0929-8266(01)00126-4

104. Simon CJ, Dupuy DE, Mayo-Smith WW. Microwave ablation: Principles and applications. In: *Radiographics.* Vol 25. Radiographics; 2005. doi:10.1148/rg.25si055501

105. Ahmed M, Brace CL, Lee FT, Goldberg SN. Principles of and advances in percutaneous ablation. *Radiology.* 2011;258(2):351–369. doi:10.1148/radiol.10081634

106. Laimer G, Schullian P, Jaschke N, et al. Minimal ablative margin (MAM) assessment with image fusion: An independent predictor for local tumor progression in hepatocellular carcinoma after stereotactic radiofrequency ablation. *Eur Radiol.* 2020;30(5):2463–2472. doi:10.1007/s00330-019-06609-7

107. Sandu RM, Paolucci I, Ruiter SJS, et al. Volumetric quantitative ablation margins for assessment of ablation completeness in thermal ablation of liver tumors. *Front Oncol.* 2021;11(March):1–11. doi:10.3389/fonc.2021.623098

108. Collins JA, Brown D, Kingham TP, Jarnagin WR, Miga MI, Clements LW. Method for evaluation of predictive models of microwave ablation via post-procedural clinical imaging. In: Webster RJ, Yaniv ZR, eds. *Proceedings Volume 9415, Medical Imaging 2015: Image-Guided Procedures, Robotic Interventions, and Modeling.*; 2015:94152F. doi:10.1117/12.2082910

109. Makuuchi M, Torzilli G, Machi J. History of intraoperative ultrasound. *Ultrasound Med Biol.* 1998;24(9):1229–1242. doi:10.1016/S0301-5629(98)00112-4

110. Zacherl J, Scheuba C, Imhof M, et al. Current value of intraoperative sonography during surgery for hepatic neoplasms. *World J Surg.* 2002;26(5):550–554. doi:10.1007/s00268-001-0266-2

111. Machi J, Oishi AJ, Furumoto NL, Oishi RH. Intraoperative ultrasound. *Surg Clin.* 2004;84(4):1085–1111. doi:10.1016/J.SUC.2004.04.001

112. Minami Y, Kudo M. Ultrasound fusion imaging technologies for guidance in ablation therapy for liver cancer. *J Med Ultrason.* 2020;47(2):257–263. doi:10.1007/s10396-020-01006-w

113. Makino Y, Imai Y, Igura T, et al. Feasibility of extracted-overlay fusion imaging for intraoperative treatment evaluation of radiofrequency ablation for hepatocellular carcinoma. *Liver Cancer.* 2016;5(4):269–279. doi:10.1159/000449338

114. Park HJ, Lee MW, Rhim H, et al. Percutaneous ultrasonography-guided radiofrequency ablation of hepatocellular carcinomas: Usefulness of image fusion with three-dimensional ultrasonography. *Clin Radiol.* 2015;70(4):387–394. doi:10.1016/j.crad.2014.12.003

115. Stippel DL, Böhm S, Beckurts KTE, Brochhagen HG, Hölscher AH. Experimental evaluation of accuracy of radiofrequency ablation using conventional ultrasound or a third-dimension navigation tool. *Langenbeck's Arch Surg 2002 3877.* 2002;387(7):303–308. doi:10.1007/S00423-002-0315-9

116. Fenster A, Downey DB, Cardinal HN. Three-dimensional ultrasound imaging. *Phys Med Biol.* 2001;46(5):67–99. doi:10.1088/0031-9155/46/5/201

117. Rankin RN, Fenster A, Downey DB, Munk PL, Levin MF, Vellet AD. Three-dimensional sonographic reconstruction: Techniques and diagnostic applications. *Am J Roentgenol.* 1993;161(4):695–702. doi:10.2214/ajr.161.4.8372741

118. Rose SC, Hassanein TI, Easter DW, et al. Value of three-dimensional US for optimizing guidance for ablating focal liver tumors. *J Vasc Interv Radiol.* 2001;12(4):507–515. doi:10.1016/S1051-0443(07)61892-2

119. Xu HX, Yin XY, Lu M, De, Xie XY, Xu ZF, Liu GJ. Usefulness of three-dimensional sonography in procedures of ablation for liver cancers: Initial experience. *J Ultrasound Med.* 2003;22(11):1239–1247. doi:10.7863/jum.2003.22.11.1239

120. Sjølie E, Langø T, Ystgaard B, Tangen GA, Hernes TAN, Ma°rvik RM. 3D ultrasound-based navigation for radiofrequency thermal ablation in the treatment of liver malignancies. *Surg Endosc.* 2003;17(6):933. doi:10.1007/s00464-002-9116-z

5 The Use of 3D Ultrasound in Gynecologic Brachytherapy

Jessica Robin Rodgers

INTRODUCTION

GYNECOLOGIC CANCERS

Globally, gynecologic cancers are among the most common cancers affecting women, and these often require multimodal treatment strategies for oncologic management.[1,2] Gynecologic cancers generally refer to cancers originating from five different anatomical sites: the endometrium or uterine corpus, uterine cervix, vagina, vulva, or ovaries. Together, this class of cancers has the fourth highest incidence and is the fourth most common cause of cancer death in females with many of the individual types among the ten most common cancers in regions with both high and low human development indices.[1,2] While endometrial or uterine cancer is the most common gynecologic cancer in developed regions, cervical cancer is one of the most common cancers in less developed regions where wide global disparities exist in the mortality rate and treatment of these cancers, leading to heavy disease burdens.

BRACHYTHERAPY

In addition to management through surgical methods or chemotherapy, effective treatment approaches of gynecologic cancers frequently include radiotherapy delivered as a combination of external beam radiation therapy and brachytherapy.[3–8] Brachytherapy is a type of radiotherapy that allows for the placement of radioactive isotopes in close proximity to the cancerous tissues, providing conformal irradiation in the region surrounding the tumor. These techniques can be employed to escalate the radiation dose locally, enabling tailored treatment volumes around the tumor and sparing nearby normal tissues.[9–12] The use of brachytherapy (also known as internal radiotherapy) is an integral part of the definitive treatment of endometrial, cervical, and vaginal cancers and is essential for managing all localized recurrences of gynecologic cancer.[4,6,8,12–17]

Placement of the radioactive sources is done using a variety of devices (applicators/implants) and techniques that allow the source to be placed inside or next to the cancerous tissues. Although the broader field of radiotherapy has embraced advanced three-dimensional (3D) imaging techniques to enable precise treatment geometries, the use of 3D imaging for visualizing gynecologic brachytherapy applicator placement and implant geometry has lagged.[11,18] The lack of a standard imaging modality for visualizing these implants during insertion and the limited use of 3D imaging perpetuates challenges in delivering high-quality gynecologic brachytherapy treatments, resulting in failures to identify unacceptable implant geometries.[11,19] These deficiencies can result in suboptimal dose distributions, affecting the efficacy of the treatment and complication rates.[11,19] The use of 3D ultrasound (US) imaging at the time of implant placement is currently being investigated to mitigate misplacements, as it has the potential to improve adoption of 3D imaging methods by providing a relatively low-cost and accessible approach.

DOI: 10.1201/9781003299462-7

TYPES OF GYNECOLOGIC BRACHYTHERAPY

Given the diversity in disease geometries and anatomical presentations, treatment of gynecologic cancers is highly complex and must balance treatment efficacy with avoidance of critical gastrointestinal and genitourinary structures in close proximity.[20] During gynecologic brachytherapy treatment, bladder, rectum, and bowel are generally considered as the organs at risk (OARs) owing to their close proximity to the treatment sites and consequently require specific dosimetric consideration. Thus, the varied nature of these cancers and nearby structures necessitates multiple brachytherapy approaches, where all of these techniques have the potential to be aided by the integration of 3D imaging into the workflow.

Dose delivery for brachytherapy can generally be performed using either low-dose-rate (LDR) or high-dose-rate (HDR) methods, which offer a number of trade-offs and often depend on the resources available at the treatment center. Permanent LDR implants have limited use for the treatment of gynecologic cancers but temporary placement of radioactive sources can be employed for both LDR and HDR approaches.[6,13] Both techniques employ isotopes that allow for a short active distance and rapid dose fall-off, such as iridium-192 (^{192}Ir) for HDR procedures, to boost dose in a highly localized region.[9,18,21] Consequently, the source placement and associated geometry is the most critical aspect influencing the balance between efficacy and toxicity.[9,18,21]

The placement of the radioactive sources used in gynecologic brachytherapy utilizes applicators that provide channels for positioning the encapsulated source. These applicators can be inserted transvaginally or transperineally depending on the characteristics of the target region and typically require the patient to be placed under general or spinal anesthesia laying their back with their legs raised in stirrups for the insertion.[6,8,11,22,23] In general, the implantation techniques can be classified as intracavitary and interstitial approaches, although modern hybrid approaches are also emerging to improve treatment flexibility.[24] Intracavitary procedures use applicators that are inserted into the vagina and/or the uterine cavity. These are typically the first choice for brachytherapy treatment unless contraindicated. The most common applicator geometries are tandem-and-ovoids or tandem-and-ring configurations, as shown in Figure 5.1.[24] In both of these applicator types, the tandem, as well as the ring or ovoids, provide source channels for radioisotope placement.[21]

In interstitial brachytherapy, hollow needles or catheters are inserted directly into the tumor and surrounding area to provide the channels for the radioactive source. Typically 5–36 of these needles are placed, depending on the disease geometry, with 10-mm increments between adjacent needles, maintaining roughly parallel trajectories.[6,11,23,26–29] Although there are a variety of

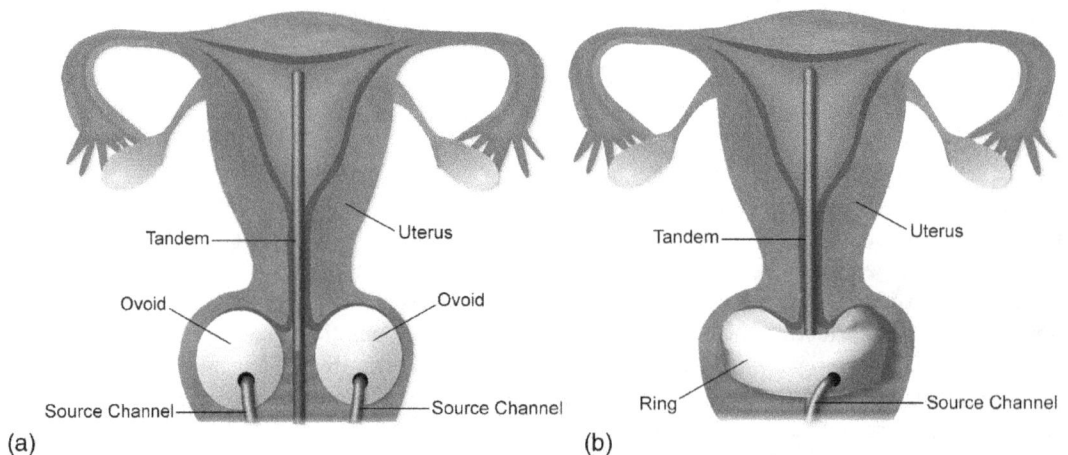

(a) Tandem — Uterus — Ovoid — Ovoid — Source Channel — Source Channel

(b) Tandem — Uterus — Ring — Source Channel

FIGURE 5.1 Most common intracavitary applicators, including (a) tandem-and-ovoids and (b) tandem-and-ring configurations.[25]

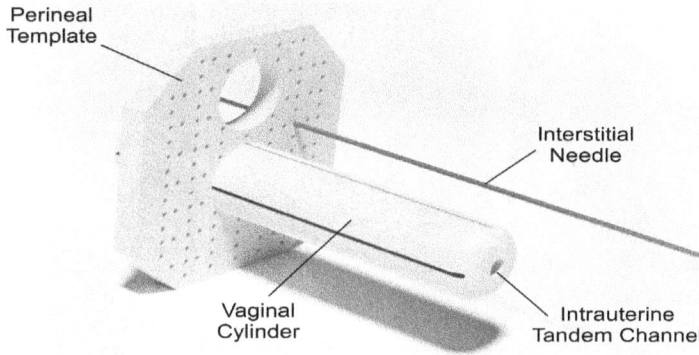

FIGURE 5.2 Illustration of a transperineal interstitial needle template with a central vaginal cylinder.[25]

techniques in clinical practice, needle insertion is typically performed through a template, as shown in Figure 5.2, and often includes a vaginal cylinder or obturator that can accommodate an intrauterine tandem while separating the vaginal walls and giving the implant greater stability.[6,23] This type of implant provides greater flexibility to tailor the dose distribution and conformality of treatment but at the cost of greater complexity compared to intracavitary approaches, requiring more clinical experience to achieve appropriate implant geometries.[8,11,27–29] Owing to this greater complexity and flexibility, interstitial implant procedures have the potential to be transformed by the incorporation of 3D imaging techniques, allowing for visualization of the implant geometry, target area, and nearby critical structures.

IMAGING IN GYNECOLOGIC BRACHYTHERAPY

Given the diverse nature of gynecologic brachytherapy procedures, there are advantages of incorporating imaging, and in particular 3D imaging, at almost every stage of the implant placement process. Preoperatively, imaging can be used to evaluate the target volume and tumor characteristics, as well as to identify the nearby structures, including the OARs. During the procedure, intraoperative and post-insertion imaging allow for assessment and verification of implant geometries, positioning relative to OARs, and treatment planning. Magnetic resonance imaging (MRI), X-ray or X-ray computed tomography (CT), and US imaging have all been investigated in the context of gynecologic brachytherapy with advantages and disadvantages at various stages of the process and uptake largely dependent on the resources available at the treating center. Recent developments have focused primarily on shifting practices toward 3D imaging-based techniques; however, adoption rate is low in resource-constrained settings and often the required equipment is inaccessible in an intraoperative environment.[18,22,30,31] Where available, the transition to 3D imaging for volumetric delineation of target volumes and OARs as part of post-insertion treatment planning has led to superior outcomes for patients, with decreased toxicity compared to conventional two-dimensional (2D) modalities.[8,22,30,32,33] These improvements resulting from the use of 3D imaging is exemplified by its application to cervical cancer brachytherapy that have demonstrated improved local control rates of 90% compared to historical rates of 60–70%.[34] Despite these advantages, 3D imaging techniques are still unavailable at many centers, particularly those with the heaviest patient burdens and fewest resources available, necessitating an accessible approach to 3D imaging in the future.

CONVENTIONAL ULTRASOUND USE

Given the advantages in portability, cost-effectiveness, accessibility, and temporal resolution, conventional US has been widely implemented for interventional procedures, commonly being used during gynecologic brachytherapy procedures for monitoring the insertion of the applicator

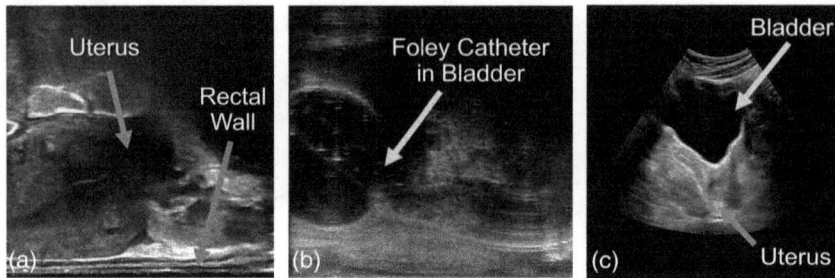

FIGURE 5.3 (a) A sagittally acquired 2D transabdominal ultrasound (TRUS) image showing the patient's uterus and rectal wall prior to applicator insertion for gynecologic brachytherapy treatment. (b) A sagittally-acquired 2D TVUS image showing the Foley catheter inserted into the patient's bladder before treatments. (c) A 2D TAUS image showing a healthy uterus and bladder.[25]

intraoperatively and assessing the placement of the implant relative to the target volumes and OARs. The immediate image feedback provided by US systems can be used to improve the technical quality of gynecologic brachytherapy implants with surveys showing use rates >60% internationally for visualization of applicator insertions.[22,35] A variety of conventional US probe configurations can play a role during gynecologic brachytherapy, with endo-cavity imaging transrectally or transvaginally and transabdominal ultrasound (TAUS) imaging most commonly employed in this application.[30] Examples of 2D US images acquired from each of these perspectives are shown in Figure 5.3.

Conventional transrectal ultrasound (TRUS) imaging is most commonly used to aid interstitial brachytherapy implant placements, allowing the positions of the needles to be assessed. Evaluation of the needle positions may require a biplanar approach with conventional 2D probes, as described by Stock et al.,[36] utilizing both axially and sagittally acquired images to assess the needle tips and trajectories.[11,30,35,37] Conversely, transvaginal ultrasound (TVUS) has limited use intraoperatively for assessment of implants, as the placement of intracavitary applicators, vaginal cylinders, and intrauterine tandems prevents the insertion of the probe transvaginally. As a result, TVUS imaging is primarily used diagnostically and for preoperative image acquisition, though it may occasionally be used for visualization of anteriorly placed interstitial needles prior to the insertion of the central obturator.[30]

The incorporation of 2D US imaging into both intracavitary and interstitial gynecologic brachytherapy procedures for immediate feedback on the positioning of the intrauterine tandem has been shown to greatly improve the quality of placement and treatment.[35,38–41] TAUS imaging is most commonly used for this purpose with the 2D images used routinely to identify misplaced or suboptimal tandem positions with the added benefit of providing visualization of needle positions relative to OARs, especially the bladder, for interstitial brachytherapy cases.[11,42] Furthermore, a number of studies have demonstrated strong correlations between dimensions of the uterus and cervix observed on TAUS images compared to MR images, which indicates the potential for TAUS-based treatment planning.[30,43–45]

Conventional 2D US is widely accessible in most regions compared to other imaging modalities, such as MRI or CT. As a result, contouring and treatment planning based on conventional US has been an active area of investigation in recent years.[30,35,46,47] Studies have demonstrated that measurements of critical structures in US images are in good agreement with measurements taken from MR images, making image-based treatment planning using conventional US images feasible.[43,48,49] Furthermore, TRUS and TAUS images independently, and in combination with other imaging modalities, have been investigated for image-based treatment planning during intracavitary gynecologic brachytherapy, demonstrating promising preliminary results; but their feasibility and application to interstitial planning is largely unexplored.[35,44,46,48,50] In general, the lack of volumetric information associated with conventional 2D US imaging limits its applicability to image-based planning, necessitating advances in 3D US to bridge this gap.[35,37,51]

ADVANCES IN 3D ULTRASOUND SYSTEMS AND DEVICES

The transition to 3D imaging modalities provides an opportunity for the investigation and implementation of 3D US technologies for gynecologic brachytherapy procedures. As elucidated throughout this book, the development of 3D US systems and devices has overcome many of the limitations typically associated with conventional 2D US imaging, such as reducing the operator dependence for image acquisition and diminishing the need to mentally collate and transform the 2D information associated with 3D structures, which is subject to uncertainty and variability.[52] Gynecologic brachytherapy applications also suffer from a variety of limitations on the placement of US probes as a result of anatomical restrictions and the inclusion of various applicators, offering another potential advantage for 3D US imaging systems that can often circumvent these restrictions to provide larger fields of view.[52] Compared to conventional 2D US images, these enlarged fields of view using 3D US techniques provide the advantage of a more complete visualization of the volumetric structures required for gynecologic brachytherapy treatment planning and evaluation of the implant positioning, warranting investigation into the application of 3D US systems in gynecologic brachytherapy. In particular, 3D US imaging enables more accurate measurements of the tumor volume compared to conventional thickness measurements associated with 2D imaging alone.[30] Despite these potential advantages, few studies have investigated the extension of 3D US systems and devices to intraoperative image-guidance during gynecologic brachytherapy. As with conventional 2D US imaging during gynecologic brachytherapy, both 3D endo-cavity US (TRUS and TVUS) imaging and 3D TAUS imaging have potential roles during the procedures. The following sections will highlight each of these geometries in the context of gynecologic brachytherapy.

3D TRUS Imaging

Despite the widespread acceptance of intraoperative 3D TRUS imaging systems for interstitial prostate brachytherapy, similar adoption has not been observed for gynecologic brachytherapy procedures owing to the unique challenges presented by this application. In particular, the large plastic components comprising intracavitary applicators and the vaginal cylinder commonly used for interstitial implants introduce complexities and limitations to acquiring high-quality 3D TRUS images. Thus, in general, the techniques developed for imaging during prostate procedures have not been adapted or evaluated in the context of gynecologic brachytherapy.

Despite these challenges, Mendez et al.[53] were able to demonstrate that using stacked axial TRUS images to create a 3D impression of the anatomy enabled clinicians to generate contours of the clinical target volume and OARs in a cervical cancer population with substantial inter-rater agreement and comparable variability to structures delineated on pre-applicator MR images. Although the results of this study are promising and demonstrate the utility of 3D TRUS images in this context, these images were generated prior to applicator insertion and further investigation is needed into post-insertion imaging, as the applicator positioning is critical for treatment planning and assessment relative to nearby structures.

In a proof-of-concept study, Nesvacil et al.[47] proposed combining 3D TRUS images with CT images for intracavitary gynecologic brachytherapy procedures to facilitate image-guided adaptive brachytherapy. This study included a single patient being treated with a tandem-and-ring intracavitary applicator and found that the combined approach was feasible, enabling treatment plans to be created that were comparable to MR-based plans.[47] Although requiring investigation in a larger patient cohort, in this proof-of-concept, the authors noted challenges using the 3D TRUS system for image acquisition to visualize all of the desired structures within the image field of view.[47] In particular, complete visualization of the applicator was difficult as the intrauterine tandem tip extended beyond the image field of view and the anterior side of the applicator ring was obscured by shadowing artifacts.[47] Furthermore, in this patient, they were unable to fully visualize the OARs as a result of the restricted field of view imposed by the 3D TRUS acquisition.[47] Despite these challenges,

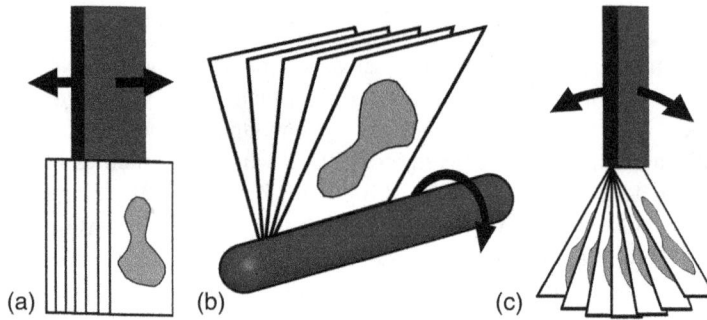

FIGURE 5.4 (a) Illustration showing the mechanical rotation of a conventional 2D endo-cavity probe to generate 3D TRUS images with the sagittal imaging plane, and (b) the fan shape of the resulting image in the axial direction.

further investigation in a broader patient cohort is warranted, as along with continued exploration into combined imaging methods including 3D US approaches.

In the study by Rodgers et al.,[54] a 3D US device was proposed for use during interstitial gynecologic brachytherapy, mimicking 3D US workflows from prostate brachytherapy. This system manipulates a conventional 2D TRUS probe with a sagittal acquisition (Figure 5.4a) through a 170° rotation to create a fan-shaped 3D image (Figure 5.4b) in 12 s. This study provides geometric validation of the system along with proof-of-concept assessment in a preliminary cohort of five patients undergoing HDR interstitial brachytherapy for treatment of vaginal masses (including three recurrent uterine cancers and two vaginal cancers).[54] To assess the visualization of the interstitial implant, the positions of the interstitial needle tips in 3D US images were compared to those seen on the post-insertion clinical CT images.[54] The resulting images, such as the one shown in Figure 5.5, allowed for visualization of the Foley catheter in the patient's bladder, as well as the vaginal cylinder associated with the perineal template and most interstitial needles (88% of needle paths and 79% of needle tips). However, some needle tips extended beyond the image field of view and some anterior needles were obscured by shadowing artifacts produced by the vaginal cylinder.[54] Of the 73 needles inserted during the study, 6 were not identified as a result of this shadowing artifact but this may be overcome in the future by modifying the material composition of the vaginal cylinder or modifying the workflow to take an intermediate 3D US image after inserting anterior needles and prior

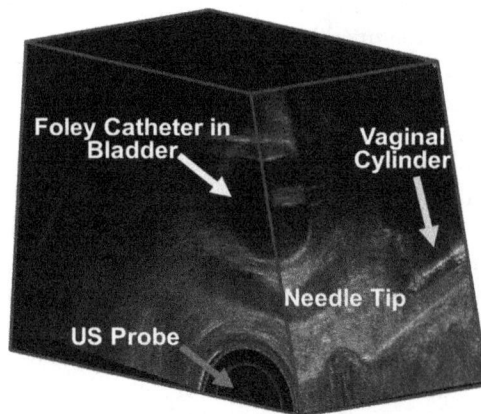

FIGURE 5.5 Example of a 3D TRUS image acquired sagittally from a patient during HDR interstitial gynecologic brachytherapy. The provided view shows an oblique slice into the 3D image volume, visualizing the Foley catheter in the patient's bladder, the vaginal cylinder of the perineal template, and one needle.

to inserting the cylinder.[54] Three more needle tips were beyond the 3D TRUS image field of view; however, further investigation is required to determine if this could be overcome with changes to the imaging protocol or whether this was a result of an anatomical restriction.[54] Comparison of the needle tip positions in 3D TRUS images relative the clinical CT images yielded mean differences of 3.82 ± 1.86 mm with the smallest differences observed in the needle insertion direction relative to the template.[54] Furthermore, as a result of the rotational acquisition, the largest errors relative to the US probe were observed in the tangential direction (i.e., the axial direction), related to reduced spatial resolution in this plane.[52,54] A small bias away from the face of the US probe was also observed, likely a result of probe pressure on the rectal wall. This study, however, was limited by the need to register two different imaging modalities from different time points for validation, which introduces changes in patient position, swelling, and internal geometries owing to the lack of fixed or rigid anatomical structures. This system has the potential to increase certainty in needle positions for HDR interstitial gynecologic brachytherapy, allowing needle placement to be assessed intraoperatively and providing 3D image guidance.

3D TVUS Imaging

Given the transvaginal insertion of applicators for intracavitary gynecologic brachytherapy, the utility of 3D TVUS imaging for these cases is currently limited to preoperative or pre-insertion imaging. Although not a 3D US method, one study embracing the principles of 3D US imaging demonstrated that a rotating US probe inserted transvaginally to the level of the cervix could be used to widen the field of view, providing increased visualization of nearby structures.[55] This imaging was performed prior to applicator placement and is limited to a single patient in this study, requiring further investigation. But it hints at the potential value of acquiring large field of view US images for assessment of the tumor and anatomy at the time of applicator insertion.[55]

Despite the advantages offered by the 3D TRUS imaging for interstitial gynecologic brachytherapy, a 3D TVUS approach may be more appropriate in cases without an intrauterine tandem, where needles may be inserted beyond the field of view of TRUS imaging or in anterior positions that are obscured by shadowing artifacts. To visualize needles inserted on all sides of the vagina, particularly through a perineal template, a 360° 3D TVUS imaging system is ideal. Previously 360° 3D TVUS imaging was investigated for examinations of the pelvic floor, as well as anorectal US imaging using specialized 3D US probes.[56–58] In 2019, Rodgers et al.[59] reported on preliminary studies using a system based on the same mechanically acquired 3D US principles shown in Figure 5.4, to create ring-shaped 3D TVUS images from six HDR interstitial gynecologic brachytherapy patients. This approach brings needles to closer proximity with the US transducer compared to the 3D TRUS method and allows needles to be visualized nearly parallel to the probe, improving clarity of visualization.[59] The adapted system, shown in Figure 5.6, also provided a series of updates to the device to facilitate workflow, featuring a counterbalanced stabilizer system that connects to the procedure room bed. This stabilizer supports the motorized scanning mechanism with a trigger-activated locking system that allows the clinicians to easily manipulate the US probe into their desired position and then secure it in place, minimizing motion during image acquisition.[59] This system also features a stepper motor for mechanical image acquisition and fine-tuning the US probe position.[59] In addition to generating 3D US images, the system can be used to align the 2D US plane with the intended template hole, allowing for real-time monitoring of needle insertion at that location. To accommodate the endo-cavity US probe transvaginally while maintaining the interstitial template geometry, a vaginal cylinder was created with the same dimensions as the clinical cylinder but it was made from sonolucent plastic with a hollow core, allowing images to be acquired by a probe inserted inside it.[59]

In the proof-of-concept study presented by Rodgers et al.,[59] anatomical features were clearly visualized in the resulting 360° 3D TVUS images, including the rectum and Foley catheter in the urethra in all patient images and the bladder in the majority of images (Figure 5.7). All interstitial

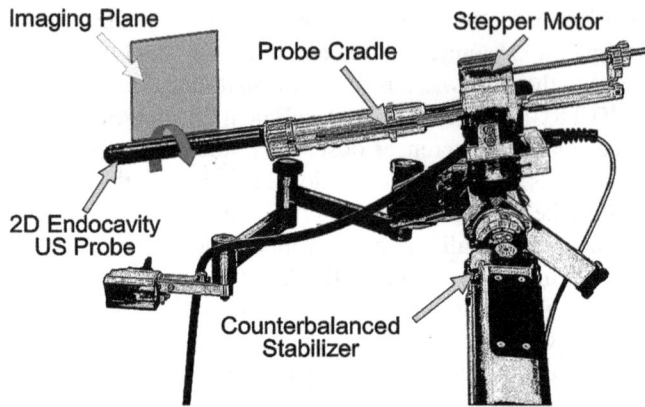

FIGURE 5.6 3D endo-cavity US system proposed by Rodgers et al.[59] for mechanically acquiring 3D US images using conventional 2D US probes with a stepper motor and including a counterbalanced stabilizer device to easily position and secure the US probe during imaging.

needles were also clearly visualized though portions of some needle paths were obstructed by small air artifacts and many needle tips extended beyond the image field of view.[59] This approach suffered from similar validation challenges as the earlier 3D TRUS study, requiring alignment with clinical CT images for verification of needle positions and introducing uncertainties as a result of differences in patient posture and swelling over time. This implementation of the system has limited applicability to tumors that are superior to the mid-vagina owing to the direction and position of the sagittal imaging plane along the endo-cavity transducer shaft. Proposed future studies involve combining these 360° sagittally acquired 3D TVUS images with 3D TVUS images acquired using an end-firing endo-cavity probe with the same mechanical system, allowing these two image types to be fused together to provide information about both needle trajectories and deeply inserted needle tips.[60] An image from a proof-of-concept study combining these two types of 3D TVUS images in a phantom is shown in Figure 5.8; however, further investigation into this approach has not yet been performed.

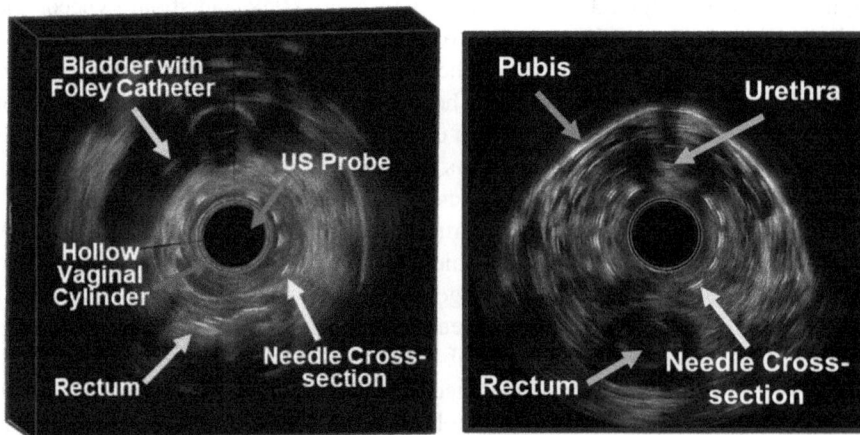

FIGURE 5.7 Examples of axial slices from two different 360° patient 3D TVUS images, showing the cross-sections of HDR interstitial brachytherapy needles, bladder with Foley catheter in place, urethra, pubis, and rectum.

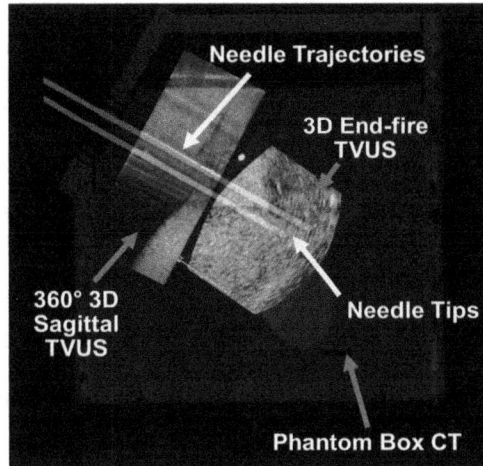

FIGURE 5.8 Image from a proof-of-concept study in a phantom combining 360° 3D sagittal TVUS images with 3D end-fire TVUS images to visualize both needle trajectories (paths) and tips for deeply inserted interstitial needles.

3D TAUS IMAGING

TAUS imaging plays a role in both intracavitary and interstitial gynecologic brachytherapy procedures, routinely implemented for verification of intrauterine tandem placement in cases where the patient has an intact uterus. Both 2D and 3D TAUS imaging techniques have been investigated in this context, particularly for intracavitary applicator placements; however, the field of view of TAUS probes can limit visualization of the parametrial extent of the disease, restricting use where heavy disease burdens exist and disease is often diagnosed at a more advanced stage.[61] To overcome this limitation, fusion approaches with other modalities have been proposed. In particular, St-Amant et al.[50] investigated the combination of 3D TAUS images with CT images. This study included five patients receiving gynecologic brachytherapy treatment with tandem-and-ovoids intracavitary applicators and found that the inclusion of the 3D TAUS images enabled superior treatment plans to be created, relative to using CT images alone for planning.[50] The study also found that the plans produced with the combined 3D TAUS and CT information were more similar to MR-based plans than with CT images alone.[50] Based on these outcomes, further investigation into combined approaches including 3D TAUS imaging is necessary.

Rodgers et al.[62] also proposed a combined approach using 3D TAUS images in combination with 3D TRUS images to create a more complete visualization of OARs and applicators during intracavitary gynecologic brachytherapy. This study explored the feasibility of using the known geometry of the applicators as the basis for registration between the two 3D US imaging geometries and provided a proof-of-concept experiment in a female pelvic phantom.[62] The experiment used the previously proposed 3D TRUS/TVUS approach[54,59] with a 3D TAUS system similar to the one developed by Gillies et al.[63] for liver scanning, creating a large field of view using both tilting and translational motions. Under the idealized phantom conditions, the combined approach demonstrated the feasibility of using the 3D TRUS image to clearly visualize the rectal wall and posterior side of both ring-and-tandem and tandem-and-ovoids intracavitary applicators while recovering shadowed information from the anterior side and providing more comprehensive visualization of the tandem tip and OARs with the broad field of view from the 3D TAUS image.[62] The authors also noted that the two applicator types that were explored presented imaging and fusion trade-offs, where the more rigid configuration of the ring-and-tandem applicator aided in the alignment of the two 3D US geometries but created more anterior shadowing, whereas the flexibility of the tandem-and-ovoids applicator made alignment of applicator components more complex but the spacing between applicator components created a "window" through the applicator, improving visualization of anterior structures.[62] This work represents an early step toward 3D US-based image-guided adaptive brachytherapy for intracavitary gynecologic procedures, warranting further investigation in a patient population.

Salas and Véliz[64] have also proposed expanding the TAUS field of view with a US-only approach to facilitate US-based 3D gynecologic brachytherapy planning. In their approach, a stepper typically used for prostate brachytherapy procedures was repurposed with a 3D printed attachment to acquire TAUS images in 5-mm increments and subsequently provide 3D volumetric information for treatment planning.[64] The authors contoured the clinical target volume and OARs based on these images from 12 patients with a total of 20 treatments using intracavitary applicators.[64] Comparing the treatment plans with those generated using 2D fluoroscopy, there were significant differences noted between the plans and these differences should be explored more extensively in future investigations, including comparison against other 3D imaging methods, such as CT or MRI.[64]

3D US-BASED IMAGE PROCESSING TOOLS

Given the lack of widespread adoption of 3D US techniques in gynecologic brachytherapy at this time, few algorithms or image-processing tools have been developed in this context. The development of such tools could greatly improve and complement uptake of 3D US methodologies as they can mitigate complexity and increase clinician confidence in image interpretation and, consequently, the positions of applicators and OARs. Furthermore, these algorithms have the potential to improve workflow efficiency and procedure times, which is critical in centers with heavy disease burdens.

Among many other tools, automated contouring has the potential for substantial impact during gynecologic brachytherapy procedures, improving accessibility and enabling the intraoperative advantages of 3D US systems to be realized. In particular, automated image segmentation to assist with the localization and evaluation of applicators and interstitial needles intraoperatively would be highly advantageous, as well as it will aid in identifying the target volumes and OARs. Although conventional US is widely used in gynecologic brachytherapy, few automated segmentation approaches have been investigated for this purpose. Building on the previously developed 360° 3D TVUS system, Rodgers et al.[65] implemented a semiautomatic segmentation approach for simultaneously identifying multiple interstitial needles in the images to aid interpretation. This method was able to accurately identify needles in these images thus giving the users a more complete understanding of the needles' positions and orientations relative to each other and nearby OARs, segmenting all needles in less than 30 s despite the presence of imaging artefacts.[65] Although this is promising to improve the image interpretability, its use has not been investigated beyond the application to the 360° 3D TVUS imaging system.

As mentioned in previous sections, registering and fusing 3D US images from various perspectives or with other imaging modalities can overcome some of the challenges commonly associated with US approaches.[47,50,62] This provides further opportunities for the development of automated image registration and fusion methods, in addition to registration across timepoints associated with the procedures, such as pre- and post-applicator insertion, which may require advanced deformable approaches.

CONCLUSION

Although largely unexplored in the literature, novel 3D US systems and tools have the potential to transform both intracavitary and interstitial gynecologic brachytherapy procedures by providing an accessible and cost-effective approach to intraoperative evaluation of applicator and needle placements, and enabling low-cost volumetric treatment planning. Clinical measurements and preliminary studies with 3D US-based techniques have demonstrated great promise for the impact resulting from clinical implementation of 3D US in gynecologic brachytherapy, including improved implant quality, which can subsequently reduce toxicities and improve local control. Despite these potential benefits, most studies about the incorporation of 3D US in gynecologic brachytherapy treatment are recent works and few have advanced beyond the proof-of-concept stage, thus, all existing approaches require more extensive investigations in broader patient cohorts to fully evaluate their utility and refine their implementation.

REFERENCES

1. Bray F, Ferlay J, Soerjomataram I, Siegel RL, Torre LA, Jemal A. Global cancer statistics 2018: GLOBOCAN estimates of incidence and mortality worldwide for 36 cancers in 185 countries. *CA Cancer J Clin.* 2018;68(6):394–424. doi:10.3322/caac.21492
2. American Cancer Society. *Global Cancer: Facts & Figures.* 3rd ed. American Cancer Society; 2015.
3. Lee L, Berkowitz RS, Matulonis UA. Ovarian and Fallopian Tube Cancer. In: Halperin EC, Wazer DE, Perez CA, Brady LW, eds. *Perez & Brady's Principles and Practice of Radiation Oncology.* 7th ed. Wolters Kluwer Health; 2018:1762–1785.
4. Alektiar KM. Endometrial Cancer. In: Halperin EC, Brady LW, Wazer DE, Perez CA, eds. *Perez & Brady's Principles and Practice of Radiation Oncology.* 6th ed. Wolters Kluwer Health/Lippincott Williams & Wilkins; 2018:1740–1761.
5. Viswanathan AN. Uterine Cervix. In: Halperin EC, Wazer DE, Perez CA, Brady LW, eds. *Perez & Brady's Principles and Practice of Radiation Oncology.* 7th ed. Wolters Kluwer Health; 2018:1651–1739.
6. Kang J, Wethington SL, Viswanathan AN. Vaginal Cancer. In: Halperin EC, Wazer DE, Perez CA, Brady LW, eds. *Perez & Brady's Principles and Practice of Radiation Oncology.* 7th ed. Wolters Kluwer Health; 2018:1786–1816.
7. Chino JP, Davidson BA, Montana GS. Carcinoma of the Vulva. In: Halperin EC, Wazer DE, Perez CA, Brady LW, eds. *Perez & Brady's Principles and Practice of Radiation Oncology.* 7th ed. Wolters Kluwer Health; 2018:1828–1844.
8. Viswanathan AN, Thomadsen B, and American Brachytherapy Society Cervical Cancer Recommendations Committee. American Brachytherapy Society consensus guidelines for locally advanced carcinoma of the cervix. Part I: General principles. *Brachytherapy.* 2012;11(1):33–46. doi:10.1016/j.brachy.2011.07.003
9. Otter SJ, Holloway C, Devlin PM, Stewart AJ. Clinical Applications of Brachytherapy : Low Dose Rate and Pulsed Dose Rate. In: Halperin EC, Wazer DE, Perez CA, Brady LW, eds. *Perez & Brady's Principles and Practice of Radiation Oncology.* 7th ed. Wolters Kluwer Health; 2018:582–606.
10. Nag S, Scruggs GR, Kalapurakal JA. Clinical Aspects and Applications of High Dose Rate Brachytherapy. In: Halperin EC, Wazer DE, Perez CA, Brady LW, eds. *Perez & Brady's Principles and Practice of Radiation Oncology.* 7th ed. Wolters Kluwer Health; 2018:626–649.
11. Viswanathan AN, Erickson BE, Rownd J. Image-Based Approaches to Interstitial Brachytherapy. In: Viswanathan AN, Kirisits C, Erickson BE, Pötter R, eds. *Gynecologic Radiation Therapy: Novel Approaches to Image-Guidance and Management.* Springer-Verlag: Berlin Heidelberg; 2011:247–259. doi:10.1007/978-3-540-68958-4_24
12. Lee LJ, Damato AL, Viswanathan AN. Gynecologic Brachytherapy. In: Devlin PM, Cormack RA, Holloway CL, Stewart AJ, eds. *Brachytherapy: Applications and Techniques.* 2nd ed. Demos Medical Publishing, LLC; 2016:139–164. doi:10.1891/9781617052613.0005
13. Ehrenpreis ED, Marsh R de W, Small W, eds. *Radiation Therapy for Pelvic Malignancy and Its Consequences.* Springer: New York; 2015. doi:10.1007/978-1-4939-2217-8
14. Chino J, Annunziata CM, Beriwal S, et al. Radiation therapy for cervical cancer: Executive summary of an ASTRO clinical practice guideline. *Pract Radiat Oncol.* 2020;10(4):220–234. doi:10.1016/j.prro.2020.04.002
15. Viswanathan AN, Beriwal S, De Los Santos JF, et al. American Brachytherapy Society consensus guidelines for locally advanced carcinoma of the cervix. Part II: High-dose-rate brachytherapy. *Brachytherapy.* 2012;11(1):47–52. doi:10.1016/j.brachy.2011.07.002
16. Han K, Milosevic M, Fyles A, Pintilie M, Viswanathan AN. Trends in the utilization of brachytherapy in cervical cancer in the United States. *Int J Radiat Oncol.* 2013;87(1):111–119. doi:10.1016/j.ijrobp.2013.05.033

17. Kamrava M, Beriwal S, Erickson B, et al. American Brachytherapy Society recurrent carcinoma of the endometrium task force patterns of care and review of the literature. *Brachytherapy.* 2017;16(6):1129–1143. doi:10.1016/j.brachy.2017.07.012

18. Viswanathan AN, Kirisits C, Erickson BE, Pötter R. *Gynecologic Radiation Therapy.* (Viswanathan AN, Kirisits C, Erickson BE, Pötter R, eds.). Springer: Berlin Heidelberg; 2011. doi:10.1007/978-3-540-68958-4

19. Viswanathan AN, Moughan J, Small W Jr, et al. The quality of cervical cancer brachytherapy implantation and the impact on local recurrence and disease-free survival in radiation therapy oncology group prospective trials 0116 and 0128. *Int J Gynecol Cancer.* 2012;22(1):123–131. doi:10.1097/IGC.0b013e31823ae3c9

20. Small W, Beriwal S, Demanes DJ, et al. American Brachytherapy Society consensus guidelines for adjuvant vaginal cuff brachytherapy after hysterectomy. *Brachytherapy.* 2012;11(1):58–67. doi:10.1016/j.brachy.2011.08.005

21. Williamson JF, Brenner DJ. Physics and Biology of Brachytherapy. In: Halperin EC, Wazer DE, Perez CA, Brady LW, eds. *Perez & Brady's Principles and Practice of Radiation Oncology.* 7th ed. Wolters Kluwer Health; 2018:530–581.

22. Viswanathan AN, Creutzberg CL, Craighead P, et al. International brachytherapy practice patterns: A survey of the Gynecologic Cancer Intergroup (GCIG). *Int J Radiat Oncol.* 2012;82(1):250–255. doi:10.1016/j.ijrobp.2010.10.030

23. Haie-Meder C, Gerbaulet A, Potter R. Interstitial Brachytherapy in Gynaecological Cancer. In: Gerbaulet A, Potter R, Mazeron J-J, Meertens H, Van Limbergen E, eds. *The GEC ESTRO Handbook of Brachytherapy.* European Society for Therapeutic Radiology and Oncology (ESTRO); 2002:417–433.

24. Taggar AS, Phan T, Traptow L, Banerjee R, Doll CM. Cervical cancer brachytherapy in Canada: A focus on interstitial brachytherapy utilization. *Brachytherapy.* 2017;16(1):161–166. doi:10.1016/j.brachy.2016.10.009

25. Rodgers JR. Development and Validation of Tools for Improving Intraoperative Implant Assessment with Ultrasound during Gynaecological Brachytherapy. Published online 2020.

26. Beriwal S, Demanes DJ, Erickson B, et al. American Brachytherapy Society consensus guidelines for interstitial brachytherapy for vaginal cancer. *Brachytherapy.* 2012;11(1):68–75. doi:10.1016/j.brachy.2011.06.008

27. Beriwal S, Bhatnagar A, Heron DE, et al. High-dose-rate interstitial brachytherapy for gynecologic malignancies. *Brachytherapy.* 2006;5(4):218–222. doi:10.1016/j.brachy.2006.09.002

28. Mendez LC, Weiss Y, D'Souza D, Ravi A, Barbera L, Leung E. Three-dimensional-guided perineal-based interstitial brachytherapy in cervical cancer: A systematic review of technique, local control and toxicities. *Radiother Oncol.* 2017;123(2):312–318. doi:10.1016/j.radonc.2017.03.005

29. Glaser SM, Beriwal S. Brachytherapy for malignancies of the vagina in the 3D era. *J Contemp Brachytherapy.* 2015;4:312–318. doi:10.5114/jcb.2015.54053

30. Viswanathan AN, Erickson BA. Seeing is saving: The benefit of 3D imaging in gynecologic brachytherapy. *Gynecol Oncol.* 2015;138(1):207–215. doi:10.1016/j.ygyno.2015.02.025

31. Viswanathan AN, Erickson B. Three-dimensional imaging in gynecologic brachytherapy: A survey of the American brachytherapy society. *Int J Radiat Oncol.* 2010;76(1):104–109. doi:10.1016/j.ijrobp.2009.01.043

32. Tanderup K, Nielsen SK, Nyvang GB, et al. From point A to the sculpted pear: MR image guidance significantly improves tumour dose and sparing of organs at risk in brachytherapy of cervical cancer. *Radiother Oncol.* 2010;94(2):173–180. doi:10.1016/j.radonc.2010.01.001

33. Rijkmans EC, Nout RA, Rutten IHHM, et al. Improved survival of patients with cervical cancer treated with image-guided brachytherapy compared with conventional brachytherapy. *Gynecol Oncol.* 2014;135(2):231–238. doi:10.1016/j.ygyno.2014.08.027

34. Sturdza A, Pötter R, Fokdal LU, et al. Image guided brachytherapy in locally advanced cervical cancer: Improved pelvic control and survival in RetroEMBRACE, a multicenter cohort study. *Radiother Oncol.* 2016;120(3):428–433. doi:10.1016/j.radonc.2016.03.011

35. van Dyk S, Schneider M, Kondalsamy-Chennakesavan S, Bernshaw D, Narayan K. Ultrasound use in gynecologic brachytherapy: Time to focus the beam. *Brachytherapy.* 2015;14(3):390–400. doi:10.1016/j.brachy.2014.12.001

36. Stock RG, Chan K, Terk M, Dewyngaert JK, Stone NN, Dottino P. A new technique for performing Syed-Neblett template interstitial implants for gynecologic malignancies using transrectal-ultrasound guidance. *Int J Radiat Oncol.* 1997;37(4):819–825. doi:10.1016/S0360-3016(96)00558-5

37. Kamrava M. Potential role of ultrasound imaging in interstitial image based cervical cancer brachytherapy. *J Contemp Brachytherapy.* 2014;6(2):223–230. doi:10.5114/jcb.2014.43778

38. Granai CO, Doherty F, Allee P, Ball HG, Madoc-Jones H, Curry SL. Ultrasound for diagnosing and preventing malplacement of intrauterine tandems. *Obstet Gynecol.* 1990;75(1):110–113.

39. Small W Jr, Strauss JB, Hwang CS, Cohen L, Lurain J. Should uterine tandem applicators ever be placed without ultrasound guidance? No: A brief report and review of the literature. *Int J Gynecol Cancer.* 2011;21(5):941–944. doi:10.1097/IGC.0b013e31821bca53

40. Sapienza LG, Jhingran A, Kollmeier MA, et al. Decrease in uterine perforations with ultrasound image-guided applicator insertion in intracavitary brachytherapy for cervical cancer: A systematic review and meta-analysis. *Gynecol Oncol.* 2018;151(3):573–578. doi:10.1016/j.ygyno.2018.10.011

41. Davidson MTM, Yuen J, D'Souza DP, Radwan JS, Hammond JA, Batchelar DL. Optimization of high-dose-rate cervix brachytherapy applicator placement: The benefits of intraoperative ultrasound guidance. *Brachytherapy.* 2008;7(3):248–253. doi:10.1016/j.brachy.2008.03.004

42. Erickson BA, Foley WD, Gillin M, Albano K, Wilson JF. Ultrasound-guided transperineal interstitial implantation of gynecologic malignancies: Description of the technique. *Endocurietherapy/ Hyperthermia Oncol.* 1995;11(2):107–113.

43. Van Dyk S, Kondalsamy-Chennakesavan S, Schneider M, Bernshaw D, Narayan K. Comparison of measurements of the uterus and cervix obtained by magnetic resonance and transabdominal ultrasound imaging to identify the brachytherapy target in patients with cervix cancer. *Int J Radiat Oncol Biol Phys.* 2014;88(4):860–865. doi:10.1016/j.ijrobp.2013.12.004

44. Van Dyk S, Narayan K, Fisher R, Bernshaw D. Conformal brachytherapy planning for cervical cancer using transabdominal ultrasound. *Int J Radiat Oncol.* 2009;75(1):64–70. doi:10.1016/j.ijrobp.2008.10.057

45. Mahantshetty U, Khanna N, Swamidas J, et al. Trans-abdominal ultrasound (US) and magnetic resonance imaging (MRI) correlation for conformal intracavitary brachytherapy in carcinoma of the uterine cervix. *Radiother Oncol.* 2012;102(1):130–134. doi:10.1016/j.radonc.2011.08.001

46. Narayan K, van Dyk S, Bernshaw D, Khaw P, Mileshkin L, Kondalsamy-Chennakesavan S. Ultrasound guided conformal brachytherapy of cervix cancer: Survival, patterns of failure, and late complications. *J Gynecol Oncol.* 2014;25(3):206. doi:10.3802/jgo.2014.25.3.206

47. Nesvacil N, Schmid MP, Pötter R, Kronreif G, Kirisits C. Combining transrectal ultrasound and CT for image-guided adaptive brachytherapy of cervical cancer: Proof of concept. *Brachytherapy.* 2016;15(6):839–844. doi:10.1016/j.brachy.2016.08.009

48. Schmid MP, Pötter R, Brader P, et al. Feasibility of transrectal ultrasonography for assessment of cervical cancer. *Strahlentherapie Onkol.* 2013;189(2):123–128. doi:10.1007/s00066-012-0258-1

49. Epstein E, Testa A, Gaurilcikas A, et al. Early-stage cervical cancer: Tumor delineation by magnetic resonance imaging and ultrasound—A European multicenter trial. *Gynecol Oncol.* 2013;128(3):449–453. doi:10.1016/j.ygyno.2012.09.025

50. St-Amant P, Foster W, Froment M-A, Aubin S, Lavallée M-C, Beaulieu L. Use of 3D transabdominal ultrasound imaging for treatment planning in cervical cancer brachytherapy: Comparison to magnetic resonance and computed tomography. *Brachytherapy.* 2017;16(4):847–854. doi:10.1016/j.brachy.2017.03.006

51. Petrič P, Pötter R, Van Limbergen E, Haie-Meder C. Adaptive Contouring of the Target Volume and Organs at Risk. In: Viswanathan AN, Kirisits C, Erickson BE, Pötter R, eds. *Gynecologic Radiation Therapy: Novel Approaches to Image-Guidance and Management.* Springer-Verlag: Berlin Heidelberg; 2011:99–118. doi:10.1007/978-3-540-68958-4

52. Fenster A, Downey DB, Cardinal HN. Three-dimensional ultrasound imaging. *Phys Med Biol.* 2001;46(5):R67–R99. doi:10.1088/0031-9155/46/5/201

53. Mendez LC, Ravi A, Martell K, et al. Comparison of CTV_{HR} and organs at risk contours between TRUS and MR images in IB cervical cancers: A proof of concept study. *Radiat Oncol.* 2020;15(1):73. doi:10.1186/s13014-020-01516-4

54. Rodgers JR, Surry K, Leung E, D'Souza D, Fenster A. Toward a 3D transrectal ultrasound system for verification of needle placement during high-dose-rate interstitial gynecologic brachytherapy. *Med Phys.* 2017;44(5):1899–1911. doi:10.1002/mp.12221

55. Petric P, Kirisits C. Potential role of TRAns cervical endosonography (TRACE) in brachytherapy of cervical cancer: Proof of concept. *J Contemp Brachytherapy.* 2016;8(3):215–220. doi:10.5114/jcb.2016.60502

56. BK Medical. Application note: 3D anorectal ultrasound. Published online 2009.

57. Santoro GA, Ratto C. Pre-Operative Staging: Endorectal Ultrasound. In: Delaini GG, ed. *Rectal Cancer: New Frontiers in Diagnosis, Treatment and Rehabilitation.* Springer-Verlag Italia; 2005:35–50. doi:10.1017/CBO9781107415324.004

58. Wright JW. Endoanal ultrasound. *Surgery.* 2008;26(6):247–249. doi:10.1016/j.mpsur.2008.04.014

59. Rodgers JR, Bax J, Surry K, et al. Intraoperative 360-deg three-dimensional transvaginal ultrasound during needle insertions for high-dose-rate transperineal interstitial gynecologic brachytherapy of vaginal tumors. *J Med Imaging.* 2019;6(2):O25001. doi:10.1117/1.JMI.6.2.025001

60. Rodgers JR, Bax J, Rascevska E, et al. 3D Ultrasound System for Needle Guidance During High-Dose-Rate Interstitial Gynecologic Brachytherapy Implant Placement Procedures. In: Fei B, Linte CA, eds. *SPIE Medical Imaging 2019: Image-Guided Procedures, Robotic Interventions, and Modeling.* SPIE; 2019:109510T. doi:10.1117/12.2512527

61. Kirisits C, Schmid MP, Beriwal S, Pötter R. High-tech image-guided therapy versus low-tech, simple, cheap gynecologic brachytherapy. *Brachytherapy.* 2015;14(6):910–912. doi:10.1016/j.brachy.2015.08.010

62. Rodgers JR, Mendez LC, Hoover DA, Bax J, D'Souza D, Fenster A. Feasibility of fusing three-dimensional transabdominal and transrectal ultrasound images for comprehensive intraoperative visualization of gynecologic brachytherapy applicators. *Med Phys.* 2021;48(10):5611–5623. doi:10.1002/mp.15175

63. Gillies DJ, Bax J, Barker K, Gardi L, Kakani N, Fenster A. Geometrically variable three-dimensional ultrasound for mechanically assisted image-guided therapy of focal liver cancer tumors. *Med Phys.* 2020;47(10):5135–5146. doi:10.1002/mp.14405

64. Pari Salas JC, Apaza Véliz DG. Description of a novel technique for ultrasound-based planning for gynaecological 3D brachytherapy and comparison between plans of this technique and 2D with fluoroscopy. *Ecancermedicalscience.* 2022;16:1415. doi:10.3332/ecancer.2022.1415

65. Rodgers JR, Hrinivich WT, Surry K, Velker V, D'Souza D, Fenster A. A semiautomatic segmentation method for interstitial needles in intraoperative 3D transvaginal ultrasound images for high-dose-rate gynecologic brachytherapy of vaginal tumors. *Brachytherapy.* 2020;19(5):659–668. doi:10.1016/j.brachy.2020.05.006

6 3D Automated Breast Ultrasound

Claire Keun Sun Park

INTRODUCTION

Globally, breast cancer is the most prevalent cancer with over 2.3 million women diagnosed and 650,000 deaths recorded in 2020 alone.[1] Breast cancer has the highest disability-adjusted life years (DALY) compared to any other type of cancer, which indicates its burden as the loss of one-year equivalent of disease-free health. Despite its high prevalence, breast cancer has a favorable relative five-year survival rate, largely attributed to the early detection of breast cancer facilitated through widespread screening programs. Accurate early detection with highly sensitive and specific imaging modalities, in conjunction with modern advancements in breast cancer treatment options such as surgery, radiotherapy, chemotherapy, and immunotherapy, has significantly improved treatment decisions, management, and outcomes of early stage, nonpalpable breast cancers.

Mammography is undoubtably an important method for early detection of breast cancer with reports up to 40% reduction of breast cancer-related mortality.[2-5] While screening mammography programs have been widely employed in predominantly developed countries, in screening populations and groups with the highest risk-factors, namely, those who are assigned female at birth and are over 40 years of age,[3] there are several other groups with an increased risk of developing breast cancer. High-risk women include those individuals with specific risk factors for breast cancer, such as personal or family history, a calculated life-time risk above approximately 25%, or carriers of genetic mutations, such as *breast cancer gene 1 (BRCA1)* or *BRCA2*.[6]

For women with increased breast density, or interchangeably mammographic density (MD), which refers to the relative amount of fibrotic (connective) and glandular tissues compared to fatty adipose tissues, as seen on an X-ray mammogram, the mammographic sensitivity substantially decreases with up to 30–48% of breast cancers being missed.[7,8] Breast density (MD) influences the appearance on mammograms due to variations in X-ray attenuation characteristics across various compositions of breast tissues. Dense breast tissues show heterogeneous or extreme distributions that appear radiopaque (white), while fatty tissues appear radiolucent (dark), as shown in Figure 6.1.[9] Since breast cancers also typically exhibit radiopaque characteristics, they are mammographically occult and often missed, due to dense tissues overlapping and obscuring the identification, localization, and characterization of these suspicious breast lesions in a two-dimensional (2D) X-ray projection.[10]

In addition to the impact on sensitivity, breast density is a strong, independent risk factor for developing breast cancer, with predominantly dense breasts facing a 4–6 times increased risk compared to those with predominantly fatty breasts.[11-13] The complex dynamic between risk factors such as age, reproductive factors, hormonal status, menopausal status, and body mass index (BMI) with increased breast density over time may increase the uncertainty of breast cancer detection at the time of imaging.[14-16] These complexities contribute to uncertainties and may increase the number of false-negative findings and interval cancers.[7] Conversely, inconclusive or false-positive findings may result in increased recall rates, unnecessary breast biopsy procedures and interventions.[17] Additionally, under the current screening recommendations, the associated exposure due to ionizing radiation may not be ideal for screening some groups of younger, high-risk women.[18,19] Breast density affects women globally with varying increased risk-factors between ethnic groups and environmental factors.[20,21] While approximately 40% of women in North America have dense

DOI: 10.1201/9781003299462-8

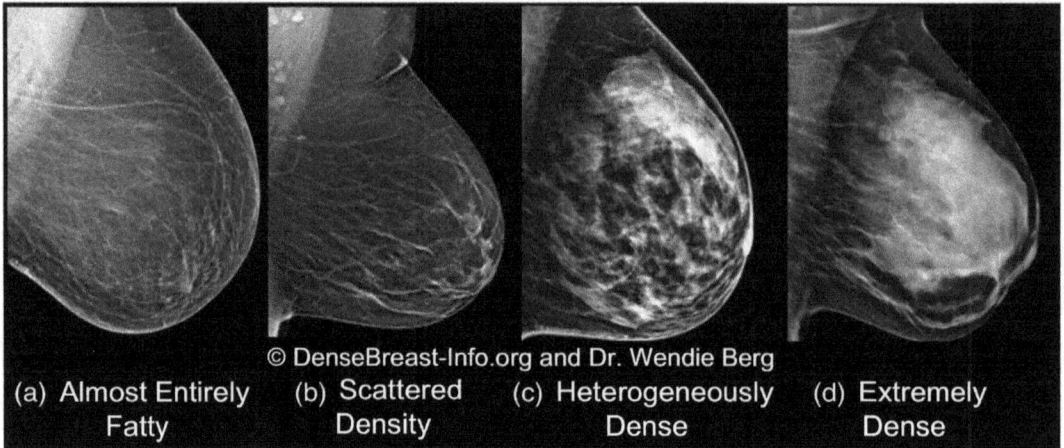

© DenseBreast-Info.org and Dr. Wendie Berg

(a) Almost Entirely Fatty   (b) Scattered Density   (c) Heterogeneously Dense   (d) Extremely Dense

FIGURE 6.1 Mammograms depicting breast density categories A–D according to the breast imaging reporting and data system (BI-RADS) developed by the American College of Radiology (ACR). Breasts in categories BI-RADS C and D are considered dense breasts. (Adapted from: DenseBreast-Info.org and Dr. Wendie A. Berg. Accessed online.[9])

breasts,[22] this proportion substantially rises up to 80% of women in some Asian populations.[23–26] Despite continuous advancements in X-ray mammography, these challenges necessitate alternative imaging techniques for effective screening in intermediate to high-risk populations, particularly those with dense breasts.

Handheld (HH) US is a widely available, nonionizing, noninvasive, and cost-effective imaging modality that has proven unparalleled importance in clinical practice worldwide. Conventional B-mode US is an effective diagnostic adjunct for supplemental screening in asymptomatic women with dense breasts (BI-RADS C and D) allowing for early detection of small, mammographically occult, invasive, and node-negative breast cancers.[6,27–30] Pivotal, large scale, and multi-institutional studies have demonstrated the utility of supplemental US for screening breast cancer. One such study, the American College of Radiology Imaging Network (ACRIN) 6666, showed that supplemental HHUS screening in high-risk women identified a significant increase in breast cancers with 5.3 additional cancers per 1000 women screened.[6] Similarly, the Japan Strategic Anti-cancer Randomized Trial (J-START), as the largest randomized clinical trial, evaluated the impact of supplemental HHUS with screening mammography in 36,752 asymptomatic women aged 40–49 years old with no history of breast cancer. The findings from J-START showed that supplemental HHUS identified 1.8 additional cancers per 1000 women, primarily comprising small, invasive, and node-negative breast cancers, which holds clinical importance for early stage detection and improved prognosis.[31] Furthermore, an extensive secondary analysis of J-START revealed that supplemental HHUS was effective for early detection of invasive breast cancers in both women with dense and non-dense breasts.[32] Therefore, HHUS shows evidential value in intermediate to high-risk populations with predominantly dense breasts, and further demonstrates clinical utility independent of breast density.

HHUS, as in many clinical applications, still suffers several limitations. Image acquisition and interpretation are time consuming, operator dependent, and lacks reproducibility, which are important for reliably monitoring temporal changes of normal breast tissues and suspicious lesions in both screening and diagnostic settings. HHUS is inherently 2D and does not permit coronal view planes parallel to the skin,[33] which restricts its ability to provide a global display of the entire breast and surrounding anatomical structures in three-dimensions (3D). Consequently, this precludes its use for whole breast assessment. Despite these limitations, HHUS shows favorable indications for screening in women with dense breasts with widespread availability and accessibility even in limited-resource settings. However, advancements in automated and whole breast US approaches are critical for their implementation as an effective screening modality.

Automated breast (AB) US imaging has been developed to address the aforementioned challenges to image the whole breast volume and overcome limitations associated with conventional HHUS approaches. ABUS employs standardized acquisition protocols, reduces operator dependence, improves reproducibility, and increases volumetric field of view.[34,35] Importantly, ABUS decouples image acquisition with interpretation, improving workflow efficiency through offline examination and assessment. While 'old-generation' ABUS scanners have existed since the 1960s,[36] advancements in ABUS technology, such as high-frequency US transducers with improved design elements, electronics, and signal processing components, have allowed for the development of modern 'new-generation' scanners. These improvements in ABUS scanner hardware, coupled with advanced software modules for computer-driven acquisition and computer-aided detection (CAD) have allowed for various approaches and techniques to improve high-resolution 3D ABUS image acquisition, multiplanar 3D reconstruction and visualization, and interpretation. Numerous studies have demonstrated the clinical value and implications of these technological advancements in clinical practice. These studies exemplify the accelerating importance for supplemental screening and diagnostic applications, particularly in intermediate to high-risk women with dense breasts.

This chapter aims to present an updated review and provide a summary of current literature related to 3D ABUS. Specifically, this chapter describes technical components and acquisition techniques of past, current, and emerging 3D ABUS systems and devices, including commercially available and experimental prototype systems. Subsequently, it will describe existing literature and evidence supporting 3D ABUS for supplemental screening and diagnostic settings, in predominantly intermediate to high-risk populations, including those with dense breasts. Finally, this chapter will conclude with discussing clinical implications of 3D ABUS, potential future directions, and advancements that can make this modality more accessible, cost-effective, and widespread in clinical practice.

3D AUTOMATED BREAST US SYSTEMS

ABUS systems can be most broadly classified by patient position, as either prone (patient lies on their stomach) or supine (patient lies on their back) acquisition types. These systems most generally contain: (a) a scan station with hardware that are either hybrid that integrate a conventional US transducer, or specialized that require a specific US transducer, (b) software for computer-driven image acquisition, and (c) a workstation for multiplanar 3D reconstruction, visualization, and interpretation. While previous comprehensive reviews have described numerous ABUS systems,[36–40] this current section provides an overview of past technologies, and presents an update on currently available and emerging technologies, including commercially available and experimental prototype systems. This section focuses on providing a technical description of these ABUS technologies, including their technique for patient positioning, breast scanning, image acquisition, visualization, and interpretation.

Prone Systems

Prone ABUS systems were first introduced in the 1970s, with the pioneering 'old-generation' prone scanner known as the multi-purpose Octoson (Australia Ultrasound Institute, Sydney, Australia) scanner. This automated full-field breast (AFB) US was composed of a large water-filled tank with eight conventional low-frequency US transducers (3–4.5 MHz operating frequency) on a motorized-drive assembly mounted at its base.[41] The patient was required to lie prone with their breasts naturally suspended in the water tank, and the scanning process involved acquiring mediolateral (transverse) and anteroposterior (sagittal) US images with scan intervals of 2–5 mm. Variations in this approach included compression using a sonolucent membrane, shorter scan intervals to 1 mm increments, or repeated scanning with different US transducers and acquisition parameters.[36] While this system was the first to achieve prone ABUS scanning, reducing operator-dependence and improving dataset acquisition reproducibility, the acquired US images were limited to 2D visualization and interpretation.

Since then, several prone ABUS systems capable of high-resolution volumetric 3D US image acquisition, visualization, and interpretation have been developed. Modern prone acquisition systems discussed in this section include the Aloka ASU-1004 (Hitachi Aloka Medical, Ltd., Tokyo, Japan), EpiSonica iABUS (EpiSonica, Hsinchu, Taiwan), Orison Embrace 3D Imaging (Orison Corp., Johnson City, Tennessee, United States), SonixEmbrace (Ultrasonix, Richmond, British Columbia, Canada), SVARA Warm Bath Ultrasound (WBU) (TechniScan, Salt Lake City, Utah, United States), SOFIA Automated Breast Ultrasound (iVu, Grapevine, Texas, United States), and SoftVue USCT (Delphinus Medical Technologies, Inc., Novi, Michigan, United States).[62]

Aloka Prosound-II ASU-1004

The Aloka ASU-1004 scanner (Prosound-II SSD-5500, Hitachi Aloka Medical, Ltd., Tokyo, Japan) is an automated full-breast (AFB) US scanner with a prone acquisition approach.[42] The ASU-1004 scanner contains a linear UST-5710 (Hitachi Aloka) transducer with a 60 mm wide footprint with a 5–10 MHz operating frequency. The US transducer is immersed inside a water tank with a thin sonolucent membrane at its surface to couple the breast. Acquisition with the Aloka ASU-1004 scanner requires the patient to be standing in a bowing posture with their breast suspended on the membrane in the water tank (Figure 6.2).[42] The US transducer is mechanically translated over a linear distance of 16.0 cm, then laterally translated by 1.0 cm partially overlap the image and acquire three parallel passes. The three-pass US images are stitched together with a registration technique based on the sum of the absolute block-mean difference (SBMD) measures,[43] resulting in a 16.0×16.0 cm whole breast US image volume. The whole breast US images are stored with DICOM standard format for review and interpretation. While these volumes are reviewed as a stack of transverse 2D US images, advanced 3D volumetric analysis can be used to characterize benign and malignant lesions.[44] Moreover, 2.5-dimensional (2.5D) analysis has been proposed to account for the anisotropic nature of the acquired volumetric dataset, demonstrating clinical evidence for diagnostic applications in breast cancer.[45]

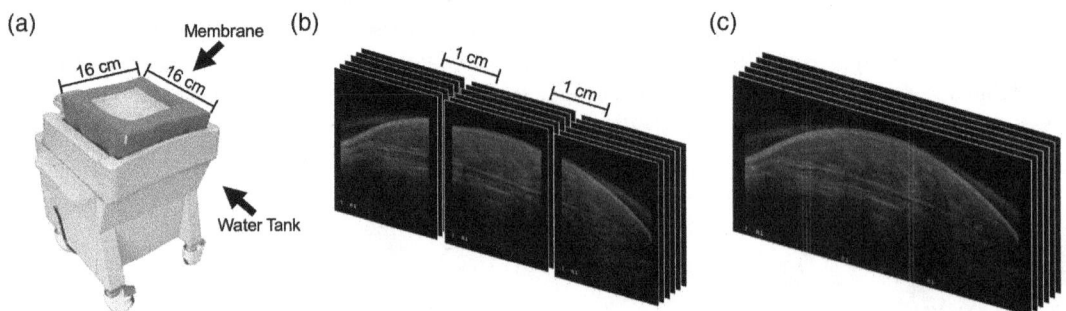

FIGURE 6.2 (a) Aloka Prosound-II ASU-1004 whole breast scanner with a thin membrane to couple the breast of a patient in a bowing position. The automated US scanning mechanism and linear US transducer is positioned inside the water tank. (b) Axial (transverse) US images of the whole breast with three partially overlapping US image passes, and (c) whole breast image slices by combining the image passes. (Adapted from H. Fujita[46] and reprinted with permission from Elsevier and Y. Ikedo et al. SPIE, reproduced with permission[47].)

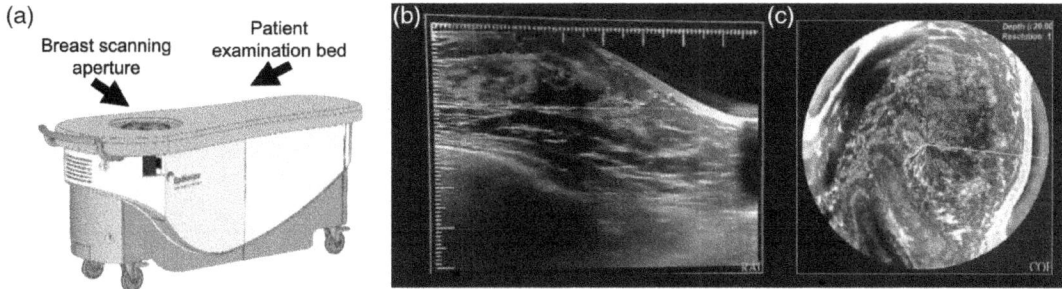

FIGURE 6.3   (a) Computer-aided design (CAD) of the EpiSonica iABUS system with a specialized patient examination bed and breast aperture for double concentric scanning. ABUS image in (b) radial (transverse) and (c) coronal views. (Adapted from Episonica.com. Accessed Online.[48])

EpiSonica iABUS

The EpiSonica Intelligent Automated Whole Breast Ultrasound Tomography System (iABUS) (EpiSonica, Hsinchu City, T'Ai-Wan, Taiwan) is a commercially available whole breast system that was received approval by the Taiwan Food and Drug Administration (TFDA).[48] The iABUS system is a prone whole breast acquisition system that uses double concentric 360° radial US scans. The system can accommodate any commercially available US transducer with a width of 50–90 mm. Double concentric scanning is performed to capture the entire breast volume, which involves a central acquisition around the nipple position and a second peripheral acquisition that includes the axillary regions.[49] The iABUS system incorporates the SonoPad™ (EpiSonica) technology that is adaptable to the natural weight and positioning of the patient allowing for improved coupling, stabilization, and minimum compression of the breast within the aperture for scanning (Figure 6.3). This design allows imaging of breasts of any size while maintaining correlation to the central nipple position, allowing for full-field whole-breast visualization and improving reproducibility between scans. The iABUS scans are reviewed with multiplanar slices, particularly radial (transverse) and coronal view planes. An advantage of this radial scanning approach offers comprehensive imaging of the ducts and lobules, as well as extension of surrounding anatomical structures in the breast. Moreover, it provides multiplanar 3D visualization for complete diagnostic interpretation for screening applications.

Embrace 3D Breast Imaging System

The Embrace 3D™ Breast Imaging System (Orison Corp., Johnson City, Tennessee, United States) is another commercially available prone-type system (Figure 6.4). The Embrace 3D system requires the patient to be sitting on a chair with their arm positioned ipsilateral on an armrest and their breast suspended naturally inside a hemispherical vessel with fluid-filled coupling medium. To ensure the breast is enclosed for whole breast imaging and accommodate for slight variations between patient body sizes and shapes, the hemispherical vessel features a custom-designed bellows seal, which was formulated based on an extensive Orison-sponsored study to map the average thoracic curve around the breast to the chest wall. The Embrace 3D acquisition approach involves a specialized rotating concave transducer to acquire a 3D ABUS image. The acquired 3D US images are then reconstructed for multiplanar 3D viewing and interpretation.

Ultrasonix SonixEmbrace

The SonixEmbrace™ (Ultrasonix Medical Corp., Richmond, British Columbia, Canada) system is another 3D ABUS system that utilizes a prone approach (Figure 6.5). With the SonixEmbrace scanner, the patient lies in a prone position on a specialized examination bed with their breast suspended and naturally compressed inside a 180 mm in diameter hemispherical aperture.

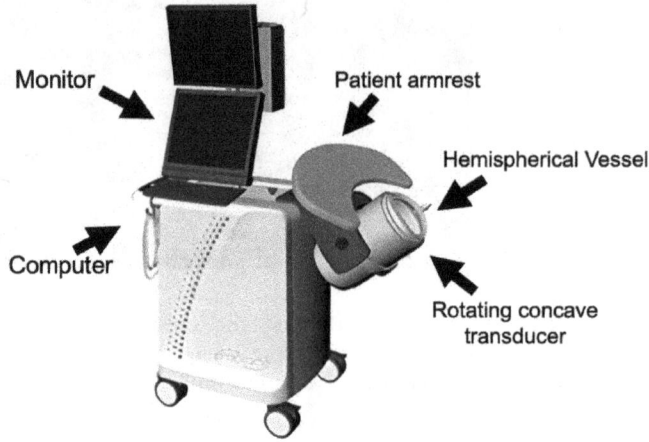

FIGURE 6.4 AComputer-aided design (CAD) of the orison embrace 3D breast imaging system with a breast aperture with a rotating concave transducer, patient armrest, and monitor to interface with the workstation. (Adapted from 2022 Nemera Insight Chicago, LLC. Accessed online.[50])

This aperture contains a special gel-based coupling pad to prove coupling, in comparison to a sonolucent membranes or fluid-filled tanks. The SonixEmbrace employs a specialized transducer with 384 elements on a 115 mm wide footprint (on a 120 mm radius of curvature) with a 10 MHz peak operating frequency. A motorized-drive mechanism rotates a concave transducer 360° to acquire 800 B-mode US images with 0.5° intervals in approximately 2 to 4 minutes. The US images are reconstructed into a 3D image, allowing for multiplanar 3D viewing in transverse, sagittal, and coronal view planes.

The SonixEmbrace 3D ABUS images are acquired with uncompressed positioning, consistent with natural anatomical orientation, facilitating the registration of these volumes with other

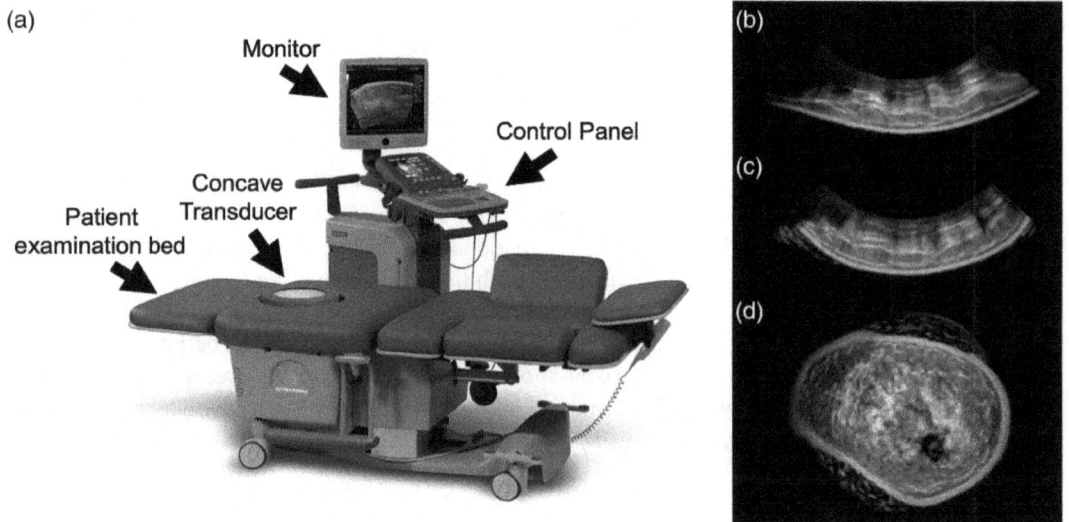

FIGURE 6.5 (a) Ultrasonix SonixEmbrace ABUS system with a specialized patient examination bed, breast aperture with a 360° concave transducer, ultrasound machine and workstation. 3D ABUS images of an uncompressed breast in (b) sagittal, (c) axial, and (d) coronal views. (Adapted from H. Tadayyon[53] and reprinted with permission from Analogic Ultrasound. Accessed online and reproduced with permission.[53])

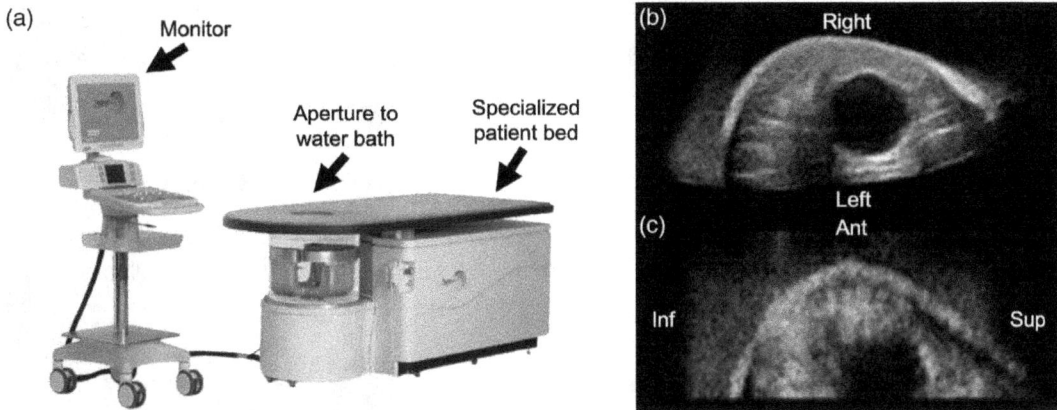

FIGURE 6.6 (a) TechniScan SVARA warm Bath ultrasound (WBU) scanner with a specialized patient examination bed and aperture to the warm water bath with scanning mechanism. WBU images in (b) axial (transverse) and (c) sagittal views. (Adapted from MedGadget, Inc. Accessed online.[54])

modalities, such as computed tomography (CT) or MRI.[51] The SonixEmbrace system can also be adapted to other commercial US transducers, allowing for the implementation of advanced imaging techniques and simultaneous acquisition of tomographic US, such as transmission, speed of sound, and acoustic attenuation, tissue elasticity (modulus) and shear wave elastography.[51] Volumetric Doppler flow imaging and photoacoustic tomography (PAT) imaging of the entire breast is also possible.[52] Currently, 3D Doppler reconstruction is not available to image the magnitude and direction of flow with this approach.[52]

SVARA Warm Bath Ultrasound

The SVARA™ Warm Bath Ultrasound (WBU) (TechniScan Inc., Salt Lake City, Utah, United States) is a 3D ABUS imaging system (Figure 6.6). The SVARA approach is performed with the patient in prone position with the breast suspended in a temperature-controlled water bath, which is maintained near skin temperature to maximize patient comfort. During image acquisition, THE US array performs a continuous scan, collecting and storing a series of 2D US images of the breast. The SVARA system utilizes reflection (B-mode) and transmission US to create refraction-corrected whole breast 3D US images. These 3D US images can be stored in DICOM format for review. The fully automated SVARA acquisition approach ensures that the whole breast image quality is not reliant on operator skills and experience. In addition to the volumetric 3D ABUS images, SVARA allows for the quantitative assessment of breast tissue properties using the recorded transmission information.

SOFIA Automated Breast Ultrasound

The SOFIA™ Automated Breast US (iVu Imaging, Grapevine, Texas, United States) is a prone 3D ABUS imaging that was previously acquired by Hitachi Medical Systems (Hitachi Ltd., Ibaraki, Japan). The SOFIA system is composed of a specialized patient examination bed, a linear array US probe, and custom workstation to interface with software.[55] While most prone systems use a suspension-based approach with the breast immersed within a coupling medium, the SOFIA system uses a specialized examination bed with a cone-like aperture to compress the breast with the body weight of the patient. During image acquisition, the patient lies in prone position with their contralateral leg slightly bent. This approach allows the breast to be compressed and flattened, allowing for a homogenous tissue distribution for improved image quality.

The SOFIA system uses a Hitachi–Aloka Noblus US (Hitachi Medical Systems, Tokyo, Japan) system with a linear EUP-L53L transducer with a 92 mm wide footprint and 5–10 MHz operating

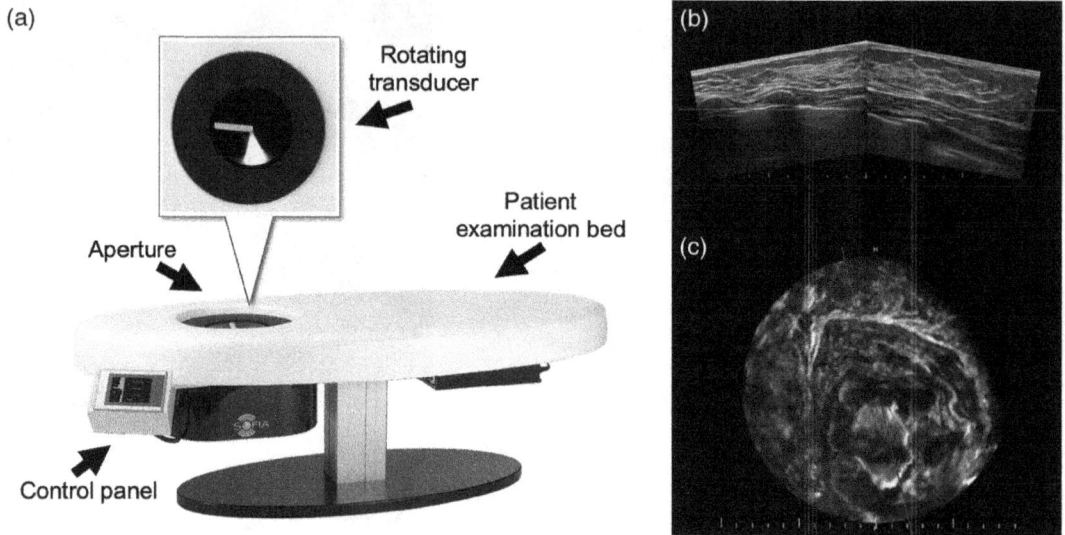

FIGURE 6.7 (a) SOFIA ABUS system with a specialized patient examination bed with a cone-like aperture for radial acquisition. 3D ABUS images of a dense breast in (b) sagittal and (c) coronal views. (Adapted from iVu Imaging Corporation (2019). Accessed online.[57])

frequency. The US transducer is positioned inside the examination bed aperture (184 mm in diameter) in a thin glass window. To acquire a 3D ABUS image, the breast is positioned on the aperture, coupled with ultrasonic lotion, with the nipple centered to ensure even distribution of the breast within the acquisition field of view. Scanning involves rotating the US transducer radially 360° clockwise from its central axis to acquire 120 US images in approximately 35 seconds, resulting in a mean total acquisition time of four minutes. Trapezoid mode can be used with the linear US transducer to expand the acquired volumetric field of view in the 3D ABUS image. The workstation performs 3D reconstruction with the transverse US images on the workstation in approximately 10 minutes, allowing for multiplanar 3D visualization on transverse, sagittal, coronal, and radial view planes. The SOFIA 3D ABUS images are compatible with DICOM standard.

For high-quality 3D ABUS images with the SOFIA approach, correct patient and nipple positioning are crucial (Figure 6.7).[56] Incorrect positioning may result in image drop-off in the peripheral regions in the reconstructed 3D US image. To mitigate this limitation, the operator can apply pressure to the patient's back to improve contact between the breast and aperture. Another limitation is that the SOFIA approach is unable to image the nipple and region immediately behind it.[49] While the SOFIA system employs a radial acquisition approach, compression of the breast and angulation of the lobules and ducts due to the patient positioning, makes the acquired datasets insufficient for ductal echography evaluation and assessment.[49] Clinical evidence supporting prone ABUS with the SOFIA system for supplemental screening are discussed in the following section.

SoftVue Whole Breast US Tomography

The SoftVue™ Automated Whole Breast Ultrasound Tomography System (Delphinus Medical Technologies, Inc., Novi, Michigan, United States) developed by the Karmanos Cancer Institute (KCI) was the first commercially available US computed tomography (CT) system. The SoftVue system received FDA approval in 2021 as an adjunct to mammography for screening women with dense breasts.[58,59] The SoftVue approach uses a unique circular US transducer array (220 mm in diameter) with a 1–3 MHz operating frequency. The transducer array consists of 2,048 elements in a uniform ring configuration. An anatomical guidance assembly (Sequr™ Breast Interface) is used to centralize the nipple position and stabilize the breast in the examination aperture for scanning, accommodating for variable breast sizes and geometries. Whole breast USCT image acquisition involves the pendulous breast inside the aperture, and performing automated scanning from the nipple to the pectoralis muscle, excluding the axilla. The resulting 3D USCT image contains approximately 30–60 images with an in-plane (sagittal and transverse) spatial resolution of approximately 0.75 mm and out-of-plane spatial resolution (elevational increments) of 2.5 mm.[59]

While other prone acquisition 3D ABUS systems use primarily conventional B-mode US images and anatomical information to characterize and differentiate between benign and malignant tissues, SoftVue 3D USCT provides quantitative measurements and simultaneous acquisition of reflection (B-mode) and transmission data, allowing for 3D reflection, acoustic attenuation, and speed-of-sound images.[60] SoftVue 3D USCT reconstruction results in five volumetric datasets for tomographic interpretation: speed-of-sound, reflection, attenuation, and waveform enhanced reflection (Wafer) image. The speed-of-sound image (units ms^{-1}) shows the propagation velocity of sound in the 3D US image. The reflection image (unitless) shows the relative intensity of the reflected US. The attenuation image (units dB mMHz^{-1}) shows the attenuation of US in the tissues. The Wafer image (unitless) allows for unique whole breast visualization, as the contrast adjusted reflection US image using the speed-of-sound data to maximize image quality.[61] Higher speed-of-sound areas are hypoechoic (darker) and lower speed-of-sound areas are hyperechoic (brighter) in these images. The relative stiffness image (unitless) shows the relative stiffness of the tissues.

Supine Systems

Another predominant method for 3D ABUS image acquisition involves a supine positioning approach, where a patient lies on their back on a standard examination bed. Early supine approaches were introduced in the 1980s by Labsonics (Australia) with a system composed of an articulated arm and suspended water-coupled instrument for whole breast acquisition.[63] The water-coupled instrument contained a water-filled, sonolucent plastic bag used for coupling, compression, and stabilization of the patient's breast in supine position. The scanner included an automated water-path 7.5 MHz polyvinylidene fluoride (PVFD) US transducer with a 4.5 MHz operating frequency. This supine acquisition approach allowed for complete 360° manipulation of the articulated arm for imaging in transverse, sagittal, and radial views. While interpretation of these early supine ABUS images was performed using conventional 2D datasets, this invention accelerated the development of modern 'new-generation' supine 3D ABUS systems.

Modern supine-type systems typically use an articulated arm with an US transducer (either conventional or specialized) with dedicated apparatus for contact with the breast, such as a rigid frame with a coupling membrane. A workstation is utilized for 3D US reconstruction, visualization, and interpretation. Supine systems discussed in this section include the SonoCiné (SonoCiné, Inc., Reno, Nevada, United States) system, Invenia Automated Breast Ultrasound (ABUS) (GE Healthcare, Chicago, Illinois, United States) (previously the SomoVu™ (Somo•V) ABUS by U-Systems Inc., San Jose, California, United States), ACUSON S2000™ Automated Breast Volume Scanner (ABVS) (Siemens Healthineers, Erlangen, Germany), SuperSonic MACH 40 Ultrasound System (Hologic, Marlborough, Massachusetts, United States), Spatially Tracked 3D US System (Robarts Research Institute, London, Ontario, Canada), ATUSA™ (iSono Health, San Francisco, California, United States), and the dedicated 3D ABUS device (Robarts Research Institute, London, Ontario, Canada).

SonoCiné Automated Breast Ultrasound

The SonoCiné™ Automated Whole Breast Ultrasound (AWBU) system (SonoCiné, Inc., Reno, Nevada, United States) was one of the first commercially available systems developed for supine ABUS imaging, which received United States FDA approval as an adjunct to screening mammography in 2008 (Figure 6.8).[37] The SonoCiné AWBU is a hybrid-type system that integrates a conventional high-resolution 2D US transducer mounted to a computer-guided, mechanically driven arm for semi-automated scanning.[64] During AWBU scanning, the patient lies supine on an examination

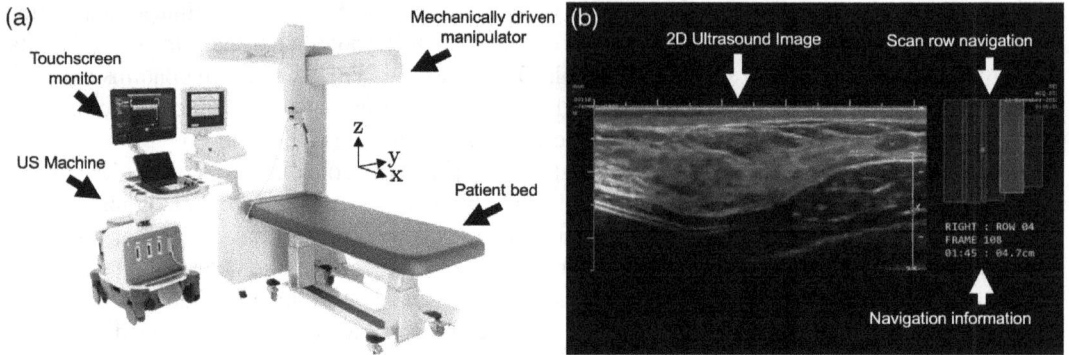

FIGURE 6.8 (a) SonoCiné automated whole breast ultrasound (AWBU) system with a scan station, mechanically driven manipulator for automated breast scanning, and patient examination bed. (b) Visual navigation software with 2D US image slice with scan row navigation and information. (Adapted from SonoCiné. Accessed online and reproduced with permission.[67])

bed, and the mechanically driven arm linearly translates the US probe across the breast in partially overlapping rows in the craniocaudal direction. An operator is responsible for maintaining contact pressure and compression with the breast during the acquisition.[65] The overall examination time is approximately 20–30 min, which includes 10 minutes for patient preparation and 10–20 min for image acquisition. Approximately 2000–5000 2D US images are acquired and stored, and dedicated software (CinéDetect) creates a cine loop for video review. While SonoCiné acquires high-resolution 2D ABUS images of the whole breast, its interpretation involves the standard review of axial 2D US images, comparable to the review to conventional HHUS in clinical practice.

Previously, SonoCiné did not have any 3D US reconstruction capabilities,[37,65,66] effectively making it a 2D ABUS approach. However, recently, software for multiplanar 3D reconstruction was announced, allowing for 3D visualization in standard transverse, sagittal, and coronal view planes of acquired datasets.[65]

Invenia ABUS

The Invenia™ Automated Breast Ultrasound (ABUS) (GE Healthcare, Chicago, Illinois, United States) is a commercially available 3D ABUS system that received FDA approval (SomoVu) in 2012 as a supplemental screening tool for screening mammography in women with dense breasts (Figure 6.9).[37] Invenia ABUS is the rebranded version of the SomoVu™ (Somo•V) ABUS (U-Systems Inc., San Jose, California, United States) prior to GE Healthcare's purchase acquisition of U-Systems in 2012. The Invenia ABUS system uses a supine acquisition protocol that consists of an articulated manipulator, specialized transducer, and touchscreen monitor (scan station) and dedicated workstation (view station) for 3D reconstruction, visualization, and image interpretation. While hybrid-type systems integrate conventional US transducers, the Invenia ABUS system (SomoVu) uses a specialized high-frequency C14-6XW (CXW) ultrawide US transducer with a 15.4 cm wide footprint and 14–16 MHz operating frequency. The specialized Invenia ABUS transducer is slightly concave to follow the natural curvature of the breast (Reverse Curve™) to improve tissue contact and coverage, while maintaining uniform compression and minimizing acquisition artifacts at the periphery.[68] The US transducer is positioned inside a sonolucent breast compression paddle and attached to a mechanical articulated arm that can be coupled to the skin with a disposable taut membrane to provide uniform compression.

The 3D ABUS image acquisition is performed by a trained operator. Positioning involves the patient lying in supine position with their ipsilateral arm raised above their head, and a sponge towel placed underneath their shoulder to flatten the breast and stabilize the anatomy.[40] A hypoallergenic

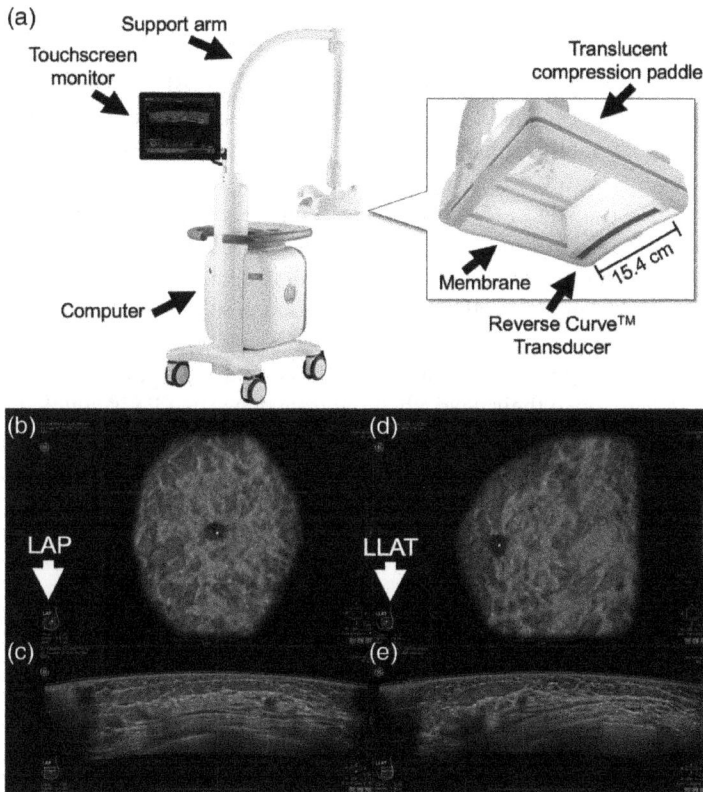

FIGURE 6.9 (a) Invenia ABUS system (GE) and compression paddle with Reverse Curve™ transducer. 3D ABUS images of an occult invasive cancer in (a) left anteroposterior (LAP) volume in the coronal and (c) axial view, and (d) left lateral (LLAT) volume in the coronal view and (e) axial view. (Adapted from General Electric (GE) Healthcare United States (2022) accessed online.[69])

lotion is applied evenly to couple the breast with the transducer paddle. The transducer paddle is positioned on the breast, and automated scanning is activated, mechanically translating the transducer 17.0 cm scan distance with 0.2 mm scan intervals (slice thickness) without overlap. Prior to acquisition, three preset compression levels can be applied prior to acquisition to reduce tissue thickness, improving coupling with the breast and detail resolution at the penetration depth. The 3D ABUS acquisition takes approximately one minute per acquisition view, with a total acquisition time of 15–20 minutes. Depth settings up to 5.0 cm are selected by the operator to ensure that deep central and peripheral breast tissues are included, resulting in a maximum $15.4 \times 17.0 \times 5.0$ cm^3 image volume. This 3D ABUS approach is reliant on the patient remaining stationary without motion and breathing slowly and smoothly during acquisition.

Acquisition involves three views, including anteroposterior (AP), lateral (LAT), and medial (MED) views in each breast. For women with larger breasts, additional image volumes can be acquired to expand the field of view with six optional views per breast: superior, inferior, axilla, upper-outer quadrant, lower-outer quadrant, and upper-inner quadrant, with the nipple position localized in each volume as an anatomical landmark. Once the 3D ABUS images are acquired, the images are processed in real-time into standard multiplanar 3D views (transverse, sagittal, and coronal views) then transferred to the workstation for offline interpretation and analysis. The 3D ABUS image quality is reliant on operator training for correct patient positioning, adequate transducer placement, and uniform compression of the breast.

Acuson S2000 ABVS

The ACUSON S2000™ Automated Breast Volume Scanner (ABVS) (Siemens Healthineers, Erlangen, Germany) is a supine 3D ABUS system. Similar to the Invenia ABUS system, the AUCSON S2000 ABVS system consists of an articulated manipulator with a specialized ultrawide US transducer and a sonolucent membrane for coupling, touchscreen monitor, and dedicated workstation (view station) for interfacing with software (Figure 6.10). The ACUSON S2000 ABVS system uses a specialized high-frequency linear 14L5BV transducer (5–14 MHz operating frequency) with 15.4 cm wide footprint. To improve uniformity in the field of view, Hanafy Lens transducer technology with variable thickness in the linear array design is integrated, improving spatial resolution and contrast.[70] Image acquisition involves collecting up to 448 images with a 16.8 cm translation with depth settings up to 6.0 cm, resulting in a maximum $15.4 \times 16.8 \times 6.0$ cm^3 volume with a 0.5 mm scan interval.

To acquire an ABVS image, the patient lies supine or supine-oblique on the examination bed with their ipsilateral arm above their head, then the transducer paddle is coupled to the breast with a disposable membrane and compression is applied. 3D ABVS imaging is performed with three predefined AP, LAT, and MED acquisition directions, which translate the transducer relative to the nipple position. In patients with smaller breasts, two views (MED and LAT) are acquired, while in patients with larger breasts, additional views are acquired, such as the superior, inferior, apex and

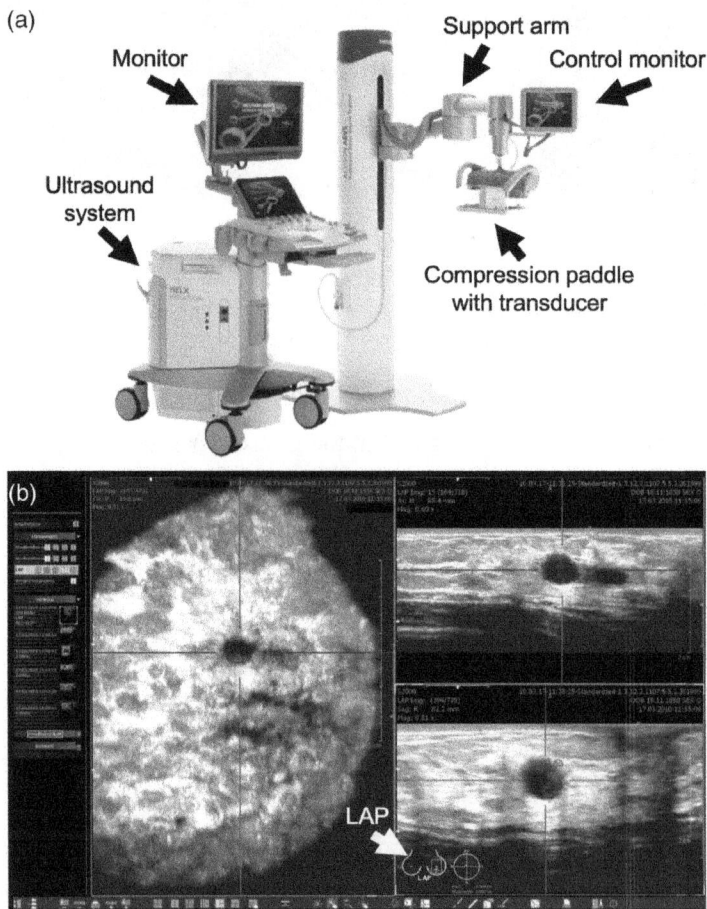

FIGURE 6.10 (a) ACUSON S2000 ABVS system with an US system and workstation, and support arm for ABUS imaging; (b) multiplanar ABVS image in the lateral anteroposterior (LAP) for review. (Adapted from Siemens Healthineers Limited (2022) accessed online[74] and S. Wojcinski (2013) BMC Medical Imaging.[71])

axillary volumes.[71] Alternative scanning patterns, such as a quadrants-based protocol with sequential scans in the upper-outer, lower-outer, lower-inner, and upper-inner regions, have been proposed to optimize whole breast coverage. This approach employs a scan direction is from the periphery of the mammary gland to the nipple.[72] The acquired images are processed by multiplanar 3D reconstruction into standard transverse, sagittal, and coronal views for interpretation.

One advantage of the ACUSON S200 ABVS system is its integration with a fully operational US machine, allowing for both 3D ABVS and 2D HHUS scanning. This may be useful for complementary evaluation in suspicious lesions or in the axilla. Advanced US imaging techniques, including, shear wave electrography and strain imaging can improve tissue characterization with 2D HHUS images. Moreover, tissue harmonic imaging, spatial compounding, and dynamic tissue contrast enhancement can be added to enhance diagnostic value in ABUS images. However, the capability to measure vascularity and blood flow with 3D Doppler to enhance characterization is currently not available with commercially available ABUS (ABVS) systems.[73]

Spatially Tracked 3D US System

The spatially tracked 3D US system, developed by the Translational Ultrasound Technologies (TRUST) lab at the Robarts Research Institute (Western University) in London, Ontario, Canada,[75] is a supine, hybrid-type system. The system is composed of a six-axis user-operated manipulator, counterbalanced stabilizer, and a quick-release 3D US scanner. While the spatially tracked 3D US system is adaptable to any commercially available US system and transducer, an Aplio i800 US system (Canon Medical Systems, Otawara, Tochigi, Japan) and a high-resolution PLT-1005BT (14L5) linear transducer with a 58 mm footprint and 10 MHz peak operating frequency were used in the prototype. The quick-release 3D US scanner is easily removable from the robotic manipulator, allowing for both bedside 2D and 3D HHUS imaging. The spatially tracked 3D US system can be mounted bedside or on a portable medical cart with a computer and monitor to interface with a dedicated workstation with custom software modules, enabling point-of-care image acquisition, reconstruction, and interpretation.

The spatially tracked 3D US system offers the advantage of a tracked six-axis approach to record the 3D spatial position and orientation of the 3D US scanner and its image for whole breast imaging (Figure 6.11). The six-axis approach includes a three-axis rotational wrist-joint, allowing the

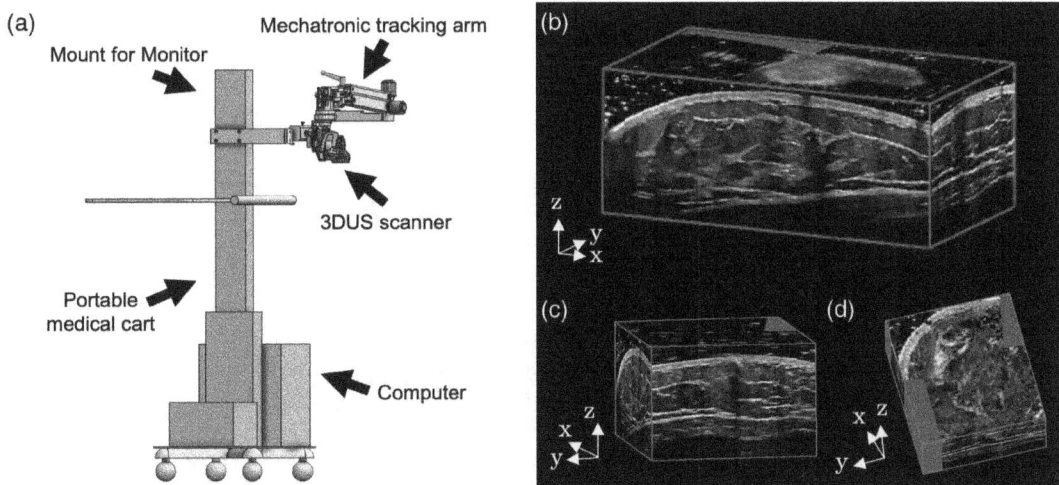

FIGURE 6.11   (a) Spatially tracked whole breast 3D US system with a mechatronic spatial tracking arm and removable 3DUS scanner mounted on a portable medical cart. (b) Spatially tracked 3DUS image volume with (b) axial, (c) sagittal, and (d) coronal views. (Adapted from C. K. S Park et al.,[75,77] reproduced with permission.)

operator to vary the applied pressure and angle of the US transducer in real-time, which is a challenge with commercially available supine ABUS (ABVS) systems due to preset automated workflows.[55,73,76] When the scanner can be positioned on the breast and coupled with US transmission gel, computer-driven, automated scanning linearly translates the 2D US transducer over a preset scan distance. While any scan distance can be calibrated, a 4.0 cm scan distance was chosen to minimize contact pressure and deformation over the natural curvature of the breast during a 3D US acquisition. Notably, this approach allows for a breast quadrants-based acquisition removing the need for compression, potentially making it an ideal choice for ABUS equipment.[49] Multiple spatially tracked 3D US images can be acquired at various positions on the breast, and multi-image registration and fusion can be performed generate a volumetric whole breast image. Multi-image fusion of 3DUS images shows potential to recover the loss of anatomical information due to posterior shadowing behind the nipple, which is an area the TRUST lab intends to explore. Although the spatially tracked system shows potential for effective 3D ABUS imaging, optimization of the whole breast acquisition protocol is required prior to its clinical implementation.

iSono Health ATUSA

The ATUSA™ System (iSono Health, San Francisco, California, United States) is a compact and portable system that allows for ABUS imaging, which received United States FDA approval in 2022.[78] It is composed of a wearable accessory that attaches to a patient like a bra, with a characteristic cone shape and an integrated US scanner, a custom software application for real-time image acquisition, display, and offline visualization. To perform image acquisition, the patient lies supine, coupling medium is added to the scanner assembly, and hands-free ABUS acquisition is initiated with a push-button, which takes approximately two minutes per breast. The acquired 3D ABUS image can be reviewed with an interactive multiplanar viewing on an online software application in transverse, sagittal, coronal, and radial views. This ATUSA system is designed for point-of-care imaging by physicians, nurses, and medical assistants. Deep-learning modules for breast lesion localization and characterization are currently being developed to enhance clinical utility and facilitate breast monitoring over time.

Dedicated 3D ABUS System

An alternative, compact, cost-effective dedicated 3D ABUS system was developed by the TRUST lab at Robarts Research Institute (Western University) in London, Ontario, Canada (Figure 6.12).[79] This system was developed to address challenges observed using the spatially tracked 3D US system, and create a more robust, cost-effective, adaptable 3D ABUS scanner.[75] The dedicated 3D ABUS system includes: (a) a personalizable rapid-prototype, 3D-printed dam to comfortably conform onto the patient and account for diverse breast sizes and geometries, (b) an adjustable compression assembly with a sonolucent TPX plate to stabilize the breast and reduce tissue thickness, and (c) a removable computer-driven motorized linear scanner. The dedicated 3D ABUS system is a hybrid-type system, is fabricated entirely with economical 3D-printed materials, and can be adapted to any commercially available US transducer to generate a 3D US image.

Acquisition involves affixing the device directly to the patient's breast, adding US transmission gel for coupling, and performing hands-free 3D ABUS scanning by activating the motorized scanner. Spatially encoded 2D US images are continually acquired at a selectable and fixed spatial interval in either the craniocaudal (CC) or mediolateral (ML) directions, as the transducer is translated, and reconstructed in real-time into a 3DUS image.[80] The US transducer can be translated laterally to acquire another 3DUS image to expand the volumetric field of view, then the images can be registered and fused into a single 3D US image.[75] The prototype acquisition settings allow for the acquisition of an image volume of approximately 10×10 cm^2 with any US depth setting.

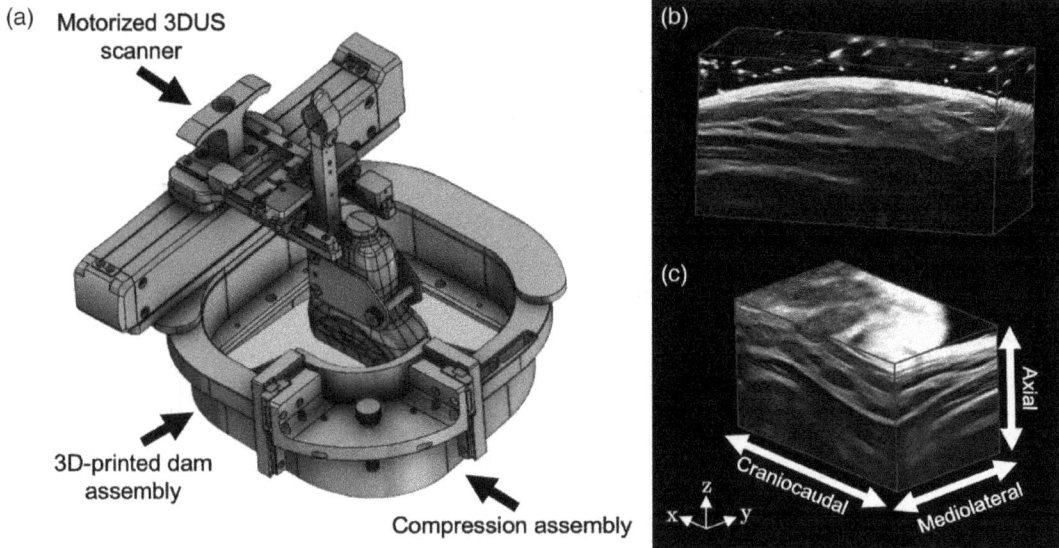

FIGURE 6.12 (a) Dedicated 3D ABUS device with a patient conforming 3D-printed dam, adjustable compression assembly, and motorized 3DUS scanner that can accommodate any linear US transducer. (b) Whole breast 3D ABUS image, illustrating (c) the craniocaudal, mediolateral, and axial directions for multiplanar 3D visualization. (Adapted from C. K. S. Park et al..[81])

COMPARISON BETWEEN PRONE AND SUPINE SYSTEMS

A comparison of characteristics between the prone and supine systems described, including compression, operator dependence, automated acquisition, and 3D reconstruction are summarized in Table 6.1. Both prone and supine 3D ABUS approaches exhibit distinct characteristics, each with advantages and disadvantages. Prone 3D ABUS systems involve the breast being naturally suspended, immersed in a coupling medium, or naturally compressed with the patient's own body weight, offering increased operator independence. Furthermore, prone approaches allow for minimal to compression-free acquisition, which may potentially alleviate discomfort and pain in patients with implants, post-surgeries, or post-neoadjuvant chemotherapies.[49] Moreover, unlike most commercially available supine ABUS systems, prone positioning systems remove the need for an external mechanical support arm (manipulator) and complex methods for coupling the breast during image acquisition. However, prone systems require a complete specialized patient examination bed with an aperture for scanning the breast, which demand correct patient positioning. Comparable to supine 3D ABUS systems, prone ABUS systems often have limited assessment in the axilla nodes, while they may be easily included with the use of 2D or 3D HHUS imaging. Consequently, the presence of an external US machine in the examination room may be required for clinical use. Lastly, most modern prone systems are typically specialized scanners, and are not capable of being adapted to any commercially available US system and transducer, which may be costly to implement into the clinic.

ALTERNATIVE 3D ABUS SYSTEMS

Upright Automated Ultrasound

Upright positioning ABUS is an alternative scanning technique to prone and supine acquisition approaches.[36] The upright scanner mimics a conventional mammography unit and consists of an upper compression scanning assembly in a rigid frame with a thin film sheet and lower compression

TABLE 6.1
Summary of Commercially Available and Experimental Prototype Automated Breast Ultrasound (ABUS) Systems

System	FDA	Type	Position	Compression	Operator Dependent	Automated	3D	Type	Frequency	Width
Aloka Prosound-II	No	Hybrid	Prone	No	No	Yes	No	Linear	5–10 MHz	50 mm
EpiSonica	2016 (Taiwan)	Hybrid	Prone	No	No	Yes	Yes	Linear	Variable	50–90 mm
Embrace 3D	No	Specialized	Prone	No	No	Yes	Yes	Concave		
SonixEmbrace	Yes	Specialized	Prone	No	No	Yes	Yes	Concave	5–14 MHz	120 mm in diameter
SVARA	No	Specialized	Prone	No	No	Yes	Yes	Specialized		
SOFIA	Yes	Specialized	Prone	Yes	No	Yes	Yes	Linear	5–10 MHz	92 mm
SoftVue	Yes	Specialized	Prone	No	No	Yes	Yes	Circular	1–3 MHz	220 mm in diameter
SonoCine	Yes	Hybrid	Supine	Yes	Yes	Semi	No	Linear	7–12 MHz	52 mm
SomoVu	Yes	Specialized	Supine	Yes	Yes	Yes	Yes	Reverse Curve	8–10 MHz	14.6 cm
Invenia	Yes	Specialized	Supine	Yes	Yes	Yes	Yes	Reverse Curve	6–16 MHz	15.4 cm
ACUSON S2000	Yes	Specialized	Supine	Yes	Yes	Yes	Yes	Linear	5–14 MHz	15.4 cm
Spatially Tracked	No	Hybrid	Supine	No	Yes	Semi	Yes	Linear	5–14 MHz	58 mm
ATUSA	Yes	Specialized	Supine	No	No	Yes	Yes	Linear	5–10 MHz	
Dedicated 3DUS	No	Hybrid	Supine	No	No	Yes	Yes	Linear	5–14 MHz	58 mm

ABUS, automated breast ultrasound; 3D, three-dimensional.

plate to stabilize the breast.[82] Patient positioning involves the patient standing upright with their breast compressed between the film sheet and compression plate for acquisition in standard CC and mediolateral oblique (MLO) view planes, analogous to mammography. During image acquisition, a motorized mechanism drives the US transducer across the breast, while an integrated lubrication system continuously supplies coupling medium between the film sheet and transducer. The efficiency of this technique is reliant on the breast tissues sufficiently extending outward into the scanning area. Therefore, it is effective in patients with larger, pendulous breasts, and limited utility in women with smaller, dense breasts.

Fusion-X-US

The FUSION-X-US prototype system that combines ACUSON S2000 3D ABVS (Siemens Healthcare, Erlangen, Germany) and digital breast tomosynthesis (DBT) (Mammomat Inspiration, Siemens Healthineers, Erlangen, Germany) acquisition into a single unit.[83] The FUSION-X-US system utilizes the high-frequency US transducer (ACUSON S2000) with a 5–14 MHz operating frequency inside a rigid frame. The contact area on the upper plate is made from woven gauze and permeable to ABVS coupling lotion to improve US coupling, while mitigating excessive X-ray absorption compared to a conventional mammography compression plate.[84] The FUSION-X-US acquisition process involves a single, integrated workflow with obtaining DBT images in one projection view, then performing ABVS in the same orientation and compression.[84–86] The DBT projection view, ML, MLO or CC, is selected depending on the clinical situation including breast geometry and position of the breast lesion. The acquired FUSION-X-US data is processed and stored on a computer in DICOM format, allowing for comparative analysis of the 3D DBT and ABVS images.

Preliminary studies with the FUSION-X-US prototype system demonstrated limitations in image quality and whole breast volume coverage.[84] However, technical advancements have improved reliability of the prototype for diagnostic workflow by providing sufficient whole breast volume coverage to 90.8% total area covered, image quality with 79.9% of cases comparable to HHUS, showing distinguishable tissue structures, and patient tolerability, with only 9.9% of patients reporting discomfort or pain.[85] Improvements in whole breast coverage and reductions in artifacts (seen in 57.7% of cases) by optimizing coupling with the membrane are required. A recent prospective feasibility study with the FUSION-X-US system was performed showing sufficient image quality with 80% whole breast coverage. This approach showed lesion detectability of 97.1% and high diagnostic accuracy of 85.0%.[86] These findings suggest that combined ABUS and DBT could prove both practical and logistical for breast cancer screening settings, as an alternative and novel dual-modality approach.

CLINICAL EVIDENCE

While various studies have demonstrated the effectiveness of HHUS for detecting small, invasive, node-negative, and mammographically occult breast cancers, several barriers exist that have limited their implementation as a widespread screening technique. To address these challenges, 3D ABUS has been developed to decouple image acquisition from interpretation, allowing for more standardized, operator-independent, reproducible, and volumetric multiplanar 3D visualization. Various studies implementing 3D ABUS in clinical practice have shown clear utility in supplemental screening and diagnostic imaging applications in predominantly intermediate to high-risk women, particularly those with dense breasts. This section summarizes the current literature demonstrating clinical evidence and implications for supplemental 3D ABUS screening and diagnostic applications.

SUPPLEMENTAL SCREENING

Various studies have investigated the use of 3D ABUS for supplemental screening to conventional mammography in asymptomatic intermediate to high-risk women. Women with dense breasts make up the largest portion of intermediate-risk population with a 20–30% increased incidence for breast cancer, making it one of the most important risk factors.[87] Therefore, supplemental screening plays an important role in the early detection of breast cancer for improving treatment decisions, management, and patient outcomes.

Supplemental screening with ABUS was first explored with semi-automated, hybrid-type 2D ABUS (AWBU) systems.[64,88] Kelly et al.[64] performed a multi-center prospective study to evaluate the diagnostic performance of mammography with AWBU (SonoCiné) with 4,419 women who were high-risk or had dense breasts. In this study, the sensitivity increased by 41% and the specificity increased by 3.5%, with an additional 3.6 cancers detected per 1,000 women (Table 6.2). This study showed significant improvements in overall cancer detection with ABUS combined with mammography, when compared with screening mammography alone. In the detection of small, invasive breast cancers smaller than 10 mm, significant improvements were reported, with a 48% increase (tripled) in sensitivity between AWBU and mammography, showing tremendous implications for early detection. When considering increased breast density, adding supplemental AWBU compared to mammography alone doubled the detection rate. This evidence establishes precedent to supplemental ABUS as a screening tool for early detection in women with dense breasts. A complementary multi-reader, multi-case (MRMC) performance study showed that adding AWBU to mammography in 102 participants significantly improved the identification asymptomatic breast cancers, improving overall confidence in recall rates in women with dense breasts.[88] Adding ABUS to screening mammography substantially improves overall breast cancer detection, recall rates, and overall confidence for call-backs in intermediate to high-risk women, with dense breasts.[64]

Giuliano et al.[13] evaluated supplemental 3D ABUS with screening mammography in 3,418 asymptomatic women with dense breasts, isolating breast density as the sole independent risk-factor. With 3D ABUS (Somo•V ABUS) added to mammography, the sensitivity and specificity was improved, with an additional 7.7 breast cancers identified per 1,000 women (Table 6.2). Overall, a 2.6-fold increase in cancer detection was observed with 3D ABUS added to mammography, compared to stand-alone mammography. This resulted in a higher theoretical breast cancer miss rate compared to the findings by Kelly et al.,[64] which the authors attribute to an increase in breast density, as the dominant risk-factor in the study. This study further explored the cost-effectiveness of implementing additional ABUS imaging, suggesting that improved effectiveness and diagnostic yield in women with dense breasts may justify its need in clinical practice.

Another retrospective, multi-reader study, the U-Systems (GE Healthcare) Pivotal Clinical Retrospective Reader Study, was performed in 164 patients with 133 non-malignant and 31 biopsy confirmed cases. Supplemental ABUS (Somo•V ABUS) showed a 24% increase in cancer detection without a substantial decrease in specificity (78.1% to 76.2%) compared to screening mammography alone in women with dense breasts.[89] The results from this study resulted in the United States of America (USA) FDA approval of 3D ABUS (Somo•V ABUS) in 2012, as a supplemental screening tool for mammography in women with dense breasts.

To date, the most extensive, observational, and multi-institutional study (conducted between 2009–2011) to assess supplemental 3D ABUS screening was the SomoInsight study.[68] This study evaluated the impact of supplemental 3D ABUS (Somo•V ABUS) with full-field digital mammography (FFDM) screening with a one-year follow up. This study included 15,318 asymptomatic women (25–94 years old) with dense breasts, as an independent risk-factor, regardless of other risk-status. Supplemental ABUS screening resulted in a 26.7% increase in sensitivity with an additional 1.9 cancers detected per 1,000 women. Almost all these additional cancers detected were small, invasive, and node-negative cancers (Table 6.3). Detection of these clinically important cancers suggests positive prognostic implications from early detection, intervention, and outcomes in asymptomatic women. While supplemental 3D ABUS showed increased detection sensitivity of 26.7%, ABUS still suffers from limitations, with a 13.4% decrease in specificity and non-optimal, elevated recall rates with 134.7 additional women recalled per 1,000 women compared with screening mammography alone. In addition, the authors found that sensitivity and specificity were dependent on the quality

TABLE 6.2

Comparison of Sensitivity, Specificity, Detection Rate, and Recall Rates between Supine and Prone Supplemental Automated Breast Ultrasound (ABUS) Screening with Mammography (MMG)

Author	Year	Modality	Patients	Sensitivity %		Specificity %		Detection Rate			Recall Rate		
				FFDM	ABUS + MMG	FFDM	ABUS + MMG	FFDM	ABUS	Difference	FFDM	ABUS	Difference
Kelly et al. [1]	2010	2D AWBU	4,419	40	81	95.2	98.7	3.6	7.2	3.6	4.2	9.6	5.4
Giuliano et al. [2]	2013	3D ABUS	3,418	76	96.7	98.2	99.7	4.6	12.3	7.7	–	–	–
Brem et al. [3]	2015	3D ABUS	15,318	73.2	100	85.4	72	5.4	7.3	1.9	15	28.4	13.4
Wilczek et al. [4]	2016	3D ABUS	1,668	63.6	100	99	98.4	4.2	6.6	2.4	1.3	2.2	0.9
Giger et al. [5]	2016	3D ABUS	185	57.5	74.1	78.1	76.1	–	–	–	–	–	–
Gatta et al. [6]	2021	Prone ABUS	1,165	58.8	93.5	94	87	3.4	6.8	3.4	14.5	26.6	12.1

ABUS, automated breast ultrasound; AWBU, automated whole breast ultrasound; MMG, mammography; 2D, two-dimensional; 3D, three-dimensional.

of interpretation and interobserver variation. Stand-alone performance of 3D ABUS image interpretation was not evaluated in this study. Moreover, the SomoInsight study provided a discussion
on participant characteristics, showing that women with elevated breast density (BI-RADS 4) were
more likely to be younger, Asian, low BMI, and premenopausal, compared to women with lower
breast density.[68] These factors emphasize the importance of population demographics on risk-based
screening, suggesting that protocols, including imaging modalities, should be well-adapted to the
population of interest.

The European Asymptomatic Screening Study (EASY) was another observational, mono-center
study in Sweden (2016) that evaluated the impact of supplemental 3D ABUS screening (Somo•V
ABUS) with FFDM in 1,668 asymptomatic women (40–74 years old) with heterogeneously
(BI-RADS 3) and extremely (BI-RADS 4) dense breasts.[90] This study showed 36.4% improvement
in sensitivity with an additional 2.4 cancers detected per 1,000 women (Table 6.3). There was an
increase in recall rate with 9.0 women recalled per 1,000 women compared to mammography alone,
which comparable with the ACRIN 6666 study.[91] Importantly, this study showed that it was feasible
to implement 3D ABUS into a breast cancer screening program at a high-volume center, allowing
for improved detection of early stage breast cancers, with an admissible recall rate within diagnostic
and screening quality assurance standards.

Supplemental 3D ABUS to screening mammography shows performance accuracy comparable
to pivotal, large-scale studies with supplemental HHUS screening. In screening settings, improving cancer detection rates is only one goal for improving population cancer-based outcomes. The
SomoInsight and EASY studies showed cancer detection rates comparable to the ACRIN 6666 and
J-START studies, with 5.3 and 1.8 additional cancers per 1000 women. [6,31,68,90] Across these four
studies, the additional detected cancers were predominantly small, invasive, and node-negative cancers. While these studies included asymptomatic screening populations, the differences in detection rates in these studies were observed due to differences in their inclusion criteria, ACRIN 6666
included 2,809 high-risk women, J-START study included 36,752 women 40–49 years old with no
history of breast cancer, and SomoInsight [68] and EASY[90] studies included 15,1318 and 1,668 women
with dense breasts, respectively. When comparing recall rates, these findings were generally higher
for 3D ABUS (13.4% in SomoInsight) than HHUS (9.4% and 8.8% for ACRIN 6666 and J-START)
which may be largely attributed to the variable reader training, experience, and performance due to
a learning curve when implementing a newer imaging modality (3D ABUS) in practice.

Reader (observer) performance and interobserver variability is pertinent when considering diagnostic accuracy. Giger et al.[92] performed a multi-reader, multi-case, sequential design reader study
that evaluated supplemental 3D ABUS screening with FFDM in 185 asymptomatic women with
dense breasts (BI-RADS 3 and BI-RADS 4). Reader performance was assessed as the area under
the curve (AUC), receiver operating characteristic (ROC) curve, sensitivity and specificity. This
study showed that adding ABUS to mammography evaluation significantly improved the detection
of breast cancer in women with dense breasts with an 29% relative increase in sensitivity and a small
2.4% decrease in specificity. Another notable finding was a statistically significant 25% relative
improvement in the AUC with added ABUS in mammographically occult cancers.[92] Another study
demonstrated consistent findings that adding supplemental 3D ABUS significantly increased the
recall rate with both screening mammography and DBT in women with dense breasts.[93] Moreover,
it was observed that double-reading of ABUS during early adoption reduces false-positive and
recall rates. Generalization of double-reading protocols to 3D ABUS for screening could play an
important role in the interpretation of early stage breast cancers.[94]

The recall rate is expected to progressively decrease over time with increased reader experience and confidence in screening settings. Arleo et al.[4] retrospectively evaluated the recall rates of
screening ABUS in women with dense breasts. This study showed that recall rates decreased from
24.7% to 12.6% from the first to third month of implementation, clearly showing the learning curve
for ABUS interpretation. Improved reader experience improves recall rates and decreases unnecessary biopsies, improving the positive predictive values, which are important when considering

TABLE 6.3

Comparison of Main Extensive Studies Adding Supplemental ABUS and HHUS with Screening Mammography

Study	Year	Modality	Number Patients	Inclusion Criteria	Time Period	Additional Cancers Detected per 1,000	Additional Recall Rate per 1,000	Mean Size (mm)	Invasive (%)	Node-Negative (%)
SomoInsight [3]	2015	3D ABUS	15,318	Asymptomatic women with dense breasts	2009–2011	1.9	13.4	12.9	93.3	92.6
EASY [4]	2015	3D ABUS	1,668	Asymptomatic women with dense breasts	2010–2012	2.4	9.0	21.8	–	50.0
ACRIN 6666 [7]	2012	HHUS	2,809	Asymptomatic women 40-49 years old	2004–2006	5.3	9.4	10	93.7	96.7
J-START [8]	2016	HHUS	36,752	Asymptomatic high-risk women	2007–2011	1.8	8.8	14.2	82	85.5

ABUS, automated breast ultrasound; MMG, mammography; 2D, two-dimensional; 3D, three-dimensional.

high-volume 3D ABUS screening populations.[38] Moreover, studies have shown that reviewing 3D ABUS in conjunction with mammography results in a higher interobserver agreement.[95] Combined 3D ABUS with mammography resulted in overall increased confidence during interpretation in women with dense breasts, suggesting that these modalities could be employed as complementary screening tools to improve diagnostic performance. These studies highlight the importance of reader experience and suggests that proper training protocols, and adequate experience to surpass the learning curve, are necessary for accurate, reliable, and confident interpretation with supplemental 3D ABUS in clinical practice.

Interpretation time has also been evaluated with the use of screening ABUS. Arslan et al.[96] evaluated the interpretation time of 3D ABUS between radiologists with different experiences and expertise. This study found significant differences in mean interpretation time between a junior radiologist and senior breast radiologist. However, the mean overall time between both expertise levels was still substantially less than HHUS. Readers with different experiences can perform ABUS interpretation with comparably short times, showing implications to improve time-efficiency in resource-limited settings, regardless of clinician experience. The performance accuracy to detect and characterize breast cancer should be equally considered to ensure quality interpretation and efficient workflow, while reducing recall rates and unnecessary biopsies. To further optimize workflow, CAD systems can be implemented to optimize ABUS interpretation by improving performance accuracy and time-efficiency.[38]

In addition to observer training and experience in interpretation, there are several technical aspects to be considered to ensure high-quality 3D ABUS images are acquired in screening settings. Improving acquisition requires proper operator training protocols and standards of performance (benchmarks) to ensure proper patient positioning, sufficient volumetric field of view for whole breast coverage, contact pressure and compression, and adequate coupling with the membrane (or medium) between the US transducer and breast tissues to limit acquisition artifacts.[97]

While supplemental 3D ABUS and HHUS screening play an important role in the detection of early stage breast cancer for women with dense breasts, the capability for operator-independence, time-efficiency for image acquisition and interpretation, dataset reproducibility, and reliability shows the superior ability of 3D ABUS. Overall, clinical evidence shows that supplemental 3D ABUS screening substantially improves the detection of mammographically occult breast lesions, that are typically small, invasive, and node-negative cancers – clinically important for improved patient prognosis. Furthermore, the efficacy and ability to be successfully employed in high-volume settings shows potential for screening applications. Standardized operator training and observer interpretation protocols are still required, but present evidence shows diagnostic value for adopting supplemental 3D ABUS screening in women with dense breasts for present and future standards. Future work requires extensive multi-institutional, randomized trials in high-volume settings to assess clinical efficiency and cost-effectiveness for widespread implementation of 3D ABUS screening for mortality risk reduction.[98,99]

Prone 3D ABUS

While supplemental 3D ABUS screening with supine approaches have shown clinical evidence for positive implications, only few studies have explored the use of prone 3D ABUS systems. Gatta et al.[56] evaluated the performance of supplemental prone 3D ABUS (SOFIA) with screening mammography with 1,165 asymptomatic women with dense breasts, independent of other risk-factors. When prone 3D ABUS was added to screening mammography, sensitivity increased by 34.7%, with a 31.8% increase for interval cancers and an additional 12.1 recalls and 0.7 cancers were detected per 1000 women (Table 6.3). This study showed that supplemental prone 3D ABUS screening is promising in women with dense breasts, with diagnostic performance comparable to supine ABUS counterparts.[56] Multi-institutional studies with an increased screening volume and number of participants are required to evaluate its utility, feasibility, and implications for supplemental screening.

Supplemental UCST (SoftVue) screening has also been explored, as an alternative prone 3D ABUS approach. Clinical evidence from a preliminary multi-reader, multi-case study, as part of the Delphinus SoftVue Prospective Case Collection studies, demonstrated that USCT can accurately distinguish between dichotomously categorized benign (BI-RADS 1 and 2) and malignant (BI-RADS 3 and 4) breast lesions, with an 20% increase in sensitivity and 8% increase in specificity, in comparison to screening mammography in women with dense breasts.[60,100–104] Moreover, whole breast stiffness imaging with UCST shows the ability for characterization of breast tissues and masses, showing potential implications for screening, diagnostic evaluation, and treatment monitoring.[59]

DIAGNOSTIC EVALUATION

Beyond supplemental screening, 3D ABUS shows effective clinical utility for various diagnostic evaluation applications, which have been explored in several studies and described in several published reviews.[34,40,65,98] While screening protocols are generally recommended for specific populations, diagnostic evaluation of palpable breast lesions or other breast cancer-related symptoms with HHUS has been widely used and proven in clinical practice. With the ability of 3D ABUS to efficiently depict the entire breast with high image quality, ABUS has opened new diagnostic avenues. This subsection discusses diagnostic applications for 3D ABUS, including its diagnostic accuracy and characterization of breast lesions, its unique interpretability with the coronal view plane, preoperative and surgical-guidance settings, second-look examination, neoadjuvant chemotherapy treatment response, and applications for molecular subtypes.

Diagnostic Accuracy and Characterization

Several studies have compared the diagnostic accuracy and detectability of 3D ABUS with conventional HHUS.[71,73,105–120] Across these studies, the reported sensitivities and specificities for detecting breast cancer were within the range of 69.2–100% and 63–100% for ABUS, and 62.5–100% and 62.5–100% for HHUS. Generally, ABUS and HHUS show comparable sensitivity, specificity, and detectability of breast lesions with no statistically significant differences.[113,117,118,120,121] Conversely, an extensive, retrospective study with 5,566 asymptomatic women (1,866 women scanned with ABUS and 3,700 women scanned with HHUS) found significantly increased diagnostic accuracy and specificity with supplemental ABUS to HHUS.[106] These results were consistent with the findings from previous studies.[113,121] Another multi-reader, multi-case study showed significantly higher diagnostic accuracy for ABUS to HHUS under the ROC curves, but no observable differences in sensitivity or specificity.[105]

Considering the large number of studies conducted to investigate the diagnostic accuracy of ABUS, several comprehensive systematic reviews and meta-analyses have been performed. Meng et al.[122] performed a literature review of studies until September 2014, resulting in a meta-analysis with thirteen studies evaluating ABUS compared to pathological results as reference standard. This meta-analysis showed a pooled sensitivity, specificity, and AUC of 92% (89.9–93.8%), 84.9% (82.4–87%) and 0.96 of ABUS, respectively. Subgroup analysis showed no obvious discrepancies in diagnostic accuracy of ABUS compared to HHUS. Wang and Qi[123] performed a literature review on studies conducted until June 2018; and a meta-analysis with nine studies including 1,775 patients and 1,985 (628 malignant and 1,357 benign) lesions was performed to compare ABUS with HHUS. This resulted in a pooled sensitivity and specificity of 90.8% (95% CI, 88.3–93.0%) and 82.2% (80.0–84.2%) for ABUS, and 90.6% (88.1–92.8%) and 81.0% (78.8–83.0%) for HHUS. The AUC of the summary ROC (SROC) was 0.93 for ABUS and 0.94 for HHUS. Zhang et al.[73] conducted another systematic literature search in November 2018 across all English and Chinese literature to compare ABUS with HHUS, resulting in a meta-analysis population of 1,376 patients across nine studies. This meta-analysis revealed a pooled sensitivity of 93% (95% CI, 91–95%) for ABUS and 90% (95% CI, 88–92%) for HHUS, and pooled specificity of 86% (95% CI, 83–88%) for ABVS and

82% (95% CI, 79–84%) for HHUS. The AUC of SROC was 0.95 for ABVS and 0.91 for HHUS. Meta-regression showed no statistically significant differences in diagnostic accuracy between 3D ABUS and HHUS. Overall, these meta-analyses show that ABUS has comparable diagnostic performance to HHUS.

Most recently, when explicitly considering the differential diagnosis between benign and malignant breast lesions, Ibraheem et al.[124] performed a systematic review between 2011–2020, resulting in 21 studies with 6,009 patients and 5,488 (4,074 benign and 1,374 malignant) lesions. Across these studies, the range of sensitivity was 72–100% for ABUS and 62–100% for HHUS; and the range of specificity was 52–98% for ABVS and 49–99% for HHUS. Across the primary clinical studies included in this meta-analysis, variabilities across the range of reported diagnostic accuracy, sensitivity, specificity, and detectability values may be attributed to differences in study design, including variable participant cohorts, inclusion criteria, ABUS system, reference standards, follow-up periods, and reader experience.[124]

Detectability of breast cancer is generally comparable or improved with 3D ABUS with reported detection rates ranging from 83.0–100% for ABUS compared to 60.6–100% for HHUS.[73,107,117,120,121,125–129] Across these studies, Zhang et al.[125] and Xiao et al.[126] showed statistically higher detection rates between ABUS and HHUS. In malignant lesions, the detection performance of ABUS has been reported as substantially improved or significantly higher compared to benign lesions.[126] Chang et al.[115] suggested that malignant lesion characteristics, such as size, shape, location, and surrounding tissue changes are more perspicuous with ABUS compared to HHUS, allowing for significantly higher sensitivity. When compared with pathologic results or one-year follow up, Niu et al.[110] showed statistically significant improvements in sensitivity and AUC with ABUS compared to HHUS, but no differences in specificity and diagnostic accuracy. Furthermore, this study showed high correlation between the maximum diameter measured with ABUS with histological results,[110] consistent with previous findings demonstrating the ability of ABUS to better estimate size.[130]

When considering breast density, Zhang et al.[129] compared 3D ABUS with HHUS in 1,973 patients, 75.7% with predominantly (BI-RADS 4 and 5) dense breasts. This study showed that 78.6% of women with dense breasts were more accurately diagnosed with precancerous lesions or cancer using ABUS, 7.2% higher than with HHUS. Conversely, in women with non-dense breasts (BI-RADS 1 and 2), ABUS and HHUS had similar false-negative rates. Moreover, the overall sensitivity and specificity of ABUS in women with dense breasts were higher than HHUS, with reported false-negative rates lower than mammography. Guldogan et al.[112] evaluated the diagnostic performance of same-day ABUS with 592 participants with dense breasts and a relatively long (20–42 month) follow-up period. This study found a comparable sensitivity and specificity between ABUS and HHUS, and a good agreement between for characterization of positive and negative BI-RADS categories. ABUS may be superior in detecting malignant breast lesions, but inferior to HHUS in detecting benign lesions. In these cases, the cost-benefit between increased detection rates and recall rates should be carefully considered and long-term clinical implications should be further explored.

In the meta-analysis, Zhang et al.[73] evaluated the detectability with 1,047 pathologically confirmed malignant lesions (>BI-RADS 4A) across thirteen studies. This meta-analysis only included studies that used pathology as the gold-standard, excluding studies that compared malignancy with other imaging modalities, such as HHUS and MRI as the reference standard.[105,114] This meta-analysis revealed a pooled detection rate of 100% (95% CI, 1.00–1.00) for both ABUS and HHUS, but notable publication bias was found.[73] Ibraheem et al.[124] showed that the diagnostic accuracy was 59–98% for ABUS and 80–99% for HHUS. Across a subset of the included studies, the mean lesion size was 2.1 mm (1.6–2.6 cm), and of those lesions identified, 94% were characterized as malignant, compared to non-cancerous lesions. These findings showed that ABUS was superior in detecting early stage (BI-RADS 0) or malignant (BI-RADS 4) cancers when compared with HHUS, which conversely, identified substantially more BI-RADS 1 and BI-RADS 2 breast lesions. This evidence is consistent with the primary studies discussed, showing implications for early detection to improve clinical decision making for recalls, biopsies, and interventions. Overall, these pertinent meta-analyses show that the diagnostic performance of ABUS is comparable to HHUS, with respect to characterization of benign and malignant lesions.

Observer concordance between ABUS and HHUS is generally comparable when comparing the characterization of breast lesions.[105,113,114,131–133] Comparison between BI-RADS assessment scores in ABUS and HHUS images showed fair to substantial agreement with reported kappa (k) scores.[105,114] Shin et al.[134] found substantial agreement was found between readers across all BI-RADS categories (k = 0.63) with an elevated agreement score (k = 0.71) when the assessments were grouped into three categories: BI-RADS 1, 2, and 3, BI-RADS 4A, 4B, and 4C, and BI-RADS 5. Substantial agreement (k = 0.63) was found for the final assessment with ABUS was compared to HHUS. Additionally, there was a high correlation in the lesion location with clock position (k = 0.75) and distance from the nipple (k = 0.89), showing excellent reliability for localization and reproducibility for 3D ABUS follow-up. Golatta et al.[131] demonstrated fair agreement (k = 0.34) between ABUS and HHUS on BI-RADS scores, but substantial agreement (k = 0.68) with a dichotomized approach with benign (BI-RADS 1 and BI-RADS 2) and suspicious (BI-RADS 4 and BI-RADS 5) categories. Golatta et al.[132] further demonstrated fair agreement (k = 0.31) with the dichotomized approach in a large-scale explorative study. While this was an acceptable agreement, the authors suggest that the ABUS examiners were at a disadvantage due to improper blinding protocols, since HHUS examiners had prior knowledge about the clinical situation.[132] Across these studies, various characteristics have showed excellent agreement in 3D ABUS images, specifically, the margin[113] and orientation[105] features, making them important predictors for malignancy.

Coronal View Plane

The coronal view plane is a unique multiplanar visualization feature of 3D ABUS that provides a global view of the breast, surrounding anatomical structures, and tissues from the nipple to the chest wall for complete visualization. This view plane allows for enhanced visibility of architectural distortions[38] and enhanced edge visualization that contributes to the improved detectability and characterization assessment of malignant breast lesions.[116,135–137] Characteristics of lesions in the coronal view plane include, the presence or absence of the retraction phenomenon, continuous or discontinuous margin contour, and indistinct or circumscribed margin.[116]

The retraction phenomenon, caused by tumor infiltration, deformation, and traction of surrounding tissues, describes the architectural distortion by convergence of surrounding tissues on a mass.[73] In a coronal ABUS image, the retraction phenomenon typically appears as cord-like hyperechoic lines radiating outwards from the surface of a breast lesion.[39,116,118] The hyperechoic rim is a complementary feature to the retraction phenomenon, where the presence of a continuous rim suggests that the lesion is benign and the presence of a discontinuous rim is strongly correlated with malignancy.[136] Both the retraction phenomenon and microlobulated margins have been proven as excellent predictors of malignancy with high diagnostic value.[136] Moreover, there exists a strong correlation between spiculation patterns and retraction phenomenon with the likelihood of malignancy, significantly improving the characterization breast lesions.[135]

Clinical evidence shows that the retraction phenomenon allows for substantial diagnostic advantage in the detection and characterization of breast lesions.[117,118] Lin et al.[117] showed a high sensitivity, specificity, and accuracy of 80%, 100% and 91.4%, respectively, when differentiating between lesions. Chen et al.[118] showed a high specificity, diagnostic accuracy, and false-positive of 100% and 96.8%, and 0%, respectively, in malignant cancers; and a specificity and accuracy of 92.8% and 95.9%, respectively, of benign cancers with the hyperechoic rim. With pathological comparison, the presence of a hyperechoic rim with retraction phenomenon has been shown to strengthen the retraction pattern, as a predictor for improved prognosis of breast cancer.[137] While the retraction phenomenon is a strong predictor for malignancy, it may be present with radial and postoperative scarring due to biopsies or other interventions.[98,117]

Preoperative Applications

The coronal view has also been termed as the surgical view plane, as it reflects the surgeon's view from the operation table. Coronal 3D ABUS images show ability for accurate assessment of preoperative extent of disease.[97,136] ABUS has been proven superior to HHUS for preoperative assessment in ductal carcinoma *in situ* (DCIS) for planning surgical interventions,[138] guiding breast conserving surgery (BCS), and predicting recurrence.[139] ABUS significantly outperforms HHUS in estimating size and volumetric measurements of breast lesions.[130,136,138,139] Improving preoperative size assessment and margin identification can improve surgical planning and resection to ensure pathologically negative margins are achieved. This evidence suggests that 3D ABUS can be effectively implemented into surgical planning workflow. Beyond preoperative applications, future directions may call 3D ABUS for image-guided interventions with the combination of robotic-assisted systems.[140]

Second-Look ABUS Examination

Second-look US examination is another application where 3D ABUS shows potential use. Despite the effectiveness of breast MRI for the evaluation for disease extent and treatment planning, MRI has a low specificity for detecting breast cancer. Second-look ABUS is an efficient method to reliably and accurately detect lesions identified on breast MRI for preoperative applications, compared to second-look HHUS.[119,127] However, ABUS and HHUS show a significantly lower ability to detect non-mass lesions compared to mass-type lesions, compared with MRI, which is a limitation of US-based modalities.[127] Histopathological measurements show that ABUS allows for improved lesion localization and estimation of lesion size (k = 0.83) for preoperative applications compared to MRI, but has a similar diagnostic accuracy as HHUS.[109] Although ABUS has limited capability or compatibility for image-guided biopsy, second-look ABUS may improve clinical decision making for image-guided breast biopsies, under US and MRI guidance,[109,119,127] or with alternative methods for high-risk women with dense breasts, such as under high-resolution positron emission mammography (PEM) and US-guidance.[141]

Neoadjuvant Chemotherapy Treatment Response

ABUS shows potential use for early assessment, management, and accurate response prediction of neoadjuvant treatment. Prediction of early pathological response to neoadjuvant chemotherapy (NAC) or neoadjuvant therapy (NAT) allows for better informed treatment decisions for personalized and patient-specific care. Wang et al.[142] predicted the pathological response with pathological complete remission (pCR) after four cycles of NAC. When measuring pCR two cycles post-NAC, ABUS had a high sensitivity and specificity of 88.1% and 81.5%, respectively. However, ABUS was found less useful in predicting adverse pathological outcomes. Similarly, a case study found that ABUS images acquired two and five cycles post-NAT cycles showed significant reduction of tumor size, disappearing after seven cycles, which was confirmed with postoperative histopathological examination, indicating complete pCR.[143] While the reduction in tumor size was evident with concentric shrinkage and 3D ABUS volumetric measurements, there were still visible microcalcifications. However, this was consistent with evidence that some patients with complete pCR have no changes in microcalcification distribution and appearance post-NAT.[143–145]

Another pilot observational study, part of the multi-institutional European RESPONDER trial, compared the tumor response by 3D ABUS and MRI during and after NAC.[146] This study showed that ABUS had similar tumor volume estimation and treatment response to MRI, with excellent interobserver variability. Compared with MRI, ABUS was more favorable for patients. More recently, D'Angelo et al.[147] compared ABUS with contrast enhanced (CE) MRI in assessment of tumour response (pCR) during NAC. This study found that 3D ABVS shows accurate response in patients with pCR post-NAC, comparable to CE–MRI. On the contrary, a comparative study comparing mammography, DBT, MRI, and ABUS in the evaluation of residual tumors after NAC in early stage breast cancers showed that ABUS had the lowest accuracy in assessing pCR with an underestimation of tumor size, in comparison to MRI and DBT, that comparably performed the best.[148] Future extensive prospective studies are required to evaluate the clinical efficiency of 3D ABUS on the response to NAC.[149]

Breast Density Estimation

Breast density estimation is crucial to ensure informed treatment decisions are made for risk management and early detection of breast cancer. Early studies have evaluated breast density estimation with prone 2D ABUS (Aloka Hitachi ASU-1004) with threshold and proportion-based methods, reporting high 87.5% and 84.4% accuracies, respectively.[150,151] This study showed that reliable quantification of breast density with ABUS was possible. Moon et al.[152] showed a strong correlation between ABUS and MRI, as a gold-standard, in estimating percent breast density and whole breast volume with 91.7% and 88.4% correlations, respectively. Since 3D ABUS may have the greatest utility at the time of mammography screening, timely estimation of breast density of the entire breast volume is crucial to assess increased risk.[28] With the use of ABUS as a primary screening tool in younger, high-risk women with dense breasts, quantifying breast density with ABUS could allow these populations to make better informed decisions for longitudinal risk management. Future directions call for near real-time density estimation methods with ABUS to make justifiable, same-session, and point-of-care ABUS a reality.

Molecular Subtypes

The correlation between 3D ABUS imaging features and morphological characteristics to predict pathological factors and molecular subtypes of breast cancer is an emerging area for exploration.[34,98] The global appearance of breast anatomy on 3D ABUS allows for unique morphological features in the images and on the coronal view plane. ABUS imaging features, including lesion shape, size, margins, acoustic features, echogenicity, calcifications, and the retraction phenomenon have been explored for molecular subtypes,[137,153,154] consistent with predictors previously discussed for characterization.[105,113] Studies show significant associations between the retraction phenomenon with small breast tumor size, lower histological grade, and positive estrogen receptor (ER) and progesterone receptor (PR) expression status, and molecular subtypes, luminal A (ER positive) and triple negative breast cancers.[137] Therefore, ABUS is able to accurately identify both low-grade cancers that are early stage, and higher grade, aggressive cancers. Wang et al.[154] observed a strong correlation between echo heterogeneity with margin appearance and malignancy in different breast cancer types and stages. Increased concordance of lesion characteristics and features with specific molecular subtypes can potentially improve reader confidence for effective, noninvasive diagnosis.

ABUS Limitations

Image quality and interpretability of 3D ABUS images are reliant on several factors during image acquisition.[39,40,99,155–157] Artifacts due to improper contact with the scanning surface including, insufficient uniform compression, improper selection of the coupling medium, improper application of the coupling medium causing air interposition, and transducer motion artifacts, could cause posterior shadowing artifacts and dropout, which impact the presentation breast tissues[39] and increase false-positive findings.[95] In women with larger or more firm breasts, peripheral dropout on the edges may increase cancer misdiagnosis with standard acquisition views. Corrugation due to respiratory motion or breathing is another common artifact, which manifest as wave-like artifacts in the image and reduces detectability.[39,40,98] Proper patient positioning, preparation, and acquisition setup, is crucial to eliminate acquisition artifacts and obtain high-quality ABUS images. Beyond these technical challenges, limitations exist when imaging patients with both smaller and larger breasts, due to preset automated acquisition protocols.[49]

While ABUS is generally well-tolerated, ABUS has been reported as uncomfortable and painful during the preparation and image acquisition protocol. A study using supplemental ABUS in 340 women with dense breasts showed that 79.1% of women experienced pain, with 10.6% of women reporting severe pain.[111] Most participants that reported severe pain were thinner-frame women with smaller breasts, which may suggest that ABUS may not be optimal in this population, despite its increased utility in women with smaller breasts.[111] This pain may be caused from anisotropic compression required to distribute dense breast tissues to increase uniformity for obtaining high-quality ABUS images.[158] Most of the women in this study (59.7%) preferred HHUS to ABUS, since its examination time was perceived as shorter, less painful, and real-time interpretation was performed.[111] However, these results may be attributed to familiarity heuristic since 3D ABUS is not yet standard of care.

When considering diagnostic evaluation, acquired ABUS images manifest several unique artifacts from HHUS that affect their interpretability.[156] While technical 3D ABUS artifacts reduce the detectability, characterization, and differentiation of breast lesions, some artifacts enhance diagnostic value, specifically in the coronal view plane. A detailed description of these ABUS artifacts that aid and hinder accurate interpretation, and assistance with resolving challenges related to artifacts for clinical diagnostic settings have been published in a recent review.[156] Clinical evidence of the coronal view plane for detectability and characterization of breast cancer is described in the earlier section.

In addition to the technical acquisition and interpretation challenges, limitations exist with cost-effectiveness and cost-benefit in clinical practice. While there is strong clinical evidence supporting 3D ABUS, feasible implementation of 3D ABUS technologies for screening applications and diagnostic evaluation requires holistic assessment of cost-effectiveness and cost-benefit. Most commercially available ABUS systems are relatively costly, requiring the expensive equipment (complete dedicated ABUS system) with specialized hardware and commercial workstations to be purchased and installed for use.[28] Increased operational costs are required for high-quality acquisition by trained technologists (operators) and interpretation by experienced readers. Furthermore, the display, viewing, and interpretation of most 3D ABUS images require a dedicated workstation. Storing 3D ABUS images with picture archiving and communication systems (PACS) and DICOM standards would improve data accessibility. Standardization would open doors to offline interpretation, which would be especially useful in resource-limited settings, coupled with hybrid ABUS systems with widely available US technologies.

Cost-effectiveness can be improved with hybrid-type 3D ABUS systems that are compatible with commercially available US equipment, including sufficiently high-frequency US transducers. Adaptability with a hybrid system would further reduce potential operational costs associated with installation and maintenance. Only few studies examining the cost-benefit and economic impact of ABUS for screening have been published.[39] Future work is necessary to assess the feasibility of its implementation across various populations, given variable demographic risk-factors and availability of resources.

CONCLUSIONS

Advancements in 3D ABUS technologies have enabled diverse approaches for effective high-resolution acquisition, multiplanar 3D reconstruction, visualization, and interpretation, while continually improving operator-independence, time-efficiency, dataset reproducibility, and reliability. Implementing 3D ABUS into supplemental screening and diagnostic applications has demonstrated clinical evidence and implications for improved detectability and characterization, demonstrating accelerating importance in intermediate to high-risk patients, particularly those with dense breasts. Supplemental 3D ABUS screening substantially improves the detection of nonpalpable, mammographically occult cancers that are small, invasive, and node-negative, which are clinically important for early detection and improved patient prognosis. Compared with HHUS, 3D ABUS shows promising diagnostic utility with a high, sensitivity, specificity, detectability, and characterization accuracy of breast lesions. Improvements in high-quality acquisition, and reader experience and confidence with 3D ABUS are still necessary. Moreover, evaluating the cost-effectiveness, cost-benefit, and long-term clinical implications of 3D ABUS across diverse patient populations and risk-factors should be explored to consider 3D ABUS as widespread modality for screening and diagnostic applications in clinical practice.

REFERENCES

1. Sung H, Ferlay J, Siegel RL, et al. Global cancer statistics 2020: GLOBOCAN estimates of incidence and mortality worldwide for 36 cancers in 185 countries. *CA Cancer J Clin*. 2021;71(3):209–249. doi:10.3322/caac.21660
2. Tabár L, Vitak B, Tonychen HH, Yen MF, Duffy SW, Smith RA. Beyond randomized controlled trials: Organized mammographic screening substantially reduces breast carcinoma mortality. *Cancer*. 2001;91(9):1724–1731. doi:10.1002/1097-0142(20010501)91:9<1724::aid-cncr1190>3.0.co;2-v
3. Oeffinger KC, Fontham ETH, Etzioni R, et al. Breast cancer screening for women at average risk: 2015 guideline update from the American Cancer Society. *J Am Med Assoc*. 2015;314(15):1599–1614. doi:10.1001/jama.2015.12783
4. Arleo EK, Hendrick RE, Helvie MA, Sickles EA. Comparison of recommendations for screening mammography using CISNET models. *Cancer*. 2017;123(19):3673–3680. doi:10.1002/cncr.30842
5. Seely JM, Alhassan T. Screening for breast cancer in 2018: What should we be doing today? *Curr Oncol*. 2018;25(6):S115–S124. doi:10.3747/co.25.3770
6. Berg WA, Blume JD, Cormack JB, et al. Combined screening with ultrasound and mammography vs mammography alone in women at elevated risk of breast cancer. *J Am Med Assoc*. 2008;299(18):2151–2163. doi:10.1001/jama.299.18.2151
7. Mandelson MT, Oestreicher N, Porter PL, et al. Breast density as a predictor of mammographic detection: Comparison of interval- and screen-detected cancers. *J Natl Cancer Inst*. 2000;92(13):1081–1087. doi:10.1093/jnci/92.13.1081
8. Boyd NF, Rommens JM, Vogt K, et al. Mammographic breast density as an intermediate phenotype for breast cancer. *Lancet Oncol*. 2005;6(10):798–808. doi:10.1016/S1470-2045(05)70390-9
9. Berg WA. DenseBreasts-Info.org. DenseBreasts-Info.org
10. Majid AS, De Paredes ES, Doherty RD, Sharma NR, Salvador X. Missed breast carcinoma: Pitfalls and pearls. *Radiographics*. 2003;23(4):881–895. doi:10.1148/rg.234025083
11. Nazari SS, Mukherjee P. An overview of mammographic density and its association with breast cancer. *Breast Cancer*. 2018;25(3):259–267. doi:10.1007/s12282-018-0857-5
12. Boyd N, Guo H, Martin LJ, Sun L, Stone J, Fishell E, Jong RA, Hislop G, Chiarelli A, Salomon Minkin MJY. Mammographic density and the risk and detection of breast cancer. *N Engl J Med*. 2007;356(3):227–236. doi:10.1056/NEJMoa062790
13. Giuliano V, Giuliano C. Improved breast cancer detection in asymptomatic women using 3D-automated breast ultrasound in mammographically dense breasts. *Clin Imaging*. 2013;37(3):480–486. doi:10.1016/j.clinimag.2012.09.018
14. Titus-Ernstoff L, Tosteson ANA, Kasales C, et al. Breast cancer risk factors in relation to breast density (United States). *Cancer Causes Control*. 2006;17(10):1281–1290. doi:10.1007/s10552-006-0071-1
15. Vachon CM, Pankratz VS, Scott CG, et al. Longitudinal trends in mammographic percent density and breast cancer risk. *Cancer Epidemiol Biomarkers Prev*. 2007;16(5):921–928. doi:10.1158/1055-9965.EPI-06-1047 [CrossRef][10.1158/1055-9965.EPI-06-1047]
16. McCormack VA, Dos Santos Silva I. Breast density and parenchymal patterns as markers of breast cancer risk: A meta-analysis. *Cancer Epidemiol Biomarkers Prev*. 2006;15(6):1159–1169. doi:10.1158/1055-9965.EPI-06-0034
17. Tosteson ANA, Fryback DG, Hammond CS, et al. Consequences of false-positive screening mammograms. *JAMA Intern Med*. 2014;174(6):954–961. doi:10.1001/jamainternmed.2014.981
18. Miglioretti DL, Lange J, Van Den Broek JJ, et al. Radiation-induced breast cancer incidence and mortality from digital mammography screening a modeling study. *Ann Intern Med*. 2016;164(4):205–214. doi:10.7326/M15-1241
19. Pauwels EKJ, Foray N, Bourguignon MH. Breast cancer induced by X-ray mammography screening? A review based on recent understanding of low-dose radiobiology. *Med Princ Pract*. 2016;25(2):101–109. doi:10.1159/000442442
20. El-Bastawissi AY, White E, Mandelson MT, Taplin S. Variation in mammographic breast density by race. *Ann Epidemiol*. 2001;11(4):257–263. doi:10.1016/S1047-2797(00)00225-8

21. Nie K, Su MY, Chau MK, et al. Age- and race-dependence of the fibroglandular breast density analyzed on 3D MRI. *Med Phys.* 2010;37(6):2770–2776. doi:10.1118/1.3426317

22. Sprague BL, Gangnon RE, Burt V, et al. Prevalence of mammographically dense breasts in the United States. *J Natl Cancer Inst.* 2014;106(10):1–6. doi:10.1093/jnci/dju255

23. Tice JA, Cummings SR, Smith-Bindman R, Ichikawa L, Barlow WE, Kerlikowske K. Using clinical factors and mammographic breast density to estimate breast cancer risk: Development and validation of a new predictive model. *Ann Intern Med.* 2008;148(5):337–347. doi:10.7326/0003-4819-148-5-200803040-00004

24. Del Carmen MG, Halpern EF, Kopans DB, et al. Mammographic breast density and race. *Am J Roentgenol.* 2007;188(4):1147–1150. doi:10.2214/AJR.06.0619

25. Jo HM, Lee EH, Ko K, et al. Prevalence of women with dense breasts in Korea: Results from a nationwide cross-sectional study. *Cancer Res Treat.* 2019;51(4):1295–1301. doi:10.4143/CRT.2018.297

26. Liao YS, Zhang JY, Hsu YC, Hong MX, Lee LW. Age-specific breast density changes in Taiwanese women: A cross-sectional study. *Int J Environ Res Public Health.* 2020;17(9):3186. doi:10.3390/ijerph17093186

27. Leconte I, Feger C, Galant C, et al. Mammography and subsequent whole-breast sonography of nonpalpable breast cancers: The importance of radiologic breast density. *Am J Roentgenol.* 2003;180(6):1675–1679. doi:10.2214/ajr.180.6.1801675

28. Thigpen D, Kappler A, Brem R. The role of ultrasound in screening dense breasts - a review of the literature and practical solutions for implementation. *Diagnostics.* 2018;8(1):1–14. doi:10.3390/diagnostics8010020

29. Kaplan SS. Clinical utility of bilateral whole-breast US in the evaluation of women with dense breast tissue. *Radiology.* 2001;221(3):641–649. doi:10.1148/radiol.2213010364

30. Kolb TM, Lichy J, Newhouse JH. Comparison of the performance of screening mammography, physical examination, and breast US and evaluation of factors that influence them: An analysis of 27,825 patient evaluations. *Radiology.* 2002;255(1):165–175. doi:10.1148/radiol.2251011667

31. Ohuchi N, Suzuki A, Sobue T, et al. Sensitivity and specificity of mammography and adjunctive ultrasonography to screen for breast cancer in the Japan Strategic Anti-cancer Randomized Trial (J-START): A randomised controlled trial. *Lancet.* 2016;387(10016):341–348. doi:10.1016/S0140-6736(15)00774-6

32. Harada-Shoji N, Suzuki A, Ishida T, et al. Evaluation of adjunctive ultrasonography for breast cancer detection among women aged 40-49 years with varying breast density undergoing screening mammography: A secondary analysis of a randomized clinical trial. *JAMA Netw Open.* 2021;4(8):e2121505. doi:10.1001/jamanetworkopen.2021.21505

33. Fenster A, Parraga G, Bax J. Three-dimensional ultrasound scanning. *Interface Focus.* 2011;1(4):503–519. doi:10.1098/rsfs.2011.0019

34. Allajbeu I, Hickman SE, Payne N, et al. Automated breast ultrasound: Technical aspects, impact on breast screening, and future perspectives. *Curr Breast Cancer Rep.* 2021;13(3):141–150. doi:10.1007/s12609-021-00423-1

35. van Zelst JCM, Mann RM. Automated three-dimensional breast US for screening: Technique, artifacts, and lesion characterization. *Radiographics.* 2018;38(3):663–683. doi:10.1148/rg.2018170162

36. Chou YH, Tiu CM, Chen J, Chang RF. Automated full-field breast ultrasonography: The past and the present. *J Med Ultrasound.* 2007;15(1):31–44. doi:10.1016/S0929-6441(08)60022-3

37. Shin HJ, Kim HH, Cha JH. Current status of automated breast ultrasonography. *Ultrasonography.* 2015;34:165–172. doi:10.14366/usg.15002

38. Vourtsis A. Three-dimensional automated breast ultrasound: Technical aspects and first results. *Diagn Interv Imaging.* 2019;100(10):579–592. doi:10.1016/j.diii.2019.03.012

39. Bene IB, Ciurea AI, Ciortea CA, Dudea SM. Pros and cons for automated breast ultrasound (ABUS): A narrative review. *J Pers Med.* 2021;11(8):703. doi:10.3390/jpm11080703

40. Nicosia L, Ferrari F, Bozzini AC, et al. Automatic breast ultrasound: State of the art and future perspectives. *Ecancermedicalscience.* 2020;14(1):1062. doi:10.3332/ECANCER.2020.1062

41. Kossoff G, Carpenter DA, Robinson DE, Radovanovich G, Garrett WJ. Octoson: A new rapid general purpose echoscope. In: *Ultrasound in Medicine.*; 1976:333–339. doi:10.1007/978-1-4613-4307-3_95

42. Ikedo Y, Fukuoka D, Hara T, et al. Development of a fully automatic scheme for detection of masses in whole breast ultrasound images. *Med Phys.* 2007;34(11):4378–4388. doi:10.1118/1.2795825

43. Chang RF, Chang-Chien KC, Takada E, et al. Rapid image stitching and computer-aided detection for multipass automated breast ultrasound. *Med Phys.* 2010;37(5):2063–2073. doi:10.1118/1.3377775

44. Lee GN, Fukuoka D, Ikedo Y, et al. Classification of benign and malignant masses in ultrasound breast image based on geometric and echo features. In: *International Workshop on Digital Mammography (IWDM)*; 2008:433–439. doi:10.1007/978-3-540-70538-3_60

45. Lee GN, Okada T, Fukuoka D, et al. Classifying breast masses in volumetric whole breast ultrasound data: A 2.5-dimensional approach. In: *International Workshop on Digital Mammography (IWDM)*. Vol 6136.; 2010:636–642. doi:10.1007/978-3-642-13666-5_86

46. Fujita H, Uchiyama Y, Nakagawa T, et al. Computer-aided diagnosis: The emerging of three CAD systems induced by Japanese health care needs. *Comput Methods Programs Biomed*. 2008;92(3):238–248. doi:10.1016/j.cmpb.2008.04.003

47. Ikedo Y, Fukuoka D, Hara T, et al. Computerized mass detection in whole breast ultrasound images: Reduction of false positives using bilateral subtraction technique. *Med Imaging 2007 Comput Diagnosis*. 2007;6514(2007):65141T. doi:10.1117/12.709225

48. EpiSonica Corporation. EpiSonica iABUS. Published 2022. https://www.episonica.com/products.html

49. Amy D. *Automatic Breast Ultrasound Scanning*.; 2018. doi:10.1007/978-3-319-61681-0_19

50. 2022 Nemera Insight Chicago L. Embrace 3D – Breast Imaging System. Published online 2022.

51. Azar RZ, Leung C, Chen TK, et al. An Automated Breast Ultrasound System for Elastography. In: *IEEE International Ultrasonics Symposium, IUS*.; 2012:1–4. doi:10.1109/ULTSYM.2012.0001

52. Kelly C, Lobo J, Honarvar M, Shao Y, Salcudean S. An Automated Breast Ultrasound Scanner with Integrated Shear Wave Elastography, Doppler Flow Imaging and Photoacoustic Tomography. In: *IEEE International Ultrasonics Symposium, IUS*.; 2018:1–4. doi:10.1109/ULTSYM.2018.8580074

53. Tadayyon H, Gangeh MJ, Vlad R, Kolios MC, Czarnota GJ. Ultrasound imaging of apoptosis: Spectroscopic detection of DNA-damage effects in vivo. *Methods Mol Biol*. 2017;1644(1):41–60. doi:10.1007/978-1-4939-7187-9_4

54. Imaging Technology News. Techniscan Medical Systems. https://www.itnonline.com/company/techniscan-medical-systems

55. Farrokh A, Erdönmez H, Schäfer F, Maass N. SOFIA: A novel automated breast ultrasound system used on patients in the prone position: A pilot study on lesion detection in comparison to handheld grayscale ultrasound. *Geburtshilfe Frauenheilkd*. 2018;78(5):499–505. doi:10.1055/a-0600-2279

56. Gatta G, Cappabianca S, La Forgia D, et al. Second-generation 3D automated breast ultrasonography (prone ABUS) for dense breast cancer screening integrated to mammography: Effectiveness, performance and detection rates. *J Pers Med*. 2021;11(9):875. doi:10.3390/jpm11090875

57. iVu Imaging Corporation, Hitachi LTD. SOFIA 3D Breast Ultrasound. Published 2019. https://social-innovation.hitachi/en-us/solutions/life_economy/sofia-3d/

58. Duric N, Sak M, Fan S, et al. Using whole breast ultrasound tomography to improve breast cancer risk assessment: A novel risk factor based on the quantitative tissue property of sound speed. *J Clin Med*. 2020;9(2):367. doi:10.3390/jcm9020367

59. Littrup PJ, Duric N, Sak M, et al. Multicenter study of whole breast stiffness imaging by ultrasound tomography (Softvue) for characterization of breast tissues and masses. *J Clin Med*. 2021;10(23):5528. doi:10.3390/jcm10235528

60. Ruiter NV, Zapf M, Hopp T, et al. 3D ultrasound computer tomography of the breast: A new era? *Eur J Radiol*. 2012;81(1):S133–134. doi:10.1016/S0720-048X(12)70055-4

61. Schmidt SP, Roy O, Li C, Duric N, Huang ZF. Modification of Kirchhoff migration with variable sound speed and attenuation for tomographic imaging of the breast. In: Medical Imaging 2011: Ultrasonic Imaging, Tomography, and Therapy.; 2011:796804. doi:10.1117/12.878210

62. Delphinus Medical Technologies I. Discover SoftVue. https://delphinusmt.com

63. Maturo VG, Zusmer NR, Gilson AJ, et al. Ultrasound of the whole breast utilizing a dedicated automated breast scanner. *Radiology*. 1980;137(2):457–463. doi:10.1148/radiology.137.2.6254110

64. Kelly KM, Dean J, Comulada WS, Lee SJ. Breast cancer detection using automated whole breast ultrasound and mammography in radiographically dense breasts. *Eur Radiol*. 2010;20(3):734–742. doi:10.1007/s00330-009-1588-y

65. Zanotel M, Bednarova I, Londero V, et al. Automated breast ultrasound: Basic principles and emerging clinical applications. *Radiol Medica*. 2018;123(1):1–12. doi:10.1007/s11547-017-0805-z

66. Guo R, Lu G, Qin B, Fei B. Ultrasound imaging technologies for breast cancer detection and management: A review. *Ultrasound Med Biol*. 2018;44(1):37–70. doi:10.1016/j.ultrasmedbio.2017.09.012

67. SonoCiné. Automated Whole Breast Ultrasound. https://www.sonocine.com/

68. Brem RF, Tabár L, Duffy SW, et al. Assessing improvement in detection of breast cancer with three-dimensional automated breast US in women with dense breast tissue: The somoinsight study. *Radiology*. 2015;274(3):663–673. doi:10.1148/radiol.14132832

69. GE HealthCare. Invenia ABUS 2.0. Published 2022. https://www.gehealthcare.com/products/ultrasound/breast-ultrasound/invenia-abus

70. Harvey CJ, Pilcher JM, Eckersley RJ, Blomley MJK, Cosgrove DO. Advances in ultrasound. *Clin Radiol*. 2002;57(3):157–177. doi:10.1053/crad.2001.0918

71. Wojcinski S, Gyapong S, Farrokh A, Soergel P, Hillemanns P, Degenhardt F. Diagnostic performance and inter-observer concordance in lesion detection with the automated breast volume scanner (ABVS). *BMC Med Imaging*. 2013;36(1):36. doi:10.1186/1471-2342-13-36

72. Tozaki M, Isobe S, Yamaguchi M, et al. Optimal scanning technique to cover the whole breast using an automated breast volume scanner. *Jpn J Radiol*. 2010;28(4):325–328. doi:10.1007/s11604-010-0424-2

73. Zhang X, Chen J, Zhou Y, et al. Diagnostic value of an automated breast volume scanner compared with a hand-held ultrasound: A meta-analysis. *Gland Surg*. 2019;8(6):698–711. doi:10.21037/gs.2019.11.18

74. Siemens Healthcare Limited. ACUSON S2000 ABVS System HELX Evolution with Touch Control. Published 2022. https://www.siemens-healthineers.com/en-ca/ultrasound/breast-care/acuson-s2000-abvs-ultrasound-machine

75. Park CKS, Xing S, Papernick S, et al. Spatially tracked whole-breast three-dimensional ultrasound system toward point-of-care breast cancer screening in high-risk women with dense breasts. *Med Phys*. 2022;49(6):3944–3962. doi:10.1002/mp.15632

76. Chang JM, Moon WK, Cho N, Park JS, Kim SJ. Breast cancers initially detected by hand-held ultrasound: Detection performance of radiologists using automated breast ultrasound data. *Acta Radiol*. 2011;52(1):8–14. doi:10.1258/ar.2010.100179

77. Park CK, Papernick S, Orlando N, et al. Toward Point-of-Care Breast Cancer Diagnosis: Validation of a Spatially Tracked Automated 3D Ultrasound System. In: *SPIE Medical Imaging Conference on Ultrasonic Imaging and Tomography*. 2022;1203804(4):7. doi:10.1117/12.2607552

78. iSono Health Inc. iSono Health ATUSA. https://isonohealth.com/

79. Fenster A, Bax J, Tessier D, Park C. Wearable 3D ultrasound-based whole breast imaging system, Provisional Patent 63/335,857. Published online 2022.

80. Fenster A, Downey DB, Cardinal HN. Three-dimensional ultrasound imaging. *Phys Med Biol*. 2001;46(5):R67–99. doi:10.1088/0031-9155/46/5/201

81. Park CK, Bax J, Tessier D, Gardi L, Fenster A. Design of a patient-specific whole-breast 3D ultrasound device for point-of-care imaging. *Med Phys*. 2022;49(8):5696–5696.

82. Summers DG, Wang S, Chen J, Anderson TC. Versatile Breast Ultrasound Scanning, Patent 2008/0269613. 2008;1(19).

83. Schaefgen B, Mati M, Sinn HP. Can routine imaging after neoadjuvant chemotherapy in breast cancer predict pathologic complete response? *Ann Surg Oncol*. 2016;23(3):789–795. doi: 10.1245/s10434-015-4918-0

84. Schaefgen B, Heil J, Barr RG, et al. Initial results of the FUSION-X-US prototype combining 3D automated breast ultrasound and digital breast tomosynthesis. *Eur Radiol*. 2018;28(6):2499–2506. doi:10.1007/s00330-017-5235-8

85. Schaefgen B, Juskic M, Hertel M, et al. First proof-of-concept evaluation of the FUSION-X-US-II prototype for the performance of automated breast ultrasound in healthy volunteers. *Arch Gynecol Obstet*. 2021;304(2):559–566. doi:10.1007/s00404-021-06081-z

86. Schafgen B, Juskic M, Radicke M, et al. Evaluation of the FUSION-x-US-II prototype to combine automated breast ultrasound and tomosynthesis. *Eur Radiol*. 2021;31(6):3712–3720. doi:10.1007/s00330-020-07573-3

87. Brem RF, Lenihan MJ, Lieberman J, Torrente J. Screening breast ultrasound: Past, present, and future. *Am J Roentgenol*. 2015;204(2):234–240. doi:10.2214/AJR.13.12072

88. Kelly KM, Dean J, Lee SJ, Comulada WS. Breast cancer detection: Radiologists' performance using mammography with and without automated whole-breast ultrasound. *Eur Radiol*. 2010;20(11):2557–2564. doi:10.1007/s00330-010-1844-1

89. Burkett BJ, Hanemann CW. A review of supplemental screening ultrasound for breast cancer: Certain populations of women with dense breast tissue may benefit. *Acad Radiol*. 2016;23(12):1604–1609. doi:10.1016/j.acra.2016.05.017

90. Wilczek B, Wilczek HE, Rasouliyan L, Leifland K. Adding 3D automated breast ultrasound to mammography screening in women with heterogeneously and extremely dense breasts: Report from a hospital-based, high-volume, single-center breast cancer screening program. *Eur J Radiol*. 2016;85(9):1554–1563. doi:10.1016/j.ejrad.2016.06.004

91. Berg WA, Blume JD, Cormack JB, et al. Combined screening with ultrasound and mammography compared to mammography alone in women at elevated risk of breast cancer: Results of the first-year screen in ACRIN 6666. *JAMA*. 2008;299(18):2151–2163. doi:10.1001/jama.299.18.2151

92. Giger ML, Inciardi MF, Edwards A, et al. Automated breast ultrasound in breast cancer screening of women with dense breasts: Reader study of mammography-negative and mammography-positive cancers. *Am J Roentgenol.* 2016;206(6):1341–1350. doi:10.2214/AJR.15.15367

93. Lee JM, Partridge SC, Liao GJ, et al. Double reading of automated breast ultrasound with digital mammography or digital breast tomosynthesis for breast cancer screening. *Clin Imaging.* 2019;55(9):119–125. doi:10.1016/j.clinimag.2019.01.019

94. Taylor-Phillips S, Stinton C. Double reading in breast cancer screening: Considerations for policy-making. *Br J Radiol.* 2020;93(1106):20190610. doi:10.1259/bjr.20190610

95. Skaane P, Gullien R, Eben EB, Sandhaug M, Schulz-Wendtland R, Stoeblen F. Interpretation of automated breast ultrasound (ABUS) with and without knowledge of mammography: A reader performance study. *Acta Radiol.* 2015;54(4):404–412. doi:10.1177/0284185114528835

96. Arslan A, Ertas G, Aribal E. 3D automated breast ultrasound system: Comparison of interpretation time of senior versus junior radiologist. *Eur J Breast Heal.* 2019;15(3):153–157. doi:10.5152/ejbh.2019.4468

97. Mendelson EB, Berg WA. Training and standards for performance, interpretation, and structured reporting for supplemental breast cancer screening. *Am J Roentgenol.* 2015;204(2):265–268. doi:10.2214/AJR.14.13794

98. Vourtsis A. Three-dimensional automated breast ultrasound: Technical aspects and first results. *Diagn Interv Imaging.* 2019;100(10):579–592. doi:10.1016/j.diii.2019.03.012

99. Allajbeu I, Hickman SE, Payne N, et al. Automated breast ultrasound: Technical aspects, impact on breast screening, and future perspectives. *Curr Breast Cancer Rep.* 2021;13(1):141–150. doi:10.1007/s12609-021-00423-1

100. Duric N, Littrup P, Schmidt S, et al. Breast imaging with the SoftVue imaging system: First results. In: *Medical Imaging 2013: Ultrasonic Imaging, Tomography.*; 2013:86750K. doi:10.1117/12.2002513

101. Duric N, Littrup P, Li C, et al. Breast imaging with SoftVue: Initial clinical evaluation. In: *Medical Imaging 2014: Ultrasonic Imaging and Tomography.*; 2014:90400V. doi:10.1117/12.2043768

102. Duric N, Littrup P, Poulo L, et al. Detection of breast cancer with ultrasound tomography: First results with the computed ultrasound risk evaluation (CURE) prototype. *Med Phys.* 2007;34(2):773–785. doi:10.1118/1.2432161

103. Ruiter NV, Göbel G, Berger L, Zapf M, Gemmeke H. Realization of an optimized 3D USCT. In: *Medical Imaging 2011: Ultrasonic Imaging, Tomography.*; 2011:796805. doi:10.1117/12.877520

104. Li C, Duric N, Littrup P, Huang L. In vivo breast sound-speed imaging with ultrasound tomography. *Ultrasound Med Biol.* 2009;35(10):1615–1628. doi:10.1016/j.ultrasmedbio.2009.05.011

105. Kim H, Cha JH, Oh HY, Kim HH, Shin HJ, Chae EY. Comparison of conventional and automated breast volume ultrasound in the description and characterization of solid breast masses based on BI-RADS features. *Breast Cancer.* 2014;21(4):423–428. doi:10.1007/s12282-012-0419-1

106. Choi WJ, Cha JH, Kim HH, et al. Comparison of automated breast volume scanning and hand-held ultrasound in the detection of breast cancer: An analysis of 5,566 patient evaluations. *Asian Pacific J Cancer Prev.* 2014;15(21):9101–9105. doi:10.7314/APJCP.2014.15.21.9101

107. Jeh SK, Kim SH, Choi JJ, et al. Comparison of automated breast ultrasonography to handheld ultrasonography in detecting and diagnosing breast lesions. *Acta Radiol.* 2016;57(2):162–169. doi:10.1177/0284185115574872

108. Hellgren R, Dickman P, Leifland K, Saracco A, Hall P, Celebioglu F. Comparison of handheld ultrasound and automated breast ultrasound in women recalled after mammography screening. *Acta Radiol.* 2017;58(5):515–520. doi:10.1177/0284185116665421

109. Schmachtenberg C, Fischer T, Hamm B, Bick U. Diagnostic performance of automated breast volume scanning (ABVS) compared to handheld ultrasonography with breast MRI as the gold standard. *Acad Radiol.* 2017;24(8):954–961. doi:10.1016/j.acra.2017.01.021

110. Niu L, Bao L, Zhu L, et al. Diagnostic performance of automated breast ultrasound in differentiating benign and malignant breast masses in asymptomatic women: A comparison study with handheld ultrasound. *J Ultrasound Med.* 2019;38(11):2871–2880. doi:10.1002/jum.14991

111. Tutar B, Esen Icten G, Guldogan N, et al. Comparison of automated versus hand-held breast US in supplemental screening in asymptomatic women with dense breasts: Is there a difference regarding woman preference, lesion detection and lesion characterization? *Arch Gynecol Obstet.* 2020;301(5):1257–1265. doi:10.1007/s00404-020-05501-w

112. Güldogan N, Yılmaz E, Arslan A, Küçükkaya F, Atila N, Arıbal E. Comparison of 3D-automated breast ultrasound with handheld breast ultrasound regarding detection and BI-RADS characterization of lesions in dense breasts: A study of 592 cases. *Acad Radiol.* 2021;29(8):1143–1148. doi:10.1016/j.acra.2021.11.022

113. Kotsianos-Hermle D, Wirth S, Fischer T, Hiltawsky KM, Reiser M. First clinical use of a standardized three-dimensional ultrasound for breast imaging. *Eur J Radiol*. 2009;71:102–108. doi:10.1016/j.ejrad.2008.04.002

114. Shin HJ, Kim HH, Cha JH, Park JH, Lee KE, Kim JH. Automated ultrasound of the breast for diagnosis: Interobserver agreement on lesion detection and characterization. *Am J Roentgenol*. 2011;197(3):747–754. doi:10.2214/AJR.10.5841

115. Chang JM, Moon WK, Cho N, Park JS, Kim SJ. Radiologists' performance in the detection of benign and malignant masses with 3D automated breast ultrasound (ABUS). *Eur J Radiol*. 2011;78(1):99–103. doi:10.1016/j.ejrad.2011.01.074

116. Wang HY, Jiang YX, Zhu QL, et al. Differentiation of benign and malignant breast lesions: A comparison between automatically generated breast volume scans and handheld ultrasound examinations. *Eur J Radiol*. 2012;81(11):3190–3200. doi:10.1016/j.ejrad.2012.01.034

117. Lin X, Wang J, Han F, Fu J, Li A. Analysis of eighty-one cases with breast lesions using automated breast volume scanner and comparison with handheld ultrasound. *Eur J Radiol*. 2012;81(5):873–878. doi:10.1016/j.ejrad.2011.02.038

118. Chen L, Chen Y, Diao XH, et al. Comparative study of automated breast 3-D ultrasound and handheld B-mode ultrasound for differentiation of benign and malignant breast masses. *Ultrasound Med Biol*. 2013;39(10):1735–1742. doi:10.1016/j.ultrasmedbio.2013.04.003

119. Chae EY, Shin HJ, Kim HJ, et al. Diagnostic performance of automated breast ultrasound as a replacement for a hand-held second-look ultrasound for breast lesions detected initially on magnetic resonance imaging. *Ultrasound Med Biol*. 2013;39(12):2246–2254. doi:10.1016/j.ultrasmedbio.2013.07.005

120. Kim SH, Kang BJ, Choi BG, et al. Radiologists' performance for detecting lesions and the interobserver variability of automated whole breast ultrasound. *Korean J Radiol*. 2013;14(2):154–163. doi:10.3348/kjr.2013.14.2.154

121. Wang ZL, Xw JH, Li JL, Huang Y, Tang J. Comparison of automated breast volume scanning to hand-held ultrasound and mammography. *Radiol Medica*. 2012;117(8):1287–1293. doi:10.1007/s11547-012-0836-4

122. Meng Z, Chen C, Zhu Y, et al. Diagnostic performance of the automated breast volume scanner: A systematic review of inter-rater reliability/agreement and meta-analysis of diagnostic accuracy for differentiating benign and malignant breast lesions. *Eur Radiol*. 2015;25(12):3638–3647. doi:10.1007/s00330-015-3759-3

123. Wang L, Qi ZH. Automatic breast volume scanner versus handheld ultrasound in differentiation of benign and malignant breast lesions: A systematic review and meta-analysis. *Ultrasound Med Biol*. 2019;45(8):1874–1881. doi:10.1016/j.ultrasmedbio.2019.04.028

124. Ibraheem SA, Mahmud R, Saini SM, Hassan HA, Keiteb AS, Dirie AM. Evaluation of diagnostic performance of automatic breast volume scanner compared to handheld ultrasound on different breast lesions: A systematic review. *Diagnostics*. 2022;12(2):1–19. doi:10.3390/diagnostics12020541

125. Zhang Q, Hu B, Hu B, Li WB. Detection of breast lesions using an automated breast volume scanner system. *J Int Med Res*. 2012;40(1):300–306. doi:10.1177/147323001204000130

126. Xiao Y, Zhou Q, Chen Z. Automated breast volume scanning versus conventional ultrasound in breast cancer screening. *Acad Radiol*. 2015;22(3):387–399. doi:10.1016/j.acra.2014.08.013

127. Kim Y, Kang BJ, Kim SH, Lee EJ. Prospective study comparing two second-look ultrasound techniques: Handheld ultrasound and an automated breast volume scanner. *J Ultrasound Med*. 2016;35(10):2103–2112. doi:10.7863/ultra.15.11076

128. Choi JJ, Kim SH, Kang BJ, Song BJ. Detectability and usefulness of automated whole breast ultrasound in patients with suspicious microcalcifications on mammography: Comparison with handheld breast ultrasound. *J Breast Cancer*. 2016;19(4):429–437. doi:10.4048/jbc.2016.19.4.429

129. Zhang X, Lin X, Tan Y, et al. A multicenter hospital-based diagnosis study of automated breast ultrasound system in detecting breast cancer among Chinese women. *Chinese J Cancer Res*. 2018;30(2):231–239. doi:10.21147/j.issn.1000-9604.2018.02.06

130. Girometti R, Zanotel M, Londero V, Linda A, Lorenzon M, Zuiani C. Automated breast volume scanner (ABVS) in assessing breast cancer size: A comparison with conventional ultrasound and magnetic resonance imaging. *Eur Radiol*. 2018;28(3):1000–1008. doi:10.1007/s00330-017-5074-7

131. Golatta M, Franz D, Harcos A, et al. Interobserver reliability of automated breast volume scanner (ABVS) interpretation and agreement of ABVS findings with hand held breast ultrasound (HHUS), mammography and pathology results. *Eur J Radiol*. 2013;82(8):332–336. doi:10.1016/j.ejrad.2013.03.005

132. Golatta M, Baggs C, Schweitzer-Martin M, et al. Evaluation of an automated breast 3D-ultrasound system by comparing it with hand-held ultrasound (HHUS) and mammography. *Arch Gynecol Obstet*. 2015;291(4):889–895. doi:10.1007/s00404-014-3509-9

133. Wenkel E, Heckmann M, Heinrich M, et al. Automated breast ultrasound: Lesion detection and BI-RADS classification: A pilot study. *RoFo Fortschritte Auf Dem Gebiet Der Rontgenstrahlen Und Der Bildgeb Verfahren*. 2008;180(9):804–808. doi:10.1055/s-2008-1027563

134. Shin HJ, Kim HH, Cha JH, Park JH, Lee KE, Kim JH. Automated ultrasound of the breast for diagnosis: Interobserver agreement on lesion detection and characterization. *Am J Roentgenol*. 2011;197(3):747–754. doi:10.2214/AJR.10.5841

135. Van Zelst JCM, Platel B, Karssemeijer N, Mann RM. Multiplanar reconstructions of 3D automated breast ultrasound improve lesion differentiation by radiologists. *Acad Radiol*. 2015;22(12):1489–1496. doi:10.1016/j.acra.2015.08.006

136. Zheng FY, Yan LX, Huang BJ, et al. Comparison of retraction phenomenon and BI-RADS-US descriptors in differentiating benign and malignant breast masses using an automated breast volume scanner. *Eur J Radiol*. 2015;84(11):2123–2129. doi:10.1016/j.ejrad.2015.07.028

137. Jiang J, Chen YQ, Xu YZ, et al. Correlation between three-dimensional ultrasound features and pathological prognostic factors in breast cancer. *Eur Radiol*. 2014;24(6):1186–1196. doi:10.1007/s00330-014-3135-8

138. Li N, Jiang YX, Zhu QL, et al. Accuracy of an automated breast volume ultrasound system for assessment of the pre-operative extent of pure ductal carcinoma in situ: Comparison with a conventional handheld ultrasound examination. *Ultrasound Med Biol*. 2013;39(12):2255–2263. doi:10.1016/j.ultrasmedbio.2013.07.010

139. Huang A, Zhu L, Tan Y, et al. Evaluation of automated breast volume scanner for breast conservation surgery in ductal carcinoma in situ. *Oncol Lett*. 2016;12(4):2481–2484. doi:10.3892/ol.2016.4924

140. Xinran Z, Haiyan D, Mingyue L, Yongde Z. Breast intervention surgery robot under image navigation: A review. *Adv Mech Eng*. 2021;13(6):1–17. doi:10.1177/16878140211028113

141. Park CKS, Bax JS, Gardi L, Knull E, Fenster A. Development of a mechatronic guidance system for targeted ultrasound-guided biopsy under high-resolution positron emission mammography localization. *Med Phys*. 2021;48(4):1859–1873. doi:10.1002/mp.14768

142. Wang X, Huo L, He Y, et al. Early prediction of pathological outcomes to neoadjuvant chemotherapy in breast cancer patients using automated breast ultrasound. *Chinese J Cancer Res*. 2016;28(5):478–485. doi:10.21147/j.issn.1000-9604.2016.05.02

143. Dang X, Zhang X, Gao Y, Song H. Assessment of neoadjuvant treatment response using automated breast ultrasound in breast cancer. *J Breast Cancer*. 2022;25(4):1–5.

144. Golan O, Amitai Y, Menes T. Does change in microcalcifications with neoadjuvant treatment correlate with pathological tumour response? *Clin Radiol*. 2016;71(5):458–463. doi:10.1016/j.crad.2016.01.009

145. Li JJ, Chen C, Gu Y, et al. The role of mammographic calcification in the neoadjuvant therapy of breast cancer imaging evaluation. *PLoS One*. 2014;9(2):88853. doi:10.1371/journal.pone.0088853

146. van Egdom LSE, Lagendijk M, Heijkoop EHM, et al. Three-dimensional ultrasonography of the breast; An adequate replacement for MRI in neoadjuvant chemotherapy tumour response evaluation? RESPONDER trial. *Eur J Radiol*. 2018;104(7):94–100. doi:10.1016/j.ejrad.2018.05.005

147. D'Angelo A, Orlandi A, Bufi E, Mercogliano S, Belli P, Manfredi R. Automated breast volume scanner (ABVS) compared to handheld ultrasound (HHUS) and contrast-enhanced magnetic resonance imaging (CE-MRI) in the early assessment of breast cancer during neoadjuvant chemotherapy: An emerging role to monitoring tumor response. *Radiol Medica*. 2021;126(4):517–526. doi:10.1007/s11547-020-01319-3

148. Park J, Chae EY, Cha JH, et al. Comparison of mammography, digital breast tomosynthesis, automated breast ultrasound, magnetic resonance imaging in evaluation of residual tumor after neoadjuvant chemotherapy. *Eur J Radiol*. 2018;108(11):261–268. doi:10.1016/j.ejrad.2018.09.032

149. Kong X, Zhang Q, Wu X, et al. Advances in imaging in evaluating the efficacy of neoadjuvant chemotherapy for breast cancer. *Front Oncol*. 2022;12(5):1–19. doi:10.3389/fonc.2022.816297

150. Chang RF, Chang-Chien KC, Takada E, et al. Breast density analysis in 3-D whole breast ultrasound images. In: *Annual International Conference of the IEEE Engineering in Medicine and Biology*; 2006:2795–2798. doi:10.1109/IEMBS.2006.260217

151. Chen JH, Huang CS, Chien KCC, et al. Breast density analysis for whole breast ultrasound images. *Med Phys*. 2009;36(11):4933–4943. doi:10.1118/1.3233682

152. Moon WK, Shen YW, Huang CS, et al. Comparative study of density analysis using automated whole breast ultrasound and MRI. *Med Phys*. 2011;38(1):382–389. doi:10.1118/1.3523617

153. Zheng FY, Lu Q, Huang BJ, et al. Imaging features of automated breast volume scanner: Correlation with molecular subtypes of breast cancer. *Eur J Radiol*. 2017;86(1):267–275. doi:10.1016/j.ejrad.2016.11.032

154. Wang XL, Tao L, Zhou XL, Wei H, Sun JW. Initial experience of automated breast volume scanning (ABVS) and ultrasound elastography in predicting breast cancer subtypes and staging. *Breast*. 2016;30(12):130–135. doi:10.1016/j.breast.2016.09.012

155. Kim SH, Kim HH, Moon WK. Automated breast ultrasound screening for dense breasts. *Korean J Radiol.* 2020;21(4):15–24. doi:10.3348/kjr.2019.0176

156. Karst I, Henley C, Gottschalk N, Floyd S, Mendelson EB. Three-dimensional automated breast us: Facts and artifacts. *Radiographics.* 2019;39(4):913–932. doi:10.1148/rg.2019180104

157. Schwaab J, Diez Y, Oliver A, et al. Automated quality assessment in three-dimensional breast ultrasound images. *J Med Imaging.* 2016;3(2):027002. doi:10.1117/1.jmi.3.2.027002

158. Mussetto I, Gristina L, Schiaffino S, Tosto S, Raviola E, Calabrese M. Breast ultrasound: Automated or hand-held? Exploring patients' experience and preference. *Eur Radiol Exp.* 2020;4(1):12. doi:10.1186/s41747-019-0136-z

7 Applications of 3D Ultrasound in Musculoskeletal Research

Carla du Toit, Robert Dima, Megan Hutter, Randa Mudathir, Samuel Papernick, Aaron Fenster, and Emily Lalone

INTRODUCTION

OSTEOARTHRITIS

Osteoarthritis (OA) is the most common type of arthritis, affecting approximately 7% of the global population.[1,2] It was previously considered a disease affecting the articular cartilage and subchondral bone caused by "wear and tear" of the joint. Recent advances in medical imaging technology have indicated that OA affects the entire joint. Therefore, its definition has been updated to a degenerative disease that affects the articular cartilage, subchondral bone, surrounding vasculature structures, and synovial membrane physiology.[3] OA patients typically experience pain, weakness, and functional disability which results in decreased overall quality of life. In addition, OA has a high comorbidity with other chronic health conditions such as depression, cardiovascular disease, and diabetes mellitus, the presence of which causes poor physical and mental health, increased hospitalizations, and high mortality.[4–6] OA is one of the leading causes of disability in older adults, and the upward trends in obesity coupled with an aging population is likely to compound the extent of its effects.[7]

DISEASE PROGRESSION

As previously mentioned, OA is a multidimensional disease, meaning there are multiple possible drivers of disease activity and progression. These drivers include abnormal biomechanics, genetics, aging, tissue-specific processes, and inflammation.[8–11] Currently, OA is categorized into primary and secondary OA. Primary OA is the most common type of OA and is idiopathic. Secondary OA occurs secondary to another condition, the most common being a previous injury (post-traumatic OA). Secondary OA can also manifest due to genetic abnormalities, hormonal problems, nerve damage, joint infection, deposits of substances (hydroxyapatite or iron) in the joint, or other inflammatory arthritis (rheumatoid arthritis).[12] Regardless of the type of OA, the disease progression is similar. Several studies have demonstrated that there are multiple key changes that occur in the synovium of patients with early OA. These changes included increased vascularity in the synovium and a subsequent increase in infiltration of inflammatory mediators, leading to synovitis (inflammation of the synovium).[13] Benito et al. reported that patients in earlier stages of knee OA progression have higher levels of macrophage infiltration and blood vessel proliferation markers than patients with more advanced OA.[14] They also demonstrated that patients with early OA expressed higher levels of inflammatory mediators (interleukin-1 and tumor necrosis factor-alpha) than patients with later stages of OA. The prolonged presence of these pro-inflammatory cytokines has been linked to the initiation and progression of articular cartilage degeneration. The articular cartilage matrix of patients with OA undergoes proteolytic degradation, which is associated with increased synthesis of altered matrix components.[15–17] This contributes to early morphological changes in the cartilage and eventually results in loss of cartilage volume.

DOI: 10.1201/9781003299462-9

IMAGING IN OSTEOARTHRITIS

Digital radiography is the current clinical imaging standard for assessing and monitoring OA. Radiography is widely available, has short acquisition times, and is associated with minimal discomfort to the patient. The progression and severity of OA are monitored using radiography by assessing cartilage degradation through measurements of joint space narrowing (JSN), presence of osteophytes, and joint subluxation, particularly in the smaller joints of the hand and wrist (Figure 7.1). Grading systems such as the Kellgren and Lawrence (Knee OA) and the Eaton and Littler (First carpometacarpal OA) are accepted as the standard methods for assessing radiographic indicators of OA.[18] However, there are several limitations associated with utilizing radiography and radiographic grading scales to monitor the progression of a multidimensional disease such as OA. The primary limitation of radiography is the lack of soft-tissue contrast, meaning it cannot be used for visualization of the soft tissue components of the joint such as the synovium and the articular cartilage which have now been established as critical components in the pathogenesis of OA.

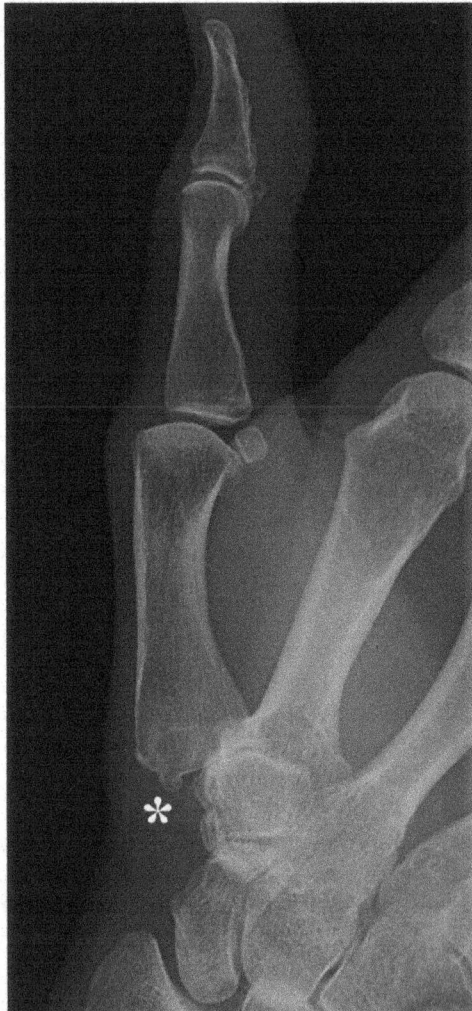

FIGURE 7.1 Radiograph of a patient with advanced first carpometacarpal osteoarthritis. This image demonstrates subluxation of the joint, a decrease in joint space, and growth of an osteophyte on the proximal metacarpal head as indicated by the asterisk.

The secondary limitation is that radiographs are acquired as a 2D projection image and subtle changes in position can change critical values such as joint space width.[19,20] Therefore, it is impossible to acquire a comprehensive image of the entire joint under observation with this modality alone.

To overcome these limitations, other three-dimensional (3D) imaging modalities, such as magnetic resonance imaging (MRI), have been used. MRI provides excellent soft-tissue contrast, enabling more comprehensive assessments of the joint structures affected by OA such as the articular cartilage, synovium, and the subchondral bone. MRI is the current imaging standard for the soft tissue structures involved in OA imaging due to its high spatial resolution, excellent soft-tissue contrast, and 3D imaging capabilities. The 3D aspect of MR images is crucial, as it allows for quantitative evaluation of key OA factors such as articular cartilage thickness and synovial tissue volume.[21,22] These measurements can be integrated into disease-monitoring and treatment effectiveness testing protocols. Quantitative measurement tools are increasingly being recognized as critical extensions of existing semi-quantitative measurement tools such as the Hand and Osteoarthritis Magnetic Resonance Scoring System (HOAMRIS) and its ultrasound counterpart Outcome Measures in Rheumatoid Arthritis Clinical Trials (OMERACT).[22] Unfortunately, MRI is associated with long wait times, high manufacturing and operating costs, making it grossly inaccessible in rural and cost-constrained healthcare systems.

APPLICATIONS OF CONVENTIONAL ULTRASOUND

Two-dimensional ultrasound (2D US) imaging is widely used at the point of care to monitor joint health status, treatment effectiveness, and as guidance for intra-articular procedures. It is increasingly integrated into clinical practice due to its portable nature, cost-effectiveness, and high resolution. Musculoskeletal applications of ultrasound include monitoring disease progression, treatment effectiveness, structural assessment, and procedure guidance (i.e., intra-articular corticosteroid injections).[23,24] Various studies have investigated the use of 2D US for OA imaging and monitoring and have found that features like articular cartilage thickness, synovitis, hypertrophy of the synovial membrane, cysts, and abnormal osteochondral growths (osteophytes) can be reliably assessed and are correlated to patient-important outcomes.[25–28] High-frequency linear transducers are most commonly used for assessing joint health in patients with OA. These transducers allow clinicians to visualize joint anatomy using multiple views and orientations to acquire images that are typically unattainable using plain radiography. Conventional ultrasound images of patients with OA are typically analyzed using semi-quantitative scoring systems such as the OMERACT and the European League Against Rheumatisms (EULAR).[29] As mentioned previously throughout this book, 2D US has several limitations. It's common practice in routine diagnostic ultrasound to estimate the volume of organs, masses, and cysts using caliper measures in three orthogonal planes. Most commercial ultrasound machines contain algorithms that provide volume measurements from these three measurements. However, this method is not routinely used in the measurement of synovitis volume as the calculation assumes an elliptical shape, whereas synovial effusion is typically an irregular and inconsistent shape. Thus, it leads to inaccurate estimations of the true volume of synovium.[30] OA treatment and research specifically require volumetric measurement of cartilage thickness and synovial tissue volume for assessments of treatment effectiveness and to aid clinical treatment planning. Also, 2D US may not be appropriate for longitudinal monitoring of disease progression due to challenges associated with standardization of transducer and patient positioning during consecutive examinations. As a 2D US image is a single-position view, clinicians are unable to visualize the status of the entire joint with a single image. Finally, 2D US lacks the field of view required to acquire comprehensive images of the whole joint, especially in larger joints such as the knee and hip. These limitations necessitate the development of point-of-care imaging devices that can provide quantitative methods for characterizing and monitoring soft tissue characteristics of OA. Examples of 2D US images acquired of the first carpometacarpal (CMC1) joint and the suprapatellar recess of the knee are shown in Figure 7.2.

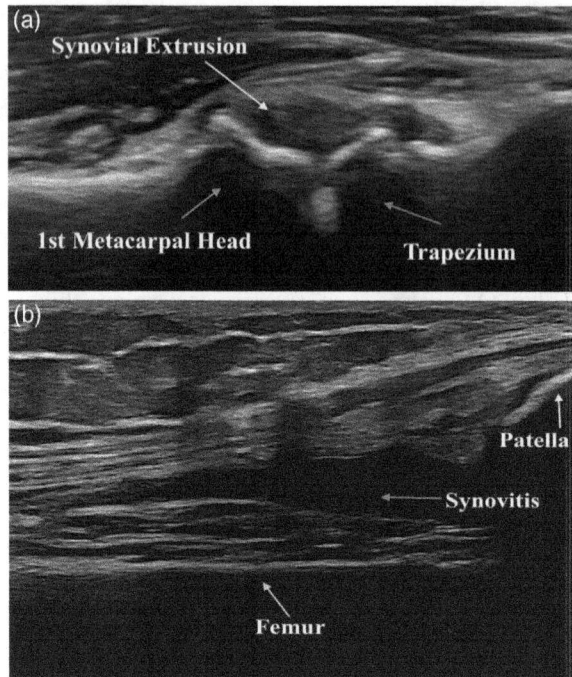

FIGURE 7.2 (a) A longitudinally acquired 2D US image of a patient's first carpometacarpal joint indicating extrusion of the synovial fluid outside of the joint line boundary. (b) A longitudinally acquired 2D US of the suprapatellar region of the knee joint indicating synovial fluid and the surround anatomical structures.

ADVANCES IN 3D US FOR MUSCULOSKELETAL HEALTH RESEARCH

As indicated throughout previous chapters in this book, various limitations of 2D US can be overcome through the development of 3D US devices for a wide variety of applications. The use of 3D US for musculoskeletal applications aids specifically in overcoming the need for operators to mentally transform 2D images into 3D impressions of joint anatomy as well as mitigating the dependency on transducer location and position.[30] In addition, 3D US techniques provide the advantage of a more complete visualization of the volumetric structures required for OA monitoring and treatment effectiveness assessment longitudinally, when imaging larger joints such as the knee and hip, warranting investigation into the application of 3D US systems in these clinical settings. The knee and CMC1 joints, two of the most common sites of OA, are currently the two main joints under investigation for the feasibility of applying 3D US to assess and monitor OA progression.[31,32] As previously stated in this chapter, current methods of imaging OA are associated with many limitations which have driven the search for cost and time-effective 3D point-of-care imaging methods.

FIRST CARPOMETACARPAL OSTEOARTHRITIS

One of the most common sites of hand OA is the first carpometacarpal joint (CMC1) located at the base of the thumb. Studies have reported that the prevalence of CMC1 OA can be as high as 33% in some populations and that post-menopausal women experience greater risk for developing the disease.[8,10,11] The CMC1 joint is a bi-concave (saddle-shaped) joint, meaning it relies heavily on the dynamic stabilizer structures (ligaments, tendons, muscles etc.) for support. This structure allows for the range of motion required to complete the wide range of thumb movements used in activities of daily living but it also unfortunately predisposes the joint to potential injury, malalignment, and the development of OA.[33] The development of CMC1 OA leads to considerable disability and pain, leading to difficulty in performing activities of daily living and ultimately decreasing the quality of life of many patients.[34]

Over the past few years, OA research has explored the role of soft-tissue structures in OA pathogenesis and how it may relate to patient symptoms. Studies have demonstrated that OA does have an inflammatory component, where chronically elevated levels of inflammatory mediators lead to synovitis which is associated with cartilage degradation and the breakdown of subchondral bone.[17,35] Despite the degree of inflammation observed in patients with OA is not as extensive as that observed in inflammatory type of arthritis, such as rheumatoid arthritis, it is now considered an important factor in OA. In the study by Du Toit et al., a 3D US device was proposed for assessing and monitoring synovial tissue volume in CMC1 OA patients. This system uses a submerged transducer mover assembly, which consisted of an 11-L tank filled with 7.25% isopropyl alcohol solution, a conventional 2D US machine, and a high-frequency linear US transducer, housed in a custom attachment. This system translates this conventional linear transducer over a 30-cm² area inside the submersion tank while acquiring a series of 2D US images at regular spatial intervals (Figure 7.3). These images are then reconstructed into a 3D US image as explained in previous chapters of this book. This study describes the linear and volumetric validation of this system and examines its reliability and validity in comparison to the imaging standard of MRI using a preliminary cohort of ten CMC1 OA patients.[36]

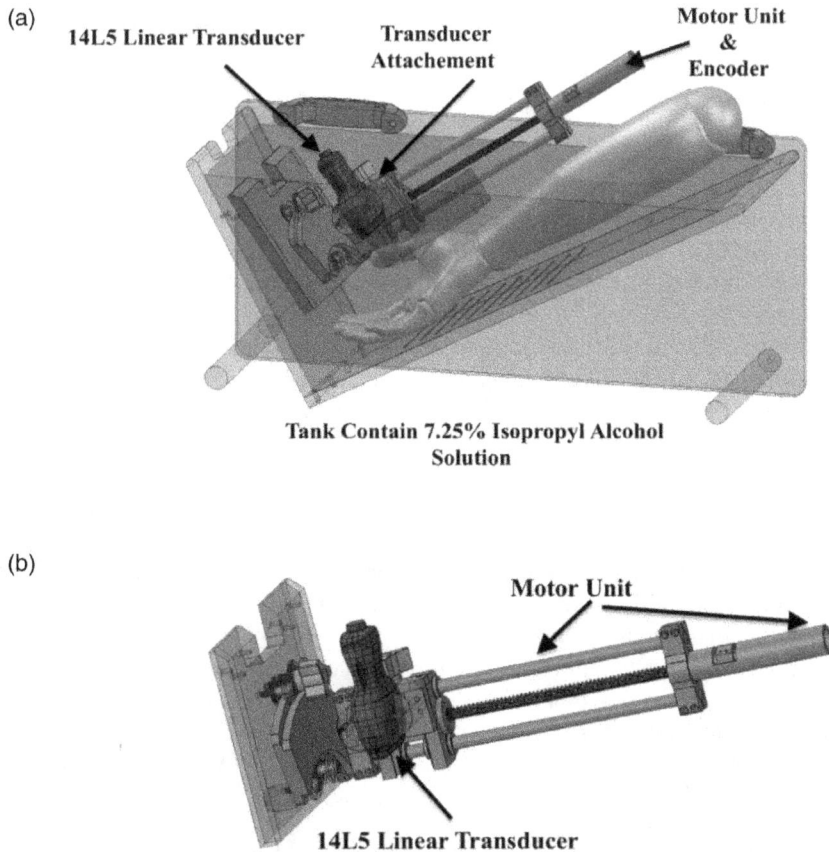

FIGURE 7.3   (a) Side view of the 3D US scanning device showing hand placement, L14-5 linear US probe, motor unit, and 7.25% isopropyl alcohol solution. (b) Side view of the mechanical components of the 3D US device.

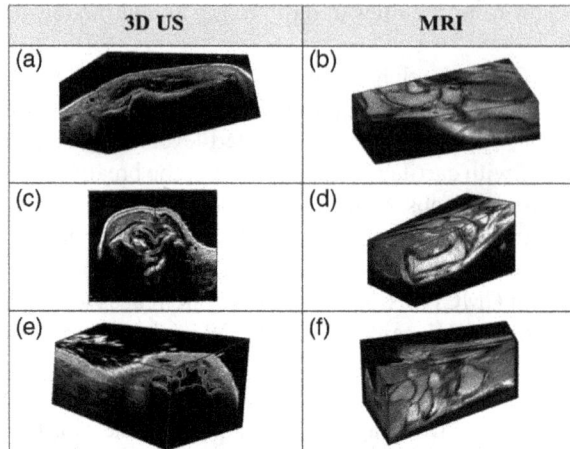

FIGURE 7.4 MRI and 3D US images of the CMC1 joint of patients with synovitis. MRI (a) and 3D US (b) of a patient with a mild amount of synovitis. MRI (c) and 3D US (d) of a patient with a moderate amount of synovitis. MRI (e) and 3D US (f) of a patient with large amounts of synovitis.

3D US images of the CMC1 joints of ten patients diagnosed with CMC1 OA were acquired through the thenar eminence on the ventral side of the hand. Complimentary MR images were acquired using a photon density fast-spin echo with fat saturation sequence for the validity analysis. Two raters with specialized training in musculoskeletal ultrasound imaging analysis identified the synovium in each of the images and manually segmented all areas containing synovitis and synovial membrane hyperplasia (thickening). Synovitis typically appears as areas of hypoechoic signal (Figure 7.4). Du Toit et al. demonstrated lower mean differences between the two raters for CMC1 synovial volume measurements in comparison to MRI and indicated that the percent difference between 3D US and MRI was 1.71% using this manual segmentation method. These observations are similar to those observed in other 3D US to MRI comparisons.[37] Visualization of smaller joints such as the joints of the hand can be more challenging due to the small size, irregular shape, and biomechanical limitations of these structures, especially in patients with OA. However, in the case of the CMC1 joint, the distinct saddle shape of the joint often leads to protrusions of the synovial fluid when inflamed. This can be observed clearly on the volar side of the joint using US which makes identification and segmentation easier (Figure 7.5). This study also demonstrated excellent inter- and intra-rater reliability for both MRI and 3D US. Interestingly, this study reported a smaller standard error of measurement for the 3D US volume measurements than the values found for MRI, indicating that there was a higher level of precision associated with the 3D US measurements than with MRI. In addition, this study demonstrated smaller minimal detectible change values for 3D US measurements compared to MRI which indicates that the 3D US imaging device was more sensitive to detecting changes in the volume measurements on test-retest. These results suggest that the use of 3D US to monitor synovial volume in OA patients should be considered in longitudinal studies with the aim of eventually introducing this technology to measure effects in clinical trials.

While developing quantitative methods for assessing and monitoring synovitis is advantageous to understanding CMC1 OA pathogenesis, it is also important to keep in mind the multifaceted nature of OA. Currently, we do not understand the exact cause of OA or what factors definitively drive the progression of the disease.[38]

Many studies have investigated different possible drivers of OA progression and how they may appear in imaging. CMC1 OA is commonly diagnosed with plain X-ray radiography; however, many studies have reported significant discrepancies between radiographic evidence of OA and patient-reported outcomes such as pain and disability.[39–41] The Multicenter Osteoarthritis Study (MOST)

Radiography	Conventional Ultrasound	3D Ultrasound
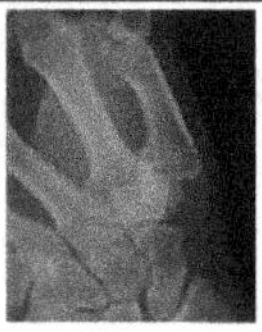	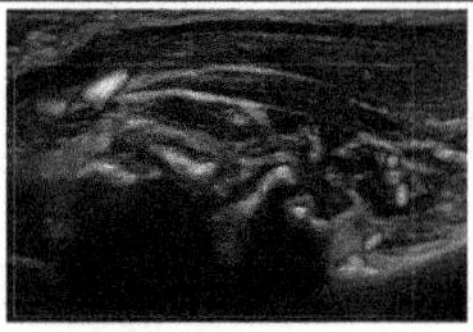	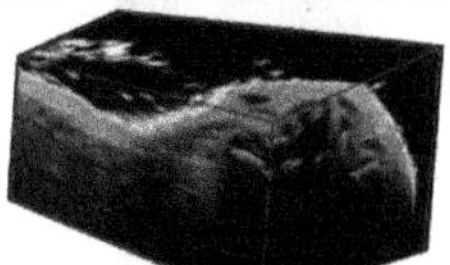
Eaton-Littler Grade: 3	OMERACT Grade: 1	Synovial Volume: 615.2 mm^3

AUSCAN Score	Pinch Grip Force
66%	6.3 kgf

FIGURE 7.5   Phenotype 1: Synovial effusion dominant cohort. Case study example of a patient allocated to the phenotype 1 cohort. This chart demonstrates the imaging features, pain interference scores, and functional grip force of this patient.

indicated that 80% of the patients who had synovitis upon imaging analysis reported at least a moderate amount of pain.[42] These reported discrepancies have been attributed to a lack of soft-tissue contrast within radiographs as well as the heterogeneous nature of OA. As previously indicated in this chapter, synovitis and synovial hypertrophy have been associated with adverse patient symptoms and has been proven to be a driver of OA pathogenesis. This makes synovitis an attractive target for disease-modifying interventions. Studies have investigated the relationship between synovitis, image-based grading scales, and patient-reported outcomes and have found conflicting results. A study conducted by Hall et al. found that despite the common observation of US abnormalities in OA patients, there is only a moderate correlation between synovitis and radiographic severity. In addition, the results of this study indicated that the relationship between these abnormal imaging features and pain is weak.[5] In contrast, Naredo et al. found that only those patients with joint effusion, meniscal protrusion, and bulging of the surrounding ligamentous structures had symptomatic OA. These patients were also at an increased risk of developing painful OA symptoms when synovial effusion was present in their images.[43] Du Toit et al. hypothesized that these discrepancies could be attributed to the lack of a standardized definition of inflammatory features in addition to a lack of sensitive quantitative measures for synovitis and the possibility of variation in pathophysiological contributors within potential CMC1 OA phenotypes. This study investigated the use of 3D US as a tool for examining synovial tissue volume and varying morphology in relation to patient-reported outcomes in a cohort of 20 CMC1 OA patients. The aims of this study were to examine what information could be gained from investigating 3D US CMC1 OA patient effusion morphology and how this information influences our current understanding of OA. Moreover, the study investigated the relationship between changes in synovial tissue volume and patient-reported outcomes.

The authors found that their patient cohort exhibited three distinct synovitis phenotypes which were distinguished by:

1. Synovial tissue volume
2. Synovial tissue margin properties
3. Histological features of effusion and hypertrophy

The first morphological group was called "synovial effusion dominant" and described the group of patients that exhibited concave extrusion of the synovial fluid beyond the joint line and in some cases synovial hypertrophy. The authors found that patients with this morphology had predominately anechoic signal present in their 3D US images indicating that the dominant inflammatory feature observed was synovial effusion rather than synovial membrane hypertrophy which presents as hypoechoic tissue (Figure 7.5). In addition, this group had moderate synovial tissue volumes

Radiography	Conventional Ultrasound	3D Ultrasound
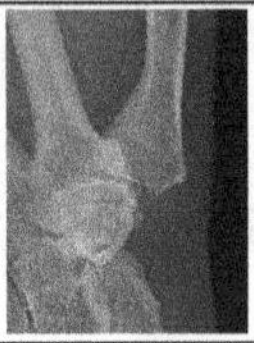	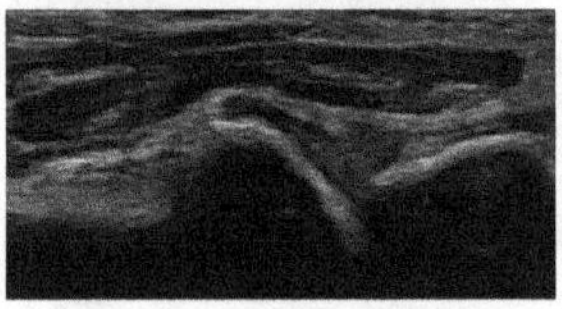	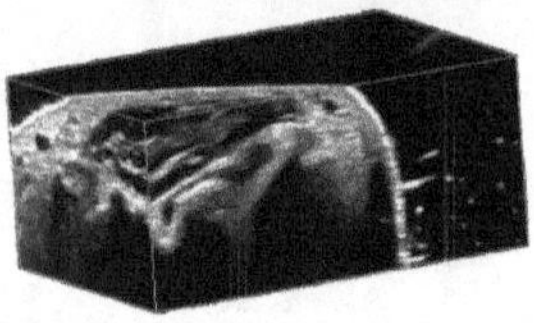
Eaton-Littler Grade: 2	OMERACT Grade: 2	Synovial Volume: 111.8 mm^3

AUSCAN Score	Pinch Grip Force
13%	19.3 kgf

FIGURE 7.6 Phenotype 2: Diffuse synovial effusion and hypertrophy cohort. Case study example of a patient allocated to the phenotype 2 cohort. This chart demonstrates the imaging features, pain interference scores, and functional grip force of this patient.

compared to the other morphological groups. The authors indicate that this group had the lowest average pain interference scores and the highest physical function.

The second morphological group was described as having diffuse synovial effusion and hypertrophy. In this group of patients, the synovial effusion and areas of membrane hypertrophy extended past the joint line and over the proximal and distal ends of the trapezium and first metacarpal, respectively (Figure 7.6). The authors indicate that patients in this subgroup presented with the greatest synovial tissue volume and reported a moderate amount of pain interference in comparison to the other two groups. The 3D US images of these patients exhibited predominantly areas of hypoechoic signal, which indicates that the tissue observed is predominantly composed of hypertrophic synovial membrane instead of fluid effusion as seen in the "effusion dominant" cohort (Figure 7.6). Interestingly, the authors reported an increase in the presence of osteophytes observed in the images of these patients. In addition, these patients performed worse functionally on the pinch grip test than those allocated to the previous group.

The third morphological group examined was the osteochondral dominant group. Patients in this group exhibited limited synovitis and synovial hypertrophy but showed advanced osteophyte growth and JSN. The authors reported minimal anechoic and hypoechoic signals in the US images indicating low amounts of synovial fluid and synovial membrane hypertrophy. Patients allocated to this group had the lowest average grip strength and the highest pain interference scores (Figure 7.7).

The morphological differences reported by Du Toit et al. indicated that there is a complex relationship between CMC1 OA disease progression, patient-related outcomes, and image-based

Radiography	Conventional Ultrasound	3D Ultrasound
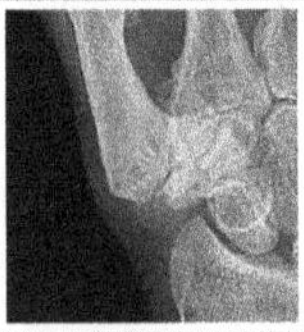	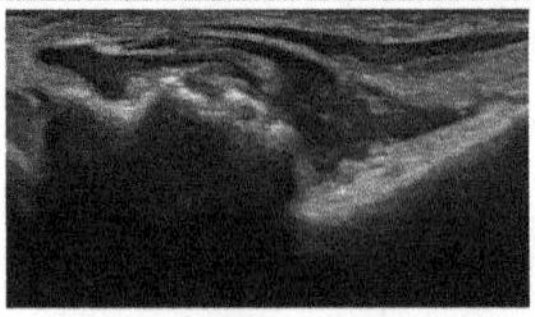	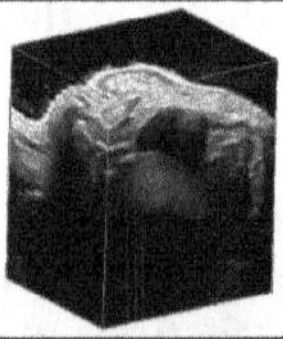
Eaton-Littler Grade: 4	OMERACT Grade: 3	Synovial Volume: 38.45 mm^3

AUSCAN Score	Pinch Grip Force
72%	5.3 kgf

FIGURE 7.7 Phenotype 3: Osteochondral dominant cohort. Case study example of a patient allocated to the phenotype 3 cohort. This chart demonstrates the imaging features, pain interference scores, and functional grip force of this patient.

evidence of OA. The authors indicate that the structures involved in CMC1 OA not only change as the disease progresses but also potentially between bouts of inflammation and between patients. The KOA literature indicates that patients have exhibited trends in earlier stages of OA, as characterized by radiographs, patients tend to have more synovial effusion, which then decreases and changes to synovial membrane hypertrophy. This eventually leads to the development of osteophytes and JSN in the later stages of the disease.[17]

The results of this study agree with these studies that theorize that synovitis seems to be the dominant characteristic and driver of OA progression in earlier stages of the disease.[35,44,45] However, this study had a small patient population and was designed as a cross-sectional feasibility study to examine 3D US characteristics of CMC1 OA. The results of this study indicate that 3D US can provide potentially valuable imaging characteristics for monitoring OA however, this research needs to be expanded to test the system's ability to monitor changes in patient images and self-reported outcomes longitudinally before introduction into clinical workflow can be considered.

3DUS FOR KNEE OSTEOARTHRITIS

The knee is the most commonly affected joint in OA, and knee osteoarthritis (KOA) has a disease prevalence of 7–17% in middle-aged adults (45 years and older).[46] Cartilage damage is one of the hallmark features of KOA and has been used as a method for characterizing disease severity through measurement of cartilage loss, where decreases in cartilage volume and quality are interpreted as increase in KOA severity.[47] In the knee, measures are specifically focused on the quality and quantity of femoral articular cartilage (FAC). Similar to the semi-quantitative rating scales used for measuring severity of CMC1 OA, grading systems for KOA like the Kellegen–Lawrence scale use tibio-fibular joint space narrowing (TF JSN) as a surrogate for femoral cartilage loss. JSN may represent FAC loss on radiographs, but radiographic grading is associated with poor sensitivity to detect FAC changes in early stage OA.[48] In addition, JSN is a composite measure of FAC and meniscal positioning and degeneration, and meniscal damage is not strongly associated with KOA severity.[49]

In a recent cross-sectional study, Papernick et al. described the development of a handheld mechanical 3D US device capable of imaging trochlear FAC (tFAC). The authors tested intra- and inter-rater reliability and concurrent validity of their device compared to MRI in a cohort of 25 healthy volunteers.[50] The study describes a high-frequency linear transducer and motorized drive mechanism attached via custom holders a seen in Figure 7.8.

3D US images were compared to MR images acquired using a 3.0 T MR system and 3D multiple echo recombined gradient echo (MERGE) sequence as recommended by the Osteoarthritis Research Society International for clinical trials.[31] The authors state that for the manual segmentations, the

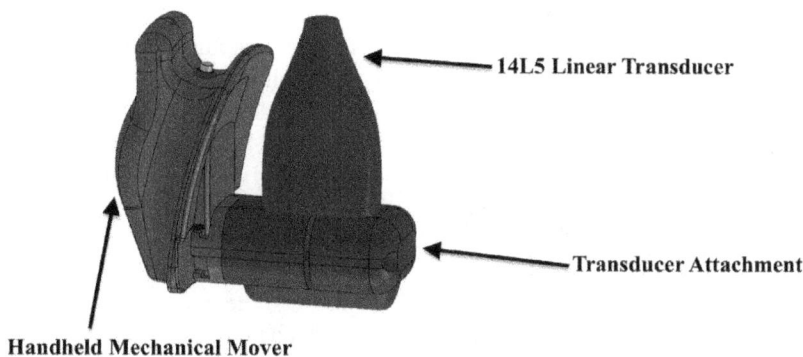

FIGURE 7.8 Schematic drawing of the handheld mechanical 3D US acquisition device for imaging knee osteoarthritis.

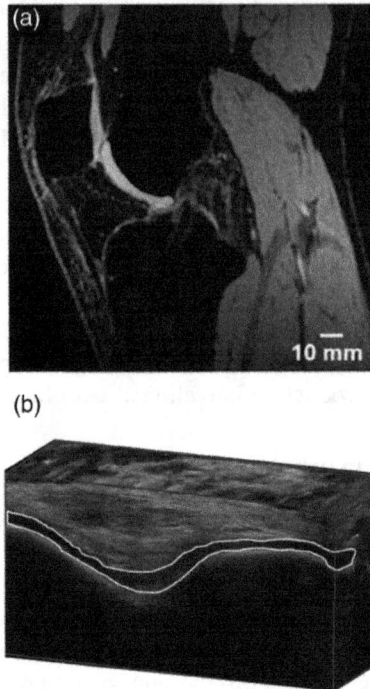

FIGURE 7.9 (a) MERGE MRI and (b) 3D US images of the trochlear articular cartilage of the knee. The articular cartilage segmentations are highlighted in yellow and indicate the area chosen for segmentation on these particular image slices.

anterior hyperechoic tFAC surface and the hyperechoic border of the cortex were defined as the boundaries for the anechoic cartilage (Figure 7.9).

Papernick et al. demonstrated excellent intra-rater reliability for both imaging modalities. For inter-rater reliability, 3D US demonstrated higher reliability than MRI. The results of this study also indicated that 3D US displayed smaller global mean surface distance and Hausdorff distances than MRI. In addition, the mean DSC was higher for 3D US than for MRI collectively suggesting that their 3D US system can quantify tFAC volume with similar reliability and precision than MRI in a healthy knee population. Despite excellent correlation between 3D US tFAC volumes and MRI tFAC volumes (Spearman's r 0.88), 3D US segmented volumes were larger on average than MRI segmentations by 16.7%. The authors attribute this difference to the high spatial resolution of US compared to the MR image acquired. MRI may not have been able to capture the true cartilage volume as effectively as 3D US due to inability to effectively capture thinning of the medial and lateral portions of the tFAC and increased challenges in delineating cartilage from thin adipose tissue. Due to the high spatial resolution associated with 3D US, the thin lateral and medial portions of the cartilage were easily visible and therefore included in the segmentations. Visualization of these sections of the cartilage is crucial to the success of clinical trials investigating joint disease, as thinner areas of cartilage are more susceptible to damage and volume loss in KOA. The device was able to visualize both the tFAC and condylar cartilage regions clearly, providing a more comprehensive model of anatomy and improved volume quantifications. This study was conducted with healthy knees rather than KOA patients. In KOA patients, there are characteristically different features to the cartilage including fissures, abrasions, and surface irregularities. Therefore, future studies should focus on investigating the measurement capabilities of these devices in a KOA patient population. Studies focused on the use of 3D US for longitudinal monitoring of KOA and potential imaging markers for earlier detection of KOA are some of the future works in progress.

FUTURE APPLICATIONS

3D US has proven to have immense potential for quantitatively characterizing joint health and for assessing OA disease status, progression, and treatment response. Future work in this area is focused on the development and implementation of automated segmentation algorithms to aid in clinical workflow by reducing the amount of time required to attain volume measurements. The deep-learning algorithms used in these studies are described in detail in other chapters. As the previous studies examined in this chapter were constructed as cross-sectional investigations, it is crucial that future studies investigate the validity and reliability of using 3D US for longitudinal monitoring of OA. This would determine the value of using 3D US as a tool for monitoring OA progression as well as treatment response. These investigations are required before 3D US can be introduced for use in clinical trials and primary clinic settings.

Finally, further research is required to broaden our understanding of OA pathology and pathogenesis. Investigating the use of 3D US to assess other potential factors and drivers of OA, such as synovitis in KOA, inflammatory responses in CMC1 OA patients, joint kinematics, and ligament recruitment in CMC1 OA patients. The effects of synovitis in KOA have been explored using various other imaging modalities such as MRI and 2D US; however, to establish the clinical feasibility of the knee 3D US system, further research is required to investigate the validity and reliability of its measurement capabilities in a patient population.

One of the current areas of interest in OA research is the role of angiogenesis in the pathophysiology of OA. Angiogenesis is the formation of new blood vessels from existing ones and has been associated with chronic inflammation.[51] Angiogenesis in the synovial membrane has been linked to synovitis and has been proposed as a potential driver.[52] The importance of angiogenesis and hypervascularization of the synovial membrane in OA pathogenesis and patient symptoms has not fully been investigated. Currently, 2D power Doppler ultrasound (2D PDUS) imaging is commonly used to evaluate fluid movement. 2D PDUS is specifically used in musculoskeletal imaging to assess the presence of synovitis, where a positive Doppler signal is indicative of active inflammation in the joint.[53,54] However, not all patients present with a positive power Doppler signal and it is theorized that this could be due to low-grade inflammation that PDUS is unable to detect. In light of

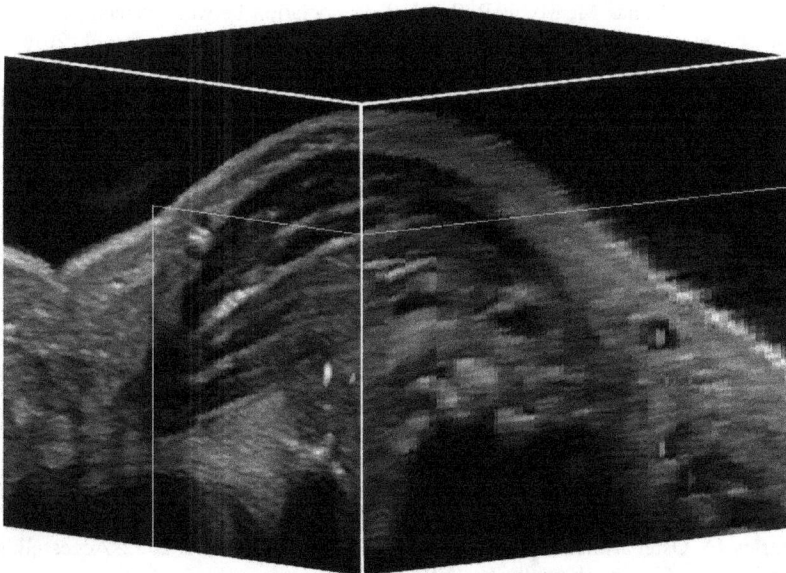

FIGURE 7.10   A 3D US image of a CMC1 OA patient presenting with a positive signal using superior microvascular imaging technology.

this limitation, we have chosen to explore the use of superb microvascular imaging (SMI), a new Doppler US technology developed by Canon Medical Systems. Previous studies have demonstrated that SMI performs better in detecting low-grade inflammation compared to DPUS.[53] Future studies aim to develop a 3D US system utilizing power Doppler and SMI technologies to visualize and quantify synovial blood flow in patients with CMC1 OA as shown in Figure 7.10.

CONCLUSION

The novel application of 3D US systems and tools has the potential to revolutionize diagnostic and monitoring protocols in musculoskeletal healthcare by providing accessible, easy-to-use approaches to evaluate joint health and to enable more effective treatment planning for future OA patients. The preliminary studies discussed in this chapter have demonstrated great potential for impacting quality of care by providing clinicians with easy-to-use quantitative methods for assessing OA disease status. Despite the potential benefits associated with these musculoskeletal 3D US systems, most of the studies mentioned are recent proof-of-concept works that require advancement beyond this stage before they can be considered for clinical implementation. Thus, extensive investigations are required in order to broaden both the technological capabilities of these systems and our understanding of OA as a multifaceted disease.

REFERENCES

1. Sharif B, Kopec J, Bansback N, et al. Projecting the direct cost burden of osteoarthritis in Canada using a microsimulation model. *Osteoarthritis Cartilage*. 2015;23(10):1654–1663. doi:10.1016/j.joca.2015.05.029
2. Badley E, Wilfong J, Zahid S, Perruccio A. *The Status of Arthritis in Canada: National Report*. ACREU for the Arthritis Society; 2019:34.
3. Kloppenburg M, Kwok WY. Hand osteoarthritis: A heterogeneous disorder. *Nat Rev Rheumatol*. 2012;8(1):22–31. doi:10.1038/nrrheum.2011.170
4. Swain S, Sarmanova A, Coupland C, Doherty M, Zhang W. Comorbidities in osteoarthritis: A systematic review and meta-analysis of observational studies. *Arthritis Care Res*. 2020;72(7):991–1000. doi:10.1002/acr.24008
5. Hall A, Stubbs B, Mamas M, Myint P, Smith T. Association between osteoarthritis and cardiovascular disease: Systematic review and meta-analysis. *Eur J Prev Cardiol*. 2016;23(9):938–946. doi:10.1177/2047487315610663
6. Louati K, Vidal C, Berenbaum F, Sellam J. Association between diabetes mellitus and osteoarthritis: Systematic literature review and meta-analysis. *RMD Open*. 2015;1(1):e000077–e000077. doi:10.1136/rmdopen-2015-000077
7. Felson DT, Naimark A, Anderson J, Kazis L, Castelli W, Meenan RF. The prevalence of knee osteoarthritis in the elderly. The framingham osteoarthritis study. *Arthritis Rheum*. 1987;30(8):914–918. doi:10.1002/art.1780300811
8. Haugen IK, Englund M, Aliabadi P, et al. Prevalence, incidence and progression of hand osteoarthritis in the general population: The Framingham osteoarthritis study. *Ann Rheum Dis*. 2011;70(9):1581–1586. doi:10.1136/ard.2011.150078
9. Sonne-Holm S, Jacobsen S. Osteoarthritis of the first carpometacarpal joint: A study of radiology and clinical epidemiology. *Osteoarthritis Cartilage*. 2006;14(5):496–500. doi:10.1016/j.joca.2005.12.001
10. Dahaghin S. Prevalence and pattern of radiographic hand osteoarthritis and association with pain and disability (the Rotterdam study). *Ann Rheum Dis*. 2005;64(5):682–687. doi:10.1136/ard.2004.023564
11. Kessler S, Stove J, Puhl W, Sturmer T. First carpometacarpal and interphalangeal osteoarthritis of the hand in patients with advanced hip or knee OA. Are there differences in the aetiology? *Clin Rheumatol*. 2003;22(6):409–413. doi:10.1007/s10067-003-0783-5
12. Sen R, Hurley JA. Osteoarthritis. In: *StatPearls*. StatPearls Publishing; 2022. Accessed: November 22, 2022. http://www.ncbi.nlm.nih.gov/books/NBK482326/
13. Pelletier JP, Martel-Pelletier J, Abramson SB. Osteoarthritis, an inflammatory disease: Potential implication for the selection of new therapeutic targets. *Arthritis Rheum*. 2001;44(6):1237–1247. doi:10.1002/1529-0131(200106)44:6<1237::AID-ART214>3.0.CO;2-F

14. Benito MJ. Synovial tissue inflammation in early and late osteoarthritis. *Ann Rheum Dis.* 2005;64(9):1263–1267. doi:10.1136/ard.2004.025270

15. Aigner T, Zien A, Gehrsitz A, Gebhard PM, McKenna L. Anabolic and catabolic gene expression pattern analysis in normal versus osteoarthritic cartilage using complementary DNA-array technology. *Arthritis Rheum.* 2001;44(12):2777–2789. doi:10.1002/1529-0131(200112)44:12<2777::aid-art465>3.0.co;2-h

16. Myers SL, Brandt KD, Ehlich JW, et al. Synovial inflammation in patients with early osteoarthritis of the knee. *J Rheumatol.* 1990;17(12):1662–1669.

17. Wenham CYJ, Conaghan PG. The role of synovitis in osteoarthritis. *Ther Adv Musculoskeletal Dis.* 2010;2(6):349–359. doi:10.1177/1759720X10378373

18. Eaton RG, Glickel SZ. Trapeziometacarpal osteoarthritis. Staging as a rationale for treatment. *Hand Clin.* 1987;3(4):455–471.

19. Angwin J, Heald G, Lloyd A, Howland K, Davy M, James MF. Reliability and sensitivity of joint space measurements in hand radiographs using computerized image analysis. *J Rheumatol.* 2001;28(8):1825–1836.

20. Huétink K, van 't Klooster R, Kaptein BL, et al. Automatic radiographic quantification of hand osteoarthritis: Accuracy and sensitivity to change in joint space width in a phantom and cadaver study. *Skeletal Radiol.* 2012;41(1):41–49. doi:10.1007/s00256-011-1110-x

21. Thoenen J, MacKay JW, Sandford HJC, Gold GE, Kogan F. Imaging of synovial inflammation in osteoarthritis, from the *AJR* special series on inflammation. *Am J Roentgenol.* 2022;218(3):405–417. doi:10.2214/AJR.21.26170

22. Haugen IK, Østergaard M, Eshed I, et al. Iterative development and reliability of the OMERACT hand osteoarthritis MRI scoring system. *J Rheumatol.* 2014;41(2):386–391. doi:10.3899/jrheum.131086

23. McNally EG. Ultrasound of the small joints of the hands and feet: Current status. *Skeletal Radiol.* 2008;37(2):99–113. doi:10.1007/s00256-007-0356-9

24. Epis O, Paoletti F, d'Errico T, et al. Ultrasonography in the diagnosis and management of patients with inflammatory arthritides. *Eur J Intern Med.* 2014;25(2):103–111. doi:10.1016/j.ejim.2013.08.700

25. Oo WM, Linklater JM, Bennell KL, et al. Are OMERACT knee osteoarthritis ultrasound scores associated with pain severity, other symptoms, and radiographic and magnetic resonance imaging findings? *J Rheumatol.* 2021;48(2):270–278. doi:10.3899/jrheum.191291

26. Wakefield RJ, Balint PV, Szkudlarek M, et al. Musculoskeletal ultrasound including definitions for ultrasonographic pathology. *J Rheumatol.* 2005;32(12):2485–2487.

27. D'Agostino MA, Conaghan P, Le Bars M, et al. EULAR report on the use of ultrasonography in painful knee osteoarthritis. Part 1: Prevalence of inflammation in osteoarthritis. *Ann Rheum Dis.* 2005;64(12):1703. doi:10.1136/ard.2005.037994

28. Walther M, Harms H, Krenn V, Radke S, Faehndrich TP, Gohlke F. Correlation of power Doppler sonography with vascularity of the synovial tissue of the knee joint in patients with osteoarthritis and rheumatoid arthritis. *Arthritis Rheum.* 2001;44(2):331–338. doi:10.1002/1529-0131(200102)44:2<331::AID-ANR50>3.0.CO;2-0

29. D'Agostino MA, Terslev L, Aegerter P, et al. Scoring ultrasound synovitis in rheumatoid arthritis: A EULAR-OMERACT ultrasound taskforce—Part 1: Definition and development of a standardised, consensus-based scoring system. *RMD Open.* 2017;3(1):e000428. doi:10.1136/rmdopen-2016-000428

30. Fenster A, Downey DB, Cardinal HN. Three-dimensional ultrasound imaging. *Phys Med Biol.* 2001;46(5):R67–R99. doi: 10.1088/0031-9155/46/5/201

31. Hunter DJ, Bierma-Zeinstra S. Osteoarthritis. *Lancet.* 2019;393(10182):1745–1759. doi:10.1016/S0140-6736(19)30417-9

32. Hunter DJ, Felson DT. Osteoarthritis. *Br Med J.* 2006;332(7542):639–642. doi:10.1136/bmj.332.7542.639

33. Menon J. The problem of trapeziometacarpal degenerative arthritis. *Clin Orthop Relat Res.* 1983;(175):155–165.

34. Van Heest AE, Kallemeier P. Thumb carpal metacarpal arthritis. *J Am Acad Orthopaed Surgeon.* 2008;16(3):140–151. doi:10.5435/00124635-200803000-00005

35. Ene R, Sinescu RD, Ene P, Cîrstoiu MM, Cîrstoiu FC. Synovial inflammation in patients with different stages of knee osteoarthritis. *Rom J Morphol Embryol.* 2015;56(1):169–173.

36. du Toit C, Dima R, Papernick S, et al. Three-dimensional ultrasound to investigate synovitis in first carpometacarpal osteoarthritis: A feasibility study. *Med Phys.* 2023; Online ahead of print. doi:10.1002/mp.16640

37. Papernick S, Gillies DJ, Appleton T, Fenster A. Three-dimensional ultrasound for monitoring knee inflammation and cartilage damage in osteoarthritis and rheumatoid arthritis. In: SPIE Medical Imaging 2020: Image-Guided Procedures, Robotic Interventions, and Modeling, Houston, TX; Vol 11315; 2020. https://doi.org/10.1117/12.2549624

38. Wieland HA, Michaelis M, Kirschbaum BJ, Rudolphi KA. Osteoarthritis—An untreatable disease? *Nat Rev Drug Discov*. 2005;4(4):331–344. doi:10.1038/nrd1693

39. Bedson J, Croft PR. The discordance between clinical and radiographic knee osteoarthritis: A systematic search and summary of the literature. *BMC Musculoskelet Disord*. 2008;9(1):116. doi:10.1186/1471-2474-9-116

40. Lawrence JS, Bremner JM, Bier F. Osteo-arthrosis. Prevalence in the population and relationship between symptoms and X-ray changes. *Ann Rheum Dis*. 1966;25(1):1–24.

41. Davis MA, Ettinger WH, Neuhaus JM, Barclay JD, Segal MR. Correlates of knee pain among US adults with and without radiographic knee osteoarthritis. *J Rheumatol*. 1992;19(12):1943–1949.

42. Roemer FW, Guermazi A, Felson DT, et al. Presence of MRI-detected joint effusion and synovitis increases the risk of cartilage loss in knees without osteoarthritis at 30-month follow-up: The MOST study. *Ann Rheum Dis*. 2011;70(10):1804. doi:10.1136/ard.2011.150243

43. Naredo E, Cabero F, Palop MJ, Collado P, Cruz A, Crespo M. Ultrasonographic findings in knee osteo-arthritis: A comparative study with clinical and radiographic assessment. *Osteoarthritis Cartilage*. 2005;13(7):568–574. doi:10.1016/j.joca.2005.02.008

44. Smith MD, Wechalekar MD. The Synovium. In: *Rheumatology*. Elsevier; 2015:27–32. doi:10.1016/B978-0-323-09138-1.00004-8

45. D. Smith M. The normal synovium. *Open Rheumatol J*. 2011;5(1):100–106. doi:10.2174/1874312901105010100

46. Cui A, Li H, Wang D, Zhong J, Chen Y, Lu H. Global, regional prevalence, incidence and risk factors of knee osteoarthritis in population-based studies. *EClinicalMedicine*. 2020;29-30:100587. doi:10.1016/j.eclinm.2020.100587

47. Felson DT. Osteoarthritis of the knee. *N Engl J Med*. 2006;354(8):841–848. doi:10.1056/NEJMcp051726

48. Amin S, LaValley MP, Guermazi A, et al. The relationship between cartilage loss on magnetic reso-nance imaging and radiographic progression in men and women with knee osteoarthritis. *Arthritis Rheum*. 2005;52(10):3152–3159. doi:10.1002/art.21296

49. Hunter DJ, Guermazi A, Roemer F, Zhang Y, Neogi T. Structural correlates of pain in joints with osteo-arthritis. *Osteoarthritis Cartilage*. 2013;21(9):1170–1178. doi:10.1016/j.joca.2013.05.017

50. Papernick S, Dima R, Gillies DJ, Appleton CT, Fenster A. Reliability and concurrent validity of three-dimensional ultrasound for quantifying knee cartilage volume. *Osteoarthritis Cartilage Open*. 2020;2(4):100127. doi: 10.1016/j.ocarto.2020.100127

51. Ashraf S, Walsh DA. Angiogenesis in osteoarthritis. *Curr Opin Rheumatol*. 2008;20(5):573–580. doi:10.1097/BOR.0b013e3283103d12

52. Walsh DA, Bonnet CS, Turner EL, Wilson D, Situ M, McWilliams DF. Angiogenesis in the synovium and at the osteochondral junction in osteoarthritis. *Osteoarthritis Cartilage*. 2007;15(7):743–751. doi:10.1016/j.joca.2007.01.020

53. Oo WM, Linklater JM, Bennell KL, et al. Superb microvascular imaging in low-grade inflammation of knee osteoarthritis compared with power Doppler: Clinical, radiographic and MRI relationship. *Ultrasound Med Biol*. 2020;46(3):566–574. doi:10.1016/j.ultrasmedbio.2019.11.017

54. Koski JM, Saarakkala S, Helle M, Hakulinen U, Heikkinen JO, Hermunen H. Power doppler ultraso-nography and synovitis: Correlating ultrasound imaging with histopathological findings and evaluat-ing the performance of ultrasound equipments. *Ann Rheum Dis*. 2006;65(12):1590–1595. doi:10.1136/ard.2005.051235

8 Local Vessel Wall and Plaque Volume Evaluation of Three-Dimensional Carotid Ultrasound Images for Sensitive Assessment of the Effect of Therapies on Atherosclerosis

Bernard Chiu, Xueli Chen, and Yuan Zhao

INTRODUCTION

Stroke is one of the main causes of mortality and disability worldwide, accounting for 5.5 million fatalities in 2016; and a major cause of stroke is atherosclerosis.[1] Medical treatments and lifestyle/dietary modifications may prevent strokes in those at high risk. The growth of atherosclerotic plaque is linked to an increased risk of stroke.[2,3] Therefore, serial monitoring of carotid atherosclerosis is essential for the stratification of stroke risk and the evaluation of anti-atherosclerotic therapies.[4] Although global measurements from ultrasound images, such as intima-media thickness (IMT), total plaque volume (TPV),[5] and vessel wall volume (VWV),[6] have been proposed, *local* plaque change may be more sensitive to anti-atherosclerotic treatments because atherosclerosis is a focal disease predominantly occurring at bends and bifurcations. We developed three-dimensional ultrasound (3DUS) imaging techniques to monitor the spatiotemporal changes of vessel-wall-plus-plaque thickness (VWT) distribution, allowing the point-by-point distance between the media-adventitia boundary (MAB) and lumen-intima boundary (LIB) and its change (ΔVWT) to be visualized and quantified consistently across patients.[7,8] Knowledge of the distribution of ΔVWT in population-based studies enables the identification of local regions that are more likely to show plaque progression/regression. Biomarkers developed to highlight these regions in ΔVWT distributions have been shown to be sensitive to the effect of medical and dietary therapies, such as atorvastatins,[9] B Vitamins,[8] and pomegranate[10] therapies.

Although ΔVWT is sensitive to changes in vessel wall and plaque thickness, it cannot detect circumferential changes. Local ΔVWT would be zero everywhere for an idealized circular artery with outward expansion from baseline but no thickness change. Excessive expansive remodeling following the formation of a plaque plays a crucial role in predicting future progression and, ultimately, the clinical significance of the plaque.[11] In around 20% of coronary arteries, plaque formation is followed by excessive expansive remodeling.[12] It reduces endothelial shear stress and accelerates plaque growth.[12] Therefore, there is a need to monitor the change in the circumferential dimension. ΔVWV, in contrast to ΔVWT, is able to monitor the collective change of thickness and circumferential changes despite its inability to localize and emphasize the changes.

DOI: 10.1201/9781003299462-10

We developed a biomarker known as voxel-based vessel-wall-and-plaque volume (VVol) that can detect *local* thickness and circumferential changes.[13] VVol measurement combines the concepts of local VWT and global VWV. The vessel wall was partitioned into a grid of cells with thickness and circumference dimensions. A grid cell correspondence workflow was proposed to match corresponding grid cells in the baseline and follow-up images. The voxel-based VVol change, denoted by $\Delta VVol$, was measured by taking the area difference between the baseline grid cells and the correspondent follow-up grid cells, where each grid cell can be considered as a voxel for the 3DUS image resliced at 1-mm intervals. Percentage VVol change, denoted by $\Delta VVol\%$ at each voxel, localizes regions where plaques are rapidly expanding from baseline. We hypothesize that biomarkers based on local $\Delta VVol$ would be more sensitive to treatment effects than existing biomarkers. In particular, we hypothesize that as $\Delta VVol$-based biomarkers account for both local thickness and circumferential changes, they would be more sensitive to therapeutic effects than ΔVWT-based biomarkers. Since the carotid geometry differs across individuals, a 3D carotid atlas was developed to represent $\Delta VVol\%$ consistently across patients so that quantitative comparisons can be made between patients and treatment groups in population-based studies. In addition, the display of $\Delta VVol\%$ in the 3D carotid atlas helps us to understand the contribution of the circumferential changes to the difference in the sensitivity between $\Delta VVol$- and ΔVWT-based biomarkers. As a major focus of this chapter is to compare the sensitivity of VVol and VWT measurements to treatment effect, we first introduce the VWT analysis workflow in Section 8.2. The description of the 2D standardized VWT map generation approach also provides necessary background for the introduction of the 3D carotid atlas generation workflow in Section 8.3. Table 8.1 lists the acronyms and notations used in this chapter.

TABLE 8.1
Acronyms and Notations Used in This Chapter

Acronym/notation	Description
3DUS	Three-dimensional ultrasound
MAB	Media-adventitia boundary
LIB	Lumen-intima boundary
CCA	Common carotid artery
ICA	Internal carotid artery
BF	Bifurcation
DL	Description length
MDL	Minimal description length
IMT	Intima-media thickness
TPV	Total plaque volume
VWV	Vessel-wall-plus-plaque volume
VWT	Vessel-wall-plus-plaque thickness
VVol	Voxel-based vessel-wall-and-plaque volume
T	Average of the two sides of the voxel-based vessel-wall-and-plaque cell in the thickness direction
C	Average of the two sides of the voxel-based vessel-wall-and-plaque cell in the circumferential direction
$\overline{\Delta VWT\%}$	Patient-based average percentage VWT change computed over the entire 2D carotid map
$\overline{\Delta T\%}$	Patient-based average percentage T change computed over the entire 3D carotid map
$\overline{\Delta C\%}$	Patient-based average percentage C change computed over the entire 3D carotid map
$\overline{\Delta VVol\%}$	Patient-based average percentage VVol change computed over the entire 3D carotid map

VESSEL-WALL-PLUS-PLAQUE THICKNESS (VWT) ANALYSIS

COMPUTATION OF 3D ΔVWT

Figure 8.1 illustrates the workflow for computing 3D ΔVWT map. VWT measures the distance between the LIB and the MAB of the carotid artery. The LIB and MAB were first segmented manually from a 3DUS image according to a previously described workflow.[6] To reduce the inter-scan measurement variability, the ultrasound volumes of the same artery acquired at baseline and follow-up scanning sections were viewed simultaneously to localize and match the bifurcation points (BF). An axis parallel to the longitudinal axis of the common carotid artery was placed centered on the bifurcation, and the 3DUS volume was resliced by a plane perpendicular to the axis with an interslice distance of 1 mm to minimize the measurement variability.[14] The LIB and MAB of the common and internal carotid arteries (CCA and ICA, respectively) on the 2D resliced images were then segmented as shown in Figure 8.1a. The adjacent 2D LIB and MAB contours in the contour stack segmented for a 3DUS volume were matched using the symmetric correspondence algorithm[7] to reconstruct the 3D LIB and MAB surfaces, shown in Figure 8.1b. To allow for point-by-point ΔVWT measurements, the 3D surfaces obtained at the follow-up imaging session was aligned to those obtained at baseline using the iterative closest point algorithm,[15] as shown in Figure 8.1c. The 3D VWT was calculated on a slice-by-slice basis. The LIB and MAB contours were matched on a point-by-point basis to compute the point-wise VWT, which was superimposed on the MAB surface to generate the 3D VWT map, as shown in Figure 8.1d. The 3D ΔVWT map in Figure 8.1e was generated by taking the VWT difference between each pair of corresponding points on the baseline and the registered follow-up VWT surfaces.[15]

GENERATION OF 2D ΔVWT MAPS

The details for generating an optimized two-dimensional (2D) ΔVWT map have been described elsewhere[9,16] and are briefly summarized here. Figure 8.2 shows the workflow of the 2D ΔVWT map generation procedures. The following steps are detailed in this subsection: (1) the 3D ΔVWT map was aligned with the standard 3D coordinate frame (Figure 8.2a); (2) the aligned ΔVWT map was projected onto an L-shaped 2D template (Figure 8.2b); (3) the L-shaped 2D map was resampled to

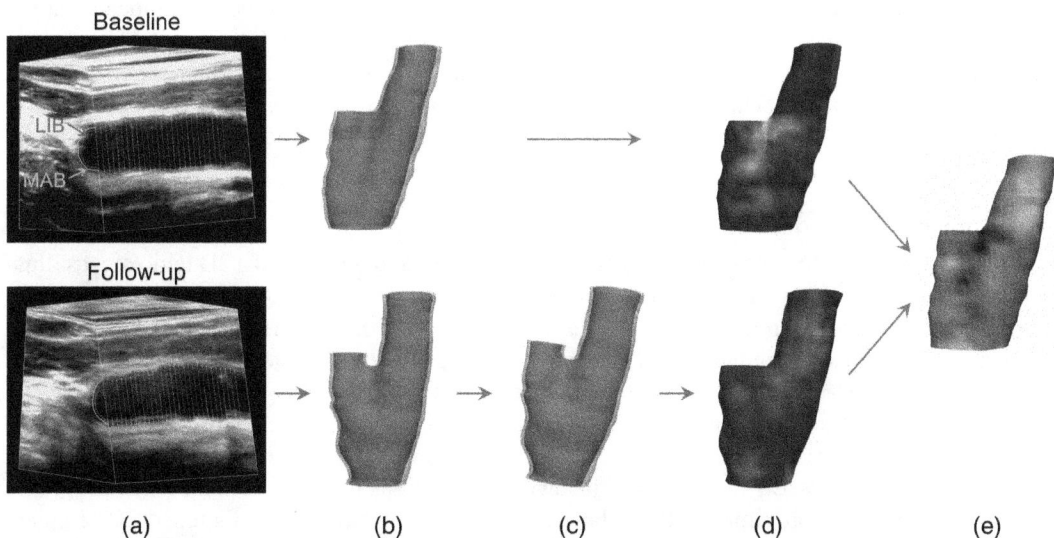

FIGURE 8.1 The workflow for 3D ΔVWT map construction: (a) Segmentation; (b) reconstruction; (c) registration, (d) VWT computation; (e) ΔVWT map computation.

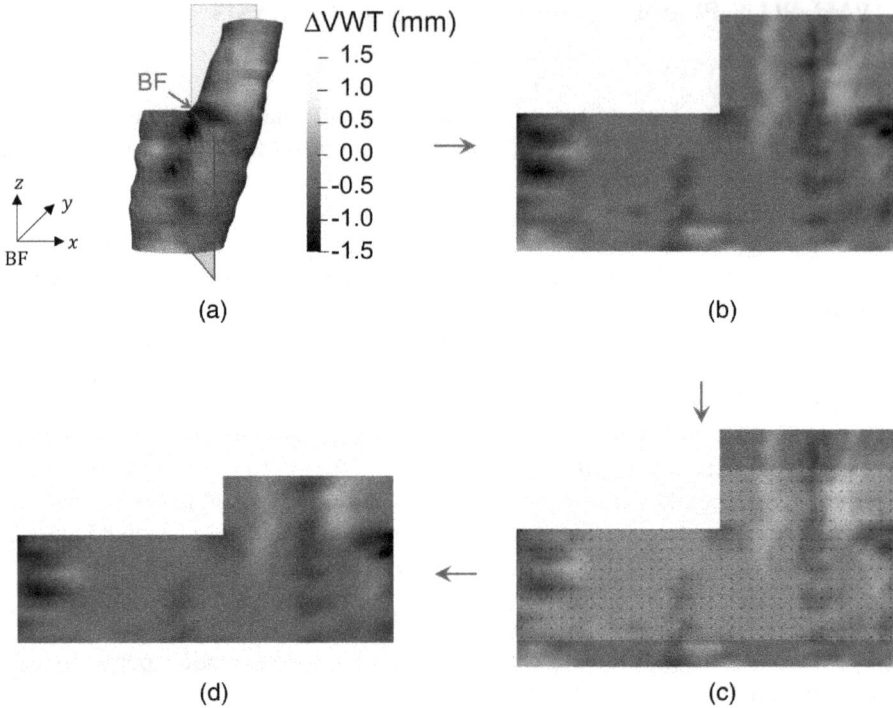

FIGURE 8.2 The schematic diagram of the 3D ΔVWT map generation workflow: (a) Alignment; (b) flattening; (c) optimization; (d) resampling.

obtain the initial 2D standardized ΔVWT map (Figure 8.2c); (4) the point correspondence in the initial 2D ΔVWT map was optimized to generate the final 2D ΔVWT map (Figure 8.2d).

Alignment of 3D VWT-change Map

The 3D VWT map was aligned to the standard 3D coordinate frame with the origin located at the carotid bifurcation (BF). The axis parallel to the longitudinal axis of the CCA, identified in manual segmentation, is aligned with the z-axis of the coordinate system, with the positive z-axis pointing in the direction of the ICA. The vector pointing from the origin to the centroid of the ICA contour on the same z-plane is aligned with the x-axis. The y-axis was defined by taking the direction of the cross-product of z- and x-axis (i.e., $\vec{z} \times \vec{x}$).

Generation of L-shaped 2D Flattened Map

The aligned 3D ΔVWT map of each artery was cut, unfolded, and mapped to a 2D template, as illustrated in Figure 8.3. The width of the top and bottom boundaries of the L-shaped carotid template corresponds to the average perimeters of ICA and CCA contours in a cohort, respectively. The plane cutting the CCA and ICA are labeled as P_{CCA} and P_{ICA}, respectively. P_{CCA} was defined by the y-axis, the BF and the centroid of the CCA contour most proximal to BF (i.e., $C_{CCA_{up}}$ in Figure 8.3a). P_{CCA} cuts the CCA surface on the negative y side. P_{ICA} was defined by the x-axis, the BF and the centroid of the ICA contour most distal from BF (i.e., $C_{ICA_{down}}$ in Figure 8.3a). The intersection between P_{ICA} and the ICA contour most distal from BF is labeled as c_1, whereas the intersection between P_{CCA} and the CCA contour most proximal to BF is labeled as c_0 in Figure 8.3a. The flattened 2D template is shown in Figure 8.3b. The intersecting line connecting BF and c_0 and that connecting BF and c_1 were projected onto the vertical edges of the 2D template. The transverse contours in the 3D ΔVWT map were unfolded in a clockwise direction, scaled, and mapped to horizontal lines in the 2D map.

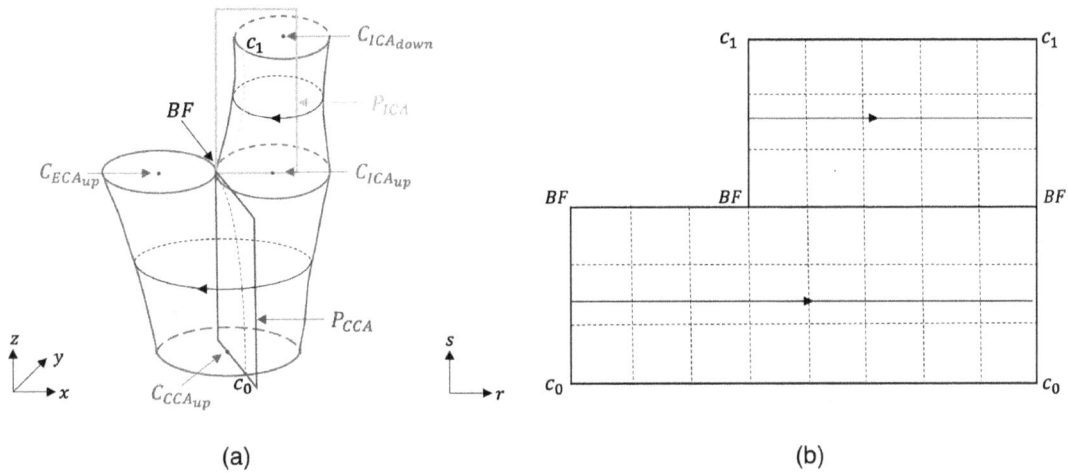

FIGURE 8.3 Construction of the 2D template. (a) 3D ΔVWT map aligned with the standard 3D coordinate frame with the origin located at BF. (b) 2D ΔVWT template.

Resampling of 2D Flattened Map

Resampling was performed to generate 2D carotid atlases with consistent dimensions across arteries. Each 2D flattened map was sampled in a 1-mm interval in both vertical and horizontal directions. While the horizontal width of the carotid atlas can be standardized based on the average perimeters of the CCA and ICA contours in a cohort, as described above, the longitudinal length of the 2D atlas cannot be standardized because the longitudinal length of the artery that can be segmented may vary across 3DUS volumes. Since the 2D carotid atlases of all arteries must be of the same size, in order for them to be comparable across subjects or between different treatment groups, resampling in the vertical direction was only performed in the overlapping region of all 2D flattened maps.

Optimization of the Initial 2D ΔVWT Map

Although 2D atlases of the same size were used to represent the ΔVWT distribution of different patients, anatomic correspondence mismatches still exist due to individual differences across patients. Anatomic correspondences were optimized across subjects using a minimal description length (MDL) algorithm developed previously.[16] Briefly, the points on the initial 2D standardized map have one-to-one mapping with the corresponding 3D surface, constituting a set of N landmarks. These landmarks were fixed on a master example arbitrarily selected. The 3D arterial surface of each subject was aligned with the master 3D arterial surface by Procrustes registration. The landmarks form a shape model that allows us to compute the DL and the gradient of DL with respect to the horizontal and vertical directions in the 2D map according to the derivation in the study by Chen et al.[16] The 2D maps were then deformed in the direction of steepest descent of DL, resulting in a new set of correspondence points. The next iteration starts with a Procrustes registration based on the new set of correspondence points followed by a new round of DL minimization described above.

VOXEL-BASED VESSEL-WALL-AND-PLAQUE VOLUME (VVol) ANALYSIS

COMPUTATION OF $\Delta VVol\%$

The MAB and LIB at the 3DUS images acquired at baseline and the follow-up sessions were manually segmented, reconstructed as surfaces, and registered with each other. Figure 8.4 illustrates the $\Delta VVol\%$ map generation process. On each axial contour, the MAB at baseline (red in Figure 8.4a)

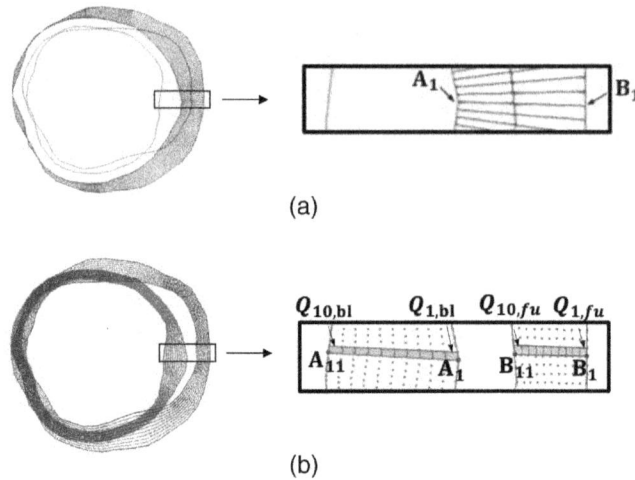

FIGURE 8.4 The slice-by-slice correspondence and sampling steps in the $\Delta VVol$ measurement workflow. (a) The MABs at baseline (red) and follow-up (green) were matched on a point-by-point basis. A_1 and B_1 are an example correspondence pair. (b) Each line connecting a corresponding point pair on the MAB–LIB pair at baseline (red contours) or the MAB–LIB pair at follow-up (green contours) was uniformly sampled at 11 locations (e.g., A_1, A_2, ..., A_{11} and B_1, B_2, ..., B_{11}). The region shaded in red enclosed by the MAB–LIB pair at baseline and the region shaded enclosed by the MAB–LIB pair at follow-up were each divided into 10 grid cells with vertices at the sampled points (i.e., $\{Q_{i,bl}\}_{i=1}^{10}$, $\{Q_{i,fu}\}_{i=1}^{10}$). $Q_{i,bl}$ was matched with $Q_{i,fu}$ for $i = 1, 2, ..., 10$. (From Chen et al.[13] with permission.)

was matched with the corresponding boundary at follow-up (Figure 8.4a) utilizing the symmetric correspondence approach established in the study by Chiu et al.[7] For example, in Figure 8.4a, point A_1 on MAB at baseline was matched with point B_1 on MAB at follow-up. The following workflow was developed to partition the vessel wall into small cells. First, the MAB–LIB pairs at baseline and follow-up were separately matched using the symmetric correspondence technique. Each corresponding line connecting the corresponding points on the MAB and LIB was sampled evenly at 11 locations, as shown in Figure 8.4b. For instance, the corresponding line starting at point A_1 was sampled at A_1, A_2, ..., A_{11}. Similarly, the corresponding line starting at point B_1 was sampled at B_1, B_2, ..., B_{11}. The region between adjacent corresponding lines, shaded in Figure 8.4b, was divided into 10 cells. For instance, the region from the baseline vessel wall shaded in red in Figure 8.4b was divided into 10 cells $\{Q_{i,bl}\}_{i=1}^{10}$ with their areas denoted by $\{S_{i,bl}\}_{i=1}^{10}$. The corresponding region from the follow-up vessel wall was divided similarly, resulting in 10 cells $\{Q_{i,fu}\}_{i=1}^{10}$ with their areas denoted by $\{S_{i,fu}\}_{i=1}^{10}$. The local $\Delta VVol$ for the cell Q_i was determined by subtracting $S_{i,bl}$ from $S_{i,fu}$ and the percentage change ($\Delta VVol\%$) was determined by dividing this difference value by $S_{i,bl}$. This change is referred to as the voxel-based $\Delta VVol$ because measurements were taken for each 1-mm reslice of the 3DUS volume and thereby the area change multiplied by 1 mm is a voxel volume measurement.

It was expected that $\Delta VVol$, accounting for the thickness and circumferential changes, would be more sensitive to treatment effect than ΔVWT, which considers thickness change only. Decomposition of $\Delta VVol$ along the thickness and circumferential directions can help to determine the contribution of thickness and circumferential changes to the sensitivity of $\Delta VVol$. Therefore, $\Delta VVol$ was decomposed into T and C, where T was defined as the average of the two sides of each cell along the direction of thickness, and C was defined as the average of the two sides along the direction of circumference. Changes of T and C (i.e., ΔT and ΔC) and their respective percentage differences (i.e., $\Delta T\%$ and $\Delta C\%$) in the corresponding voxels in the baseline and follow-up images

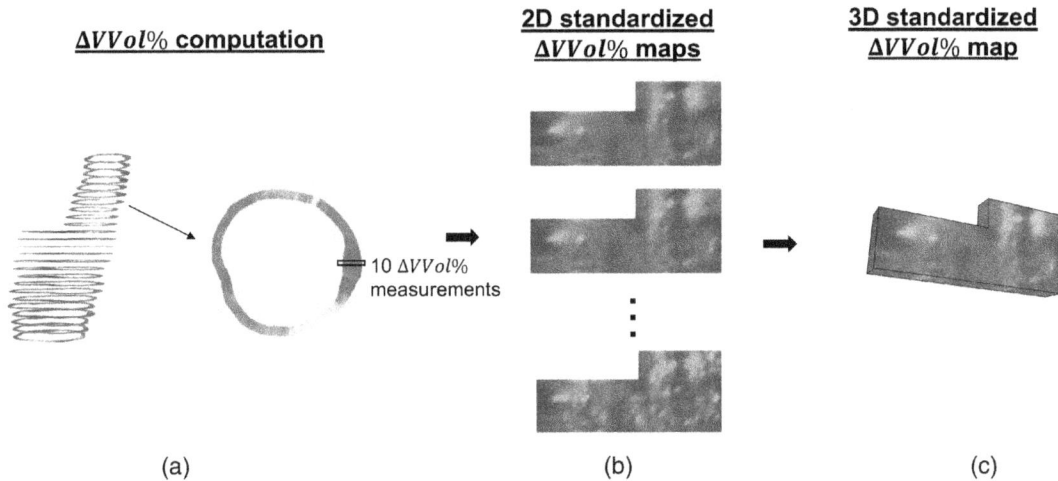

FIGURE 8.5 Construction of the 3D standardized carotid $\Delta VVol\%$ map. (From Chen et al.[13] with permission.)

were then calculated in the same manner as $\Delta VVol$. Visualizing the spatial distributions of thickness and circumferential changes can shed light on the differences between pomegranate and placebo patients in vascular remodeling.

GENERATION OF 3D STANDARDIZED $\Delta VVol\%$ MAPS

Quantification of $\Delta VVol\%$ across patients and treatment groups requires the measurements to be represented in maps with standardized geometry. The 3D standardized carotid $\Delta VVol$ map we developed was generated by the following three steps:

1. The MAB surfaces of the ICA and CCA were cut by planes P_{ICA} and P_{CCA}, respectively, unfolded, and mapped to a standardized domain as described above, generating an L-shaped 2D carotid map. The resampling was then performed for the 2D carotid map, as described above.
2. Anatomic correspondence mismatches were adjusted for a group of 2D carotid maps using the MDL correspondence optimization technique, as described above.
3. Each point on the 2D carotid map was equipped with 10 measurements for $\Delta VVol\%$ at the 10 grid cells, as described in Section 3.1 and illustrated in Figure 8.4(b). The 10 measurements were superimposed on 10 consecutive 2D carotid maps, as shown in Figure 8.5b, which are stacked to form a 3D carotid atlas, as shown in Figure 8.5c. $\Delta T\%$ and $\Delta C\%$ were represented in a 3D carotid atlas in the same manner.

QUANTITATIVE ASSESSMENT OF TREATMENT EFFECT BASED ON VWT AND VVol

STUDY SUBJECTS AND ULTRASOUND IMAGE ACQUISITION

Subjects were recruited at the Stroke Prevention & Atherosclerosis Research Centre, Robarts Research Institute (London, Ontario, Canada) for a clinical trial (registration number ISRCTN30768139). This study involved 120 patients randomized into 2 groups, with 66 patients receiving pomegranate extract in a tablet and/or juice form and 54 patients receiving a placebo tablet and juice once daily for a year. Baseline characteristics of the cohort are provided in study by Zhao et al.[10]

High-resolution 3DUS images were obtained by translating an ultrasound transducer (L12-5, Philips, Bothel, WA) mounted on a mechanical assembly at a uniform speed along the neck for about 4 cm. Ultrasound frames acquired using an ultrasound machine (ATL HDI 5000, Philips, Bothel, WA) were digitized at a rate of 30 Hz and reconstructed into a 3D image. Participants were scanned at the baseline before the study and a follow-up session after approximately a year (375 ± 17 days, range: 283–428 days).

STATISTICAL METHODS

The patient-based average measurements of $\Delta VVol\%$, $\Delta T\%$, and $\Delta C\%$ over the entire 3D map are denoted by $\overline{\Delta VVol\%}$, $\overline{\Delta T\%}$, and $\overline{\Delta C\%}$, respectively. The sensitivities of these measurements were quantified by the p-values of t tests or U tests and were compared with those of the TPV and VWV changes, denoted by ΔTPV and ΔVWV, respectively, and their corresponding percentage changes, denoted by $\Delta TPV\%$ and $\Delta VWV\%$. The patient-based average measurements of ΔVWT and $\Delta VWT\%$ over the entire 2D map are denoted by $\overline{\Delta VWT}$ and $\overline{\Delta VWT\%}$, respectively. In contrast with VWT, only the mean percentage change was calculated for VVol, since the mean change for each patient in VVol is proportional to and has the same sensitivity as ΔVWV.

The change measurements are normalized to their annual rates by linear scaling since follow-up duration differs between participants from 283 to 428 days. The Shapiro–Wilk test was first performed to test the normality of the measurements. Then, p-values were calculated using two-sample t tests for normally distributed data and Mann–Whitney U tests for non-normally distributed data. Sample size calculation estimates the number of participants required per group to achieve a significant difference at a specified significance level and statistical power. The sample size required per group for each normally distributed measurement to detect a specific effect size δ was estimated using the following equation:[17]

$$N = \frac{\left(z_{\alpha/2} + z_\beta\right)^2 \left(\sigma_0^2 + \sigma_1^2\right)}{\delta^2},$$

where Z is a normally distributed random variable and $Pr\left(Z > z_\beta\right) = \beta$. σ_0 and σ_1 are the standard deviations of the placebo and pomegranate groups, respectively. The significance level α and the statistical power $1 - \beta$ were set as 5% and 90%, respectively, in this study. The sample size required per group for each non-normally distributed measurement was estimated using the following equation:[18]

$$N = \frac{\left(z_{\alpha/2} + z_\beta\right)^2}{12t(1-t)(p-0.5)^2}$$

where $p = (1/mn)\sum_{i=1}^{m}\sum_{j=1}^{n}\left(I\left(x_i > y_j\right)\right) + 0.5I\left(x_i = y_j\right)$ and $t = n/(m+n)$. x_i and y_j are the parameters for the ith patient of the placebo group and the jth patient of the pomegranate group, respectively. m and n are the number of placebo and pomegranate subjects, respectively. $I(\cdot)$ is the indicator function.

RESULTS

No statistically significant difference was observed between the background medications taken by the pomegranate and placebo groups, according to supplementary Table 8.2 in Zhao et al.[10] P-values for the biomarkers were shown in Table 8.2. $\overline{\Delta VVol\%}$ was the most sensitive to the treatment effect. $\overline{\Delta VWT\%}$ and $\overline{\Delta T\%}$ share the same sensitivity to treatment effect as they both quantified the change in VWT. The difference between pomegranate and placebo patients in $\overline{\Delta C\%}$ was larger than in $\overline{\Delta T\%}$. The sample size per group required for various effect sizes is shown in Table 8.3.

TABLE 8.2

The Means and Standard Deviations of Nine Biomarkers Computed for the Placebo and Pomegranate Groups and the *p*-values Associated with Two-Sample *t*-Tests or Mann–Whitney U Tests. The Asterisks Represent *p*-values Obtained Using Mann–Whitney U Tests. The Remaining *p*-values were Associated with *t* Tests

Measurement	Placebo (*n* = 54)	Pomegranate (*n* = 66)	*p*-value
$\Delta TPV \left(mm^3 \right)$	25.0 ± 76.5	16.1 ± 83.9	0.24*
$\Delta TPV\%$	$18.3 \pm 47.6\%$	$13.3 \pm 48.2\%$	0.19*
$\Delta VWV \left(mm^3 \right)$	112.4 ± 118.5	58.9 ± 104.9	0.01
$\Delta VWV\%$	$19.1 \pm 20.6\%$	$9.0 \pm 14.6\%$	0.008*
$\Delta VWT \left(mm \right)$	0.045 ± 0.12	0.010 ± 0.12	0.03*
$\Delta VWT\%$	$18.6 \pm 16.2\%$	$11.9 \pm 14.3\%$	0.02
$\Delta T\%$	$21.7 \pm 16.9\%$	$15.1 \pm 15.3\%$	0.03
$\Delta C\%$	$13.6 \pm 10.2\%$	$8.1 \pm 10.7\%$	0.005
$\Delta VVol\%$	$36.9 \pm 23.1\%$	$22.3 \pm 16.2\%$	0.0002

Source: From Chen et al.[13] with permission.

Interestingly, we observe from Table 8.2 that the sum of the group mean $\overline{\Delta C\%}$ and $\overline{\Delta T\%}$ approximated the group mean $\overline{\Delta VVol\%}$. This observation agrees with the analysis illustrated in Figure 8.6. It can be established that $\Delta VVol\% \approx \Delta T\% + \Delta C\%$ with an approximation $\Delta C\Delta T \approx 0$. This approximation can be made because $\Delta C\Delta T$ in a small cell is much smaller than $C\Delta T$ and $T\Delta C$. The analysis provides a basis explaining the higher sensitivity of $\overline{\Delta VVol\%}$, as compared to $\overline{\Delta VWT\%}$.

To allow visual comparison between the 3D $\Delta VVol\%$, $\Delta T\%$ and $\Delta C\%$ maps and the 2D $\Delta VWT\%$ map, each of the 3D maps was flattened to 2D by taking the mean of ten 2D maps along the radial

TABLE 8.3

Sample Sizes Per Group Required for Various Effect Sizes in a One-Year Study, with a 90% Statistical Power and a Significant Level of 5% (Two-Tailed). The Effect Sizes Are Expressed as the Percentage of the Effect Exhibited in the Current Placebo-Controlled Study. For Non-normally Distributed Measurements, Only Sample Sizes with 100% Effect Size Were Available

Measurement	100%	75%	50%
ΔTPV	2363	/	/
$\Delta TPV\%$	1496	/	/
ΔVWV	92	164	368
$\Delta VWV\%$	211	/	/
ΔVWT	324	/	/
$\Delta VWT\%$	110	196	441
$\Delta T\%$	127	225	507
$\Delta C\%$	75	134	302
$\Delta VVol\%$	39	70	157

Source: Chen et al.[13] with permission.

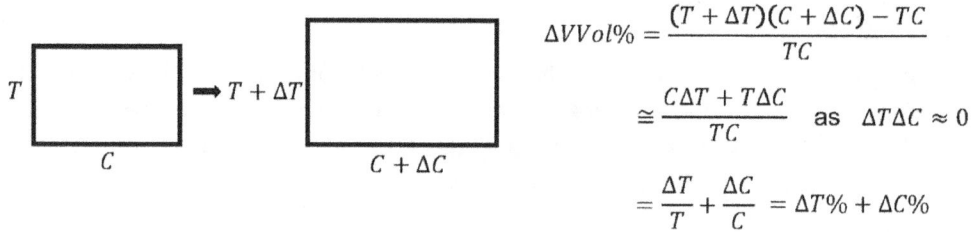

$$\Delta VVol\% = \frac{(T + \Delta T)(C + \Delta C) - TC}{TC}$$

$$\cong \frac{C\Delta T + T\Delta C}{TC} \quad \text{as} \quad \Delta T \Delta C \approx 0$$

$$= \frac{\Delta T}{T} + \frac{\Delta C}{C} = \Delta T\% + \Delta C\%$$

FIGURE 8.6 Relationship of $\Delta VVol\%$, the thickness and circumferential changes ($\Delta T\%$ and $\Delta C\%$, respectively) in an idealized regular cell. (From Chen et al.[13] with permission.)

direction, resulting in the mean 2D $\Delta VVol\%_{\bar{r}}$, $\Delta T\%_{\bar{r}}$ and $\Delta C\%_{\bar{r}}$ maps shown in Figures 8.7 and 8.8. The $\Delta T\%_{\bar{r}}$ and $\Delta VWT\%$ are almost identical, as expected. The increased sensitivity afforded by $\Delta VVol\%$ is attributable to the larger circumferential expansion of placebo subjects at regions close to the bifurcation, as pointed to by arrows in Figures 8.7 and 8.8. Figure 8.9 shows the circumferential expansion in four example arteries. $\Delta VVol\%$ highlights both plaque progression and outward expansion, while $\Delta VWT\%$ can only capture thickness change but not outward expansion. Plaque growth marked by asterisks was highlighted in both the $\Delta VVol\%_{\bar{r}}$ and $\Delta VWT\%$ maps, whereas regions with outward expansion and no local thickness change were highlighted only on the $\Delta VVol\%_{\bar{r}}$ maps, as pointed to by arrows in Figure 8.9. In addition, it is observed that circumferential expansion sometimes occurred near plaque progression, as shown in Figure 8.9a,c.

As previously mentioned, $\overline{\Delta VWV}$ is a scaled version of the subject-based mean of voxel-based $\Delta VVol$. $\Delta VVol\%$, the average of the voxel-based $\Delta VVol$ percentage change, is more able to discriminate between the pomegranate and placebo groups than $\overline{\Delta VWV}$, as shown in Table 8.2. Similarly, $\overline{\Delta VWT\%}$ is more sensitive to treatment effect than $\overline{\Delta VWT}$. The average percentage change was found to be more sensitive than the average magnitude change because percentage changes emphasize rapid plaque change, more commonly exhibited in placebo subjects than pomegranate subjects. Figure 8.10 shows two example arteries with similar $\overline{\Delta VWV}$ values but different $\Delta VVol\%$ values. The artery shown in Figure 8.10a had a larger $\Delta VVol\%$ than the artery in Figure 8.10b because it had more regions with rapid local plaque change.

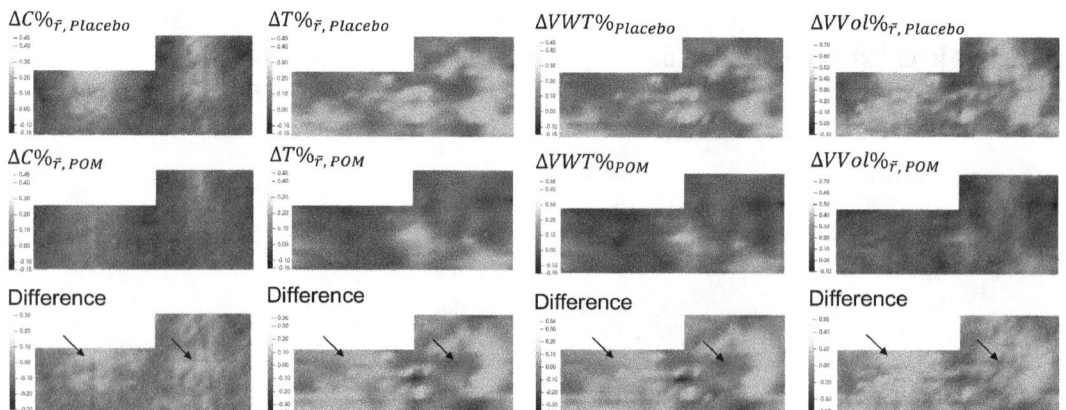

FIGURE 8.7 Groupwise average $\Delta C\%_{\bar{r}}$, $\Delta T\%_{\bar{r}}$, $\Delta VWT\%$ and $\Delta VVol\%_{\bar{r}}$ of the left carotid artery of the placebo and pomegranate groups. The arrows indicate the locations where the circumferential expansion of the arteries of placebo subjects is larger than that of pomegranate subjects, detected by $\Delta VVol\%$ but not $\Delta VWT\%$. (From Chen et al.[13] with permission.)

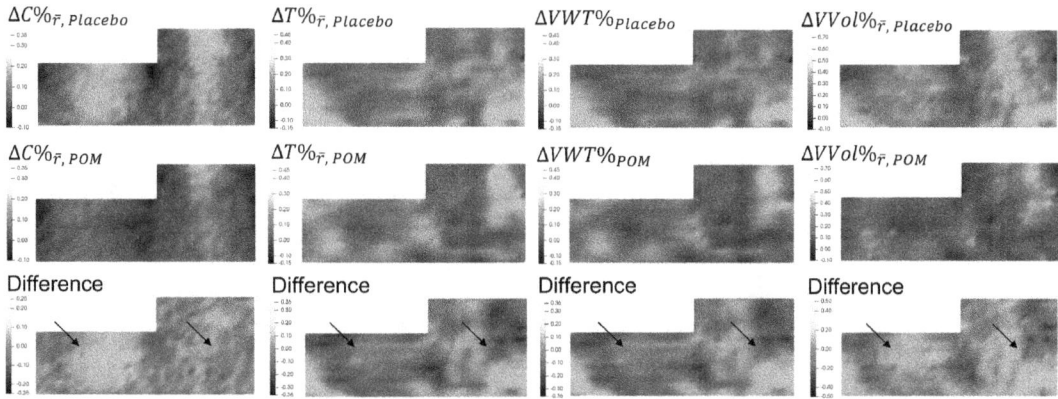

FIGURE 8.8 Groupwise average $\Delta C\%_{\bar{r}}$, $\Delta T\%_{\bar{r}}$, $\Delta VWT\%$ and $\Delta VVol\%_{\bar{r}}$ of the right carotid artery of the placebo and pomegranate groups. The arrows indicate the locations where the circumferential expansion of the arteries of placebo subjects is larger than that of pomegranate subjects, detected by $\Delta VVol\%$ but not $\Delta VWT\%$. (From Chen et al.[13] with permission.)

DISCUSSION AND FUTURE PERSPECTIVE

We developed a biomarker quantifying the percentage change of vessel-wall-and-plaque volume ($\overline{\Delta VVol\%}$) and a 3D carotid atlas to represent the voxel-based $\Delta VVol\%$ consistently across patients with variable carotid geometry. $\overline{\Delta VVol\%}$ was sensitive to the difference exhibited in the pomegranate and placebo groups. Although TPV was sensitive to the impact of intensive atorvastatin therapy,[5] it was incapable of detecting the effect of pomegranate in this study because pomegranate is a dietary supplement and has a weaker effect than high-dose atorvastatin.

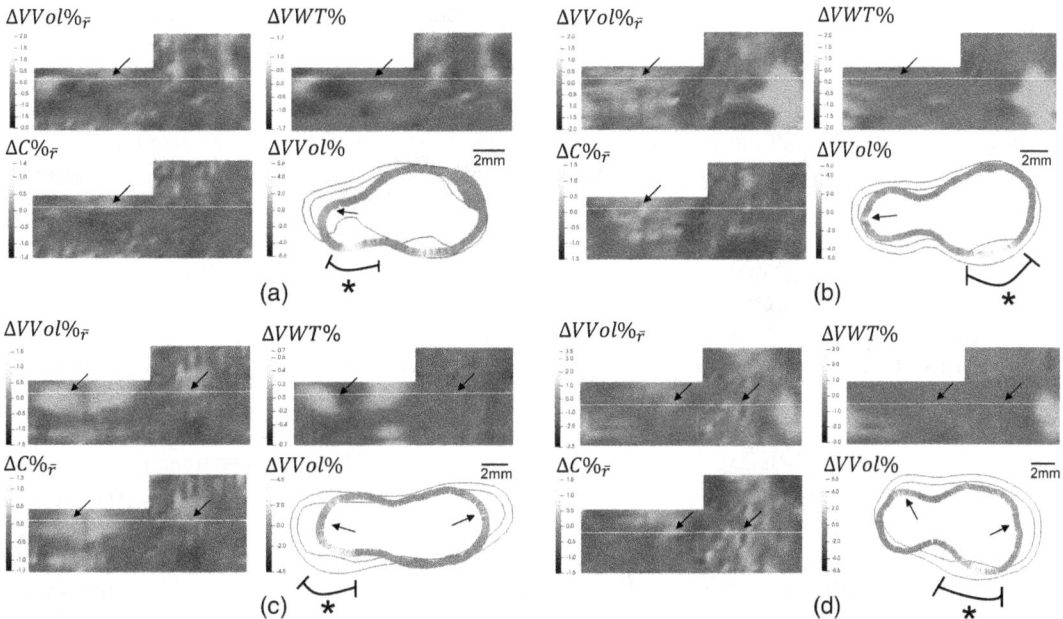

FIGURE 8.9 Four example arteries demonstrating excessive expansive remodeling without local vessel wall thickness change. The bottom right illustration in each subfigure shows an axial reslice at the location marked by the white lines shown in the flattened maps. The locations with excessive expansion are pointed to by the arrows. (From Chen et al.[13] with permission.)

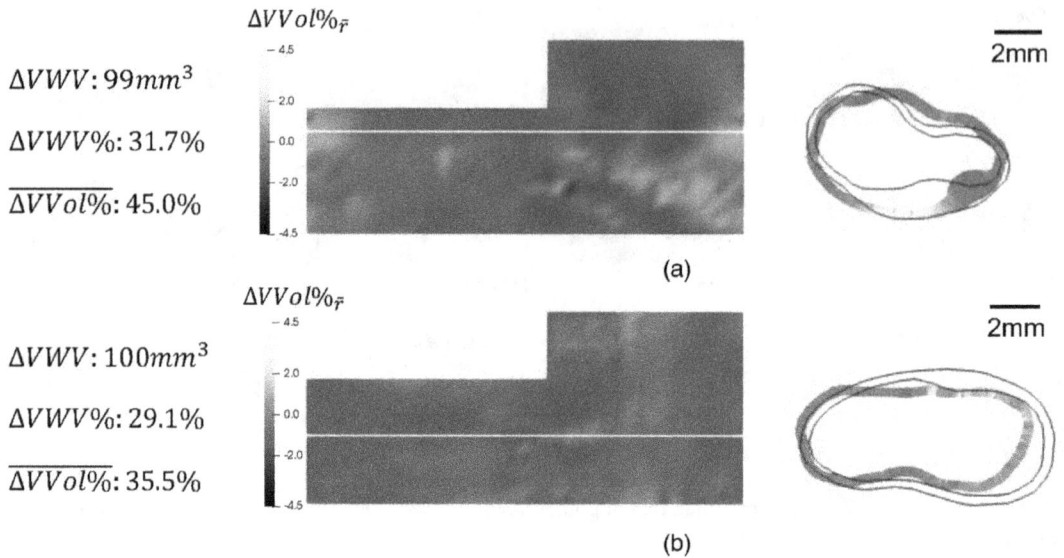

$\Delta VWV: 99mm^3$

$\Delta VWV\%: 31.7\%$

$\overline{\Delta VVol\%}: 45.0\%$

(a)

$\Delta VWV: 100mm^3$

$\Delta VWV\%: 29.1\%$

$\overline{\Delta VVol\%}: 35.5\%$

(b)

FIGURE 8.10 Two examples illustrating the sensitivity of $\overline{\Delta VVol\%}$ in highlighting rapid plaque change, in comparison to ΔVWV and $\Delta VWV\%$. For each example artery, the $\Delta VVol\%_{\bar{r}}$ map and an axial reslice of the $\Delta VVol\%$ at the location marked by the white line are illustrated. The ΔVWV and $\Delta VWV\%$ values in the two examples are comparable, but $\overline{\Delta VVol\%}$ was larger in part (a) because more regions in example (a) were associated with rapid plaque change. (From Chen et al.[13] with permission.)

We found that $\overline{\Delta VVol\%}$ was more sensitive to treatment effect than $\overline{\Delta VWT\%}$ because it can detect change along both the thickness and the circumferential directions. This was established quantitatively in Table 8.2 and analytically in Figure 8.6. The circumferential change was found to be more able to discriminate between the pomegranate and placebo groups than thickness change, as shown in Table 8.2. A comparison of the $\Delta C\%_{\bar{r}}$ and $\Delta T\%_{\bar{r}}$ maps in Figures 8.7 and 8.8 shows that a larger degree of circumferential expansion occurred in placebo subjects than in pomegranate patients. Compensatory enlargement has been known as a mechanism to maintain lumen size and endothelial shear stress in the presence of atherosclerotic plaques.[19] However, compensatory enlargement would have resulted in a large increase in thickness but a small change in circumference. The circumferential expansion observed in placebo subjects was over-compensatory. Over-compensatory remodeling was first suggested by Glagov et al.[19] and was confirmed by serial intravascular ultrasonography (IVUS) measurements of the coronary arteries.[20] According to another IVUS study,[21] over-compensatory remodeling was found in about 20% of coronary arteries.

It was thought that compensatory enlargement as a response to plaque progression was beneficial by preserving the lumen size.[19] However, over-compensatory remodeling would result in a reduction of endothelial shear stress, promoting local lipid accumulation, inflammation, and further extracellular matrix degradation (ECM).[12] It is unknown what factors determine whether the expansive remodeling response is excessive or compensatory. Earlier research suggested a connection between vascular expansion and matrix-metalloproteinases (MMPs), major enzymes involved in ECM degradation.[20,22] Pomegranate is a polyphenol-rich dietary supplement associated with the suppression of MMP-2 and MMP-9 production.[23,24] The finding of the current study suggests that pomegranate may help to slow ECM degradation and weaken the arterial wall structure.

In this study, we demonstrated that for both VVol and VWT measurements, the patient-based average percentage change was more sensitive to treatment effect than the average magnitude change. According to previous studies,[3,25,26] patients with rapid carotid plaque progression are at an increased risk for cerebrovascular and coronary events. Therefore, mean percentage change may be more precise in identifying high-risk individuals in addition to having increased sensitivity to treatment effects. As high-risk patients are anticipated to respond more to aggressive medical therapy, the increased accuracy will enhance the cost-effectiveness of these treatments.

A VWT-based biomarker[10] and a plaque-texture-based biomarker[27] were previously established to evaluate the therapeutic effects of pomegranate from the same trial. Eighty-three and 45 subjects per group were needed to be monitored for one year at the same level of significance and power as in this investigation for the VWT-based and plaque-texture-based biomarkers, respectively, compared to 39 patients per group for $\overline{\Delta VVol\%}$. The number of subjects required for $\overline{\Delta VVol\%}$ was less than half of that required for the weighted ΔVWT biomarker proposed in Zhao et al.[10] In addition, the weighted ΔVWT biomarker proposed by Zhao et al.[10] requires training on another cohort to obtain the weights for computing the weighted average ΔVWT for each subject, whereas $\overline{\Delta VVol\%}$ does not require training. Although most regions exhibiting progression or regression in the arteries could be highlighted by the trained weight map, thereby reducing the number of subjects required to achieve significance in comparison to the unweighted $\overline{\Delta VWT\%}$ (sample size: 83 for the weighted average ΔVWT vs. 110 for $\overline{\Delta VWT\%}$), the trained weight map missed some regions with elevated progression or regression, as acknowledged in Zhao et al.[10] The miss is attributable to the small difference between the VWT distribution patterns between cohorts. $\Delta VVol$ does not have this issue as training is not required.

Compared to the plaque texture biomarker proposed by Chen et al.,[27] $\overline{\Delta VVol\%}$ requires a smaller sample size to show the significance of the treatment effect. An additional advantage of $\overline{\Delta VVol\%}$ over the plaque texture biomarker is that only MAB and LIB segmentations are required for $\Delta VVol$ measurements, whereas plaque segmentation is required for analyzing plaque texture. Plaque segmentation is more challenging than MAB and LIB segmentation, due to its longer training and implementation times and higher observer variability.[28] Although manual segmentation of MAB and LIB was performed in this study focusing on the evaluation of a novel biomarker, automated deep-learning-based segmentation algorithms have been developed to segment MAB and LIB from 3DUS images[29–31]. These methods can be easily incorporated into the proposed workflow to improve the efficiency and reproducibility of the analysis.

We acknowledge some limitations of this study. First, we use manual segmentation of MAB and LIB for the 3D $\Delta VVol\%$ measurements. The manual segmentation should be replaced by an automated one for time shortening before the utility of the proposed biomarker in clinical research and practice. A deep-learning segmentation technique recently developed to segment the MAB and LIB at the CCA and ICA[29] was ideally suited for incorporation into the $\Delta VVol\%$ measurement pipeline. However, despite VWV measurements from MAB and LIB automatically segmented by this method highly correlated with the corresponding measurements from manual segmentation (Pearson's correlation was 0.94 for CCA and 0.95 for ICA as reported in Jiang et al.[29]), a similar correlation study will be required in the future to confirm that the algorithm and manual measurements of $\overline{\Delta VVol\%}$ also highly correlate. Second, the imaging information used in this study was collected from a single clinical site. The proposed biomarker should be tested in a trial involving multiple sites, multiple ultrasound systems, and a larger population before it can be deemed a clinically reliable tool.

Besides volumetric information, it is also beneficial to utilize intensity and textural information available in the 3D US images since it may provide information about the composition of the vessel wall and plaque. Plaque composition determines its vulnerability, and thereby, the risk of cardiovascular events. Plaque texture parameters can be used as a proxy for assessing plaque composition. We anticipate that texture within the vessel wall region is discriminative as plaque texture has been shown to be capable of detecting treatment effect.[27] An advantage of analyzing vessel wall texture instead of plaque texture is that segmentation of MAB and LIB is easier than plaque segmentation and can be automated. Without plaque boundaries enclosing the plaque, a challenge for vessel wall textural analysis is to appropriately highlight regions with a relatively larger plaque progression and regression for textural analysis to avoid dilution of plaque features with texture at regions with small changes. Regions with larger changes can be highlighted according to $\Delta VVol\%$, $\Delta T\%$ and $\Delta C\%$, and larger weights can be given to the texture features extracted in these important regions. This approach allows regions to be highlighted on a voxel basis without the need for training in another cohort as in Zhao et al.[10]

Advanced automatic analysis techniques based on deep neural networks have been proposed for carotid ultrasound image analysis. Applications include 2D longitudinal carotid ultrasound image segmentation to obtain biomarkers IMT[32] and TPA,[33] 3D carotid ultrasound volume segmentation to generate biomarker VWV,[29,34] and plaque component segmentation to identify vulnerable plaques.[35] However, the development has been limited to segmentation and there is a lack of deep-learning algorithms for generating new sensitive biomarkers. In future, we will develop sensitive deep-learning-based VWT-based and VVol-based biomarkers that can be generated directly from the VWT maps and VVol maps. We expect these biomarkers to be more sensitive to therapeutic treatments than simple average values of ΔVWT and $\Delta VVol$ measured in this study. Vessel wall and plaque texture features can also contribute in the development of novel sensitive biomarkers through fusion with VWT and VVol features.

ACKNOWLEDGMENT

Bernard Chiu is grateful for the funding support from the Research Grant Council of the HKSAR, China (Project Nos. CityU 11205917 and 11203218), and City University of Hong Kong Strategic Research Grant No. 7005441.

REFERENCE

1. Johnson CO, Nguyen M, Roth GA, et al. Global, regional, and national burden of stroke, 1990–2016: A systematic analysis for the Global Burden of Disease Study 2016. *Lancet Neurol.* 2019;18(5):439–458.
2. Wannarong T, Parraga G, Buchanan D, et al. Progression of carotid plaque volume predicts cardiovascular events. *Stroke.* 2013;44(7):1859–1865.
3. Hirano M, Nakamura T, Kitta Y, et al. Short-term progression of maximum intima-media thickness of carotid plaque is associated with future coronary events in patients with coronary artery disease. *Atherosclerosis.* 2011;215(2):507–512.
4. Spence JD, Hackam DG. Treating arteries instead of risk factors: A paradigm change in management of atherosclerosis. *Stroke.* 2010;41(6):1193–1199.
5. Ainsworth CD, Blake CC, Tamayo A, Beletsky V, Fenster A, Spence JD. 3D ultrasound measurement of change in carotid plaque volume: A tool for rapid evaluation of new therapies. *Stroke.* 2005;36(9):1904–1909.
6. Krasinski A, Chiu B, Spence JD, Fenster A, Parraga G. Three-dimensional ultrasound quantification of intensive statin treatment of carotid atherosclerosis. *Ultrasound Med Biol.* 2009;35(11):1763–1772.
7. Chiu B, Egger M, Spence JD, Parraga G, Fenster A. Quantification of carotid vessel wall and plaque thickness change using 3D ultrasound images. *Med Physics.* 2008;35(8):3691–3710.
8. Cheng J, Ukwatta E, Shavakh S, et al. Sensitive three-dimensional ultrasound assessment of carotid atherosclerosis by weighted average of local vessel wall and plaque thickness change. *Med Physics.* 2017;44(10):5280–5292.
9. Chiu B, Li B, Chow TW. Novel 3D ultrasound image-based biomarkers based on a feature selection from a 2D standardized vessel wall thickness map: A tool for sensitive assessment of therapies for carotid atherosclerosis. *Physics Med Biol.* 2013;58(17):5959.
10. Zhao Y, Spence JD, Chiu B. Three-dimensional ultrasound assessment of effects of therapies on carotid atherosclerosis using vessel wall thickness maps. *Ultrasound Med Biol.* 2021;47(9):2502–2513.
11. Schoenhagen P, Ziada KM, Kapadia SR, Crowe TD, Nissen SE, Tuzcu EM. Extent and direction of arterial remodeling in stable versus unstable coronary syndromes: An intravascular ultrasound study. *Circulation.* 2000;101(6):598–603.
12. Chatzizisis YS, Coskun AU, Jonas M, Edelman ER, Feldman CL, Stone PH. Role of endothelial shear stress in the natural history of coronary atherosclerosis and vascular remodeling: Molecular, cellular, and vascular behavior. *J Am Col Cardiol.* 2007;49(25):2379–2393.

13. Chen X, Zhao Y, Spence JD, Chiu B. Quantification of local vessel wall and plaque volume change for assessment of effects of therapies on carotid atherosclerosis based on 3-D ultrasound imaging. *Ultrasound Med Biol.* 2022;doi:10.1016/j.ultrasmedbio.2022.10.017

14. Landry A, Spence JD, Fenster A. Quantification of carotid plaque volume measurements using 3D ultrasound imaging. *Ultrasound Med Biol.* 2005;31(6):751–762.

15. Chiu B, Shamdasani V, Entrekin R, Yuan C, Kerwin WS. Characterization of carotid plaques on 3-dimensional ultrasound imaging by registration with multicontrast magnetic resonance imaging. *J Ultrasound Med.* 2012;31(10):1567–1580.

16. Chen Y, Chiu B. Correspondence optimization in 2D standardized carotid wall thickness map by description length minimization: A tool for increasing reproducibility of 3D ultrasound-based measurements. *Med Physics.* 2016;43(12):6474–6490.

17. Rosner B. *Fundamentals of Biostatistics.* Cengage learning; 2015.

18. Noether GE. Sample size determination for some common nonparametric tests. *J Am Stat Assoc.* 1987;82(398):645–647.

19. Glagov S, Weisenberg E, Zarins CK, Stankunavicius R, Kolettis GJ. Compensatory enlargement of human atherosclerotic coronary arteries. *N Engl J Med.* 1987;316(22):1371–1375.

20. Sipahi I, Tuzcu EM, Schoenhagen P, et al. Paradoxical increase in lumen size during progression of coronary atherosclerosis: Observations from the REVERSAL trial. *Atherosclerosis.* 2006;189(1):229–235.

21. Feldman CL, Coskun AU, Yeghiazarians Y, et al. Remodeling characteristics of minimally diseased coronary arteries are consistent along the length of the artery. *Am J Cardiol.* 2006;97(1):13–16.

22. Galis ZS, Khatri JJ. Matrix metalloproteinases in vascular remodeling and atherogenesis: The good, the bad, and the ugly. *Circulation Res.* 2002;90(3):251–262.

23. Mazani M, Fard AS, Baghi AN, Nemati A, Mogadam RA. Effect of pomegranate juice supplementation on matrix metalloproteinases 2 and 9 following exhaustive exercise in young healthy males. *J Pak Med Assoc.* 2014;64(7):785–790.

24. Stoclet J-C, Chataigneau T, Ndiaye M, et al. Vascular protection by dietary polyphenols. *Eur J Pharmacol.* 2004;500(1-3):299–313.

25. Spence JD, Eliasziw M, DiCicco M, Hackam DG, Galil R, Lohmann T. Carotid plaque area: A tool for targeting and evaluating vascular preventive therapy. *Stroke.* 2002;33(12):2916–2922.

26. Hirano M, Nakamura T, Kitta Y, et al. Rapid improvement of carotid plaque echogenicity within 1 month of pioglitazone treatment in patients with acute coronary syndrome. *Atherosclerosis.* 2009;203(2):483–488, doi:10.1016/j.atherosclerosis.2008.07.023

27. Chen X, Lin M, Cui H, et al. Three-dimensional ultrasound evaluation of the effects of pomegranate therapy on carotid plaque texture using locality preserving projection. *Comput Methods Prog Biomed.* 2020;184:105276.

28. Egger M, Spence JD, Fenster A, Parraga G. Validation of 3D ultrasound vessel wall volume: An imaging phenotype of carotid atherosclerosis. *Ultrasound Med Biol.* 2007;33(6):905–914.

29. Jiang M, Zhao Y, Chiu B. Segmentation of common and internal carotid arteries from 3D ultrasound images based on adaptive triple loss. *Med Physics.* 2021;48(9):5096–5114.

30. Tan H, Shi H, Lin M, Spence JD, Chan K-L, Chiu B. *Vessel Wall Segmentation of Common Carotid Artery via Multi-Branch Light Network.* SPIE; 2020:228–233.

31. Zhou R, Guo F, Azarpazhooh MR, et al. A voxel-based fully convolution network and continuous max-flow for carotid vessel-wall-volume segmentation from 3D ultrasound images. *IEEE Tran Med Imaging.* 2020;39(9):2844–2855.

32. Shin J, Tajbakhsh N, Hurst RT, Kendall CB, Liang J. Automating carotid intima-media thickness video interpretation with convolutional neural networks. In: IEEE Conference on Computer Vision and Pattern Recognition, 2016:2526–2535.

33. Zhou R, Azarpazhooh MR, Spence JD, et al. Deep learning-based carotid plaque segmentation from B-mode ultrasound images. *Ultrasound Med Biol.* 2021;47(9):2723–2733.

34. Zhou R, Fenster A, Xia Y, Spence JD, Ding M. Deep learning-based carotid media-adventitia and lumen-intima boundary segmentation from three-dimensional ultrasound images. *Med Physics.* 2019;46(7):3180–3193.

35. Lekadir K, Galimzianova A, Betriu À, et al. A convolutional neural network for automatic characterization of plaque composition in carotid ultrasound. *IEEE J Biomed Health Informat.* 2016;21(1):48–55.

9 3D Ultrasound for Biopsy of the Prostate

Jake Pensa, Rory Geoghegan, and Shyam Natarajan

INTRODUCTION

One in eight men get diagnosed with prostate cancer in their lifetime, accounting for 27% of all cancer diagnoses in men: approximately 270,000 new cases annually and 35,000 annual deaths [1]. Accurate disease diagnosis is vital for prostate cancer management, as overtreatment of low-grade cancer can result in severe decreases in quality of life for the patient, including sexual and urinary function, and increased healthcare expenditures; while conversely, undertreatment of aggressive disease can be fatal [2]. The standard diagnostic pathway is an ultrasound (US)-guided biopsy (Figure 9.1a) following an elevated value on a prostate-specific antigen (PSA) screening test (above 4 ngl/mL) or abnormal digital rectal exam (DRE) [3, 4]. While US allows for gland visualization and measurement, it has poor sensitivity and specificity for detection of prostate cancer with many tumors appearing isoechoic under US and many hypoechoic regions being nonmalignant upon biopsy [5]. As such, the current recommended diagnostic strategy is a pre-biopsy, multi-parametric magnetic resonance imaging (mpMRI) scan which has demonstrated improved sensitivity (89%) and specificity (73%) over standard US for imaging prostate cancer [6–9]. To fully utilize mpMRI imaging for prostate cancer detection, radiologist-identified suspicious lesions are fused to three-dimensional (3D) US scans to allow for real-time visualization during biopsy [4, 10]. While it is possible to do in-bore MRI-guided biopsy, these procedures are resource-intensive, time-consuming, complex, often cost-prohibitive, and do not offer a substantial diagnostic benefit over less expensive, and more time efficient, MRI-US fusion biopsies [11]. In general, use of US-guided MRI targeted prostate biopsies are superior to traditional systematic US-guided biopsy (Figure 9.1b,c), with higher rates of identification of clinically significant prostate cancer and fewer diagnosis of clinically insignificant cancer [9].

In this chapter, we highlight the current state of 3D US for prostate biopsies, including methods for generating 3D US volumes, various registration techniques for combining 3D US sweeps with other imaging modalities, and the advantages of fusion biopsies over traditional 2D US biopsies.

ULTRASOUND-GUIDED PROSTATE BIOPSY

Since the development of transrectal ultrasound (TRUS) in the 1970s, US has had a vital role in prostate gland visualization and guiding of systematic 12-core biopsies [12–14]. Prior to the advent of PSA testing in the late 1980s, prostate cancer screening relied heavily on DREs for diagnosis, resulting in cancers that were later stage and larger at time of biopsy than they are today, many of which were visible on US [5, 15]. Following the FDA approval of PSA testing in 1986, prostate cancer is generally identified at earlier stages and smaller than it was in prior decades. Consequently, US has poor sensitivity and specificity for prostate cancer with biopsy results often identifying benign tissue in hypoechoic regions while cancer is found within otherwise inconspicuous isoechoic regions [5]. However, US scans are still vital for prostate cancer screening as they help guide biopsy cores for systematic sampling and allow physicians to accurately measure the volume of the gland to calculate PSA density, a metric that is reported to have similar sensitivity, but superior specificity to PSA for predicting clinically significant prostate cancer [16–19]. Recently, in the past decade,

DOI: 10.1201/9781003299462-11

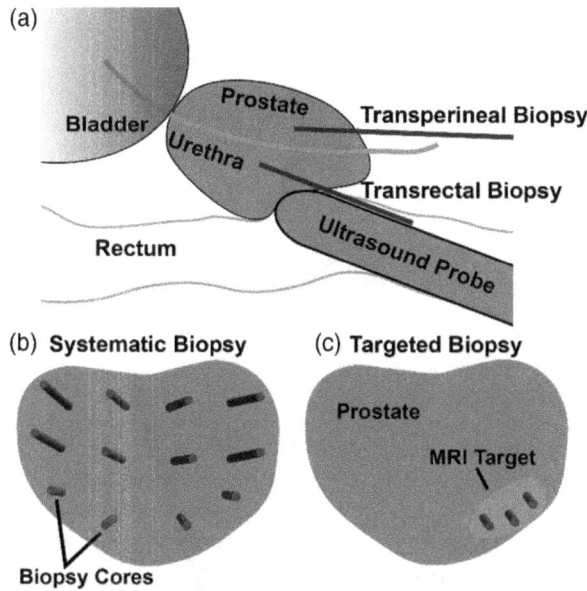

FIGURE 9.1 Anatomic position of the prostate and the relative orientation of transrectal and transperineal biopsies; both are guided via transrectal US. Example of systematic (left) and targeted (right) biopsy core maps of a prostate visualized from the coronal perspective. Systematic biopsy samples the prostate at standardized locations to mitigate the number of cores taken while retaining confidence for prostate cancer identification. Targeted biopsy samples any pre-biopsy suspicious lesion identified on MRI. A combination of these strategies is often used to identify prostate cancer during biopsy.

3D US fusion devices have revolutionized prostate biopsies, allowing for precision targeting of suspicious lesions and improved detection rates of clinically significant cancer [20].

3D ULTRASOUND ACQUISITION

Generally, two types of US probes are used for acquiring 3D US volumes during prostate biopsy (Figure 9.2):

1. One-dimensional (1D) array transducers with position tracking: these are traditional US probes that create standard 2D images. The depth of the reflection of the US beam is recorded along the dimension of the transducer, providing 1D positional data with a time component, resulting in a beam-formed 2D image. The location and orientation of the probe in 3D space is tracked while the 2D images are collected to allow for a 3D reconstruction of the imaging sweep. Multiple arrays can be embedded within a probe to image different planes simultaneously. Embedded motors may robotically move the array or probe to generate a smooth 3D reconstruction.
2. 2D array (matrix) transducers: these are more advanced US transducers that create native 3D images. The depth of the reflection of US beam is recorded along the transducer matrix, resulting in 2D positional data with a time component, which provides a 3D image.

The primary advantage of array transducers is the ablity to image the entire 3D volume without moving the probe. In contrast to 1D transducers, this enables biopsy needle insertion under direct visualization without having to navigate the probe to the desired biopsy site. This simplifies user interaction and also improves registration with MRI as each time the probe is moved, the prostate

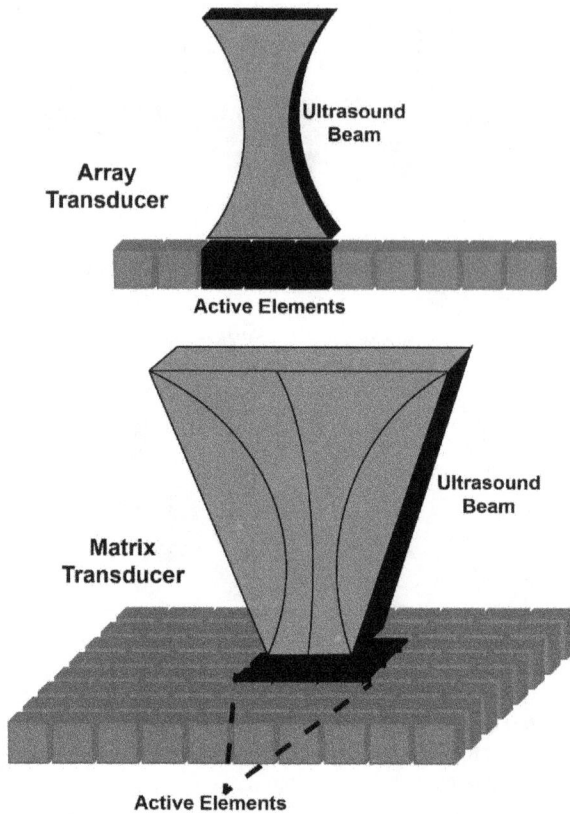

FIGURE 9.2 One-dimensional array transducers (top) create US beams from a subset of elements in the array, and the resulting echo is read by the full array to create a 2D planar image. Two-dimensional array (matrix) transducers (bottom) also create US beams from a subset of elements and read the echo from elements in a two-dimensional area to create a 3D volumetric image.

undergoes new deformation which needs to be accounted for. Imaging is acquired transrectally, therefore, minimizing probe size is of paramount importance for patient comfort. However, 2D transducer arrays have not been commercialized for prostate biopsy. Consequently, the cost and complexity of 2D probe design is prohibitive given that 1D probes already achieve adequate levels of co-registration with MRI [4, 21–24]. Today, all commercial 3D US systems for prostate biopsy rely on 1D array transducers that employ various tracking techniques during the interrogation of the prostate (Table 9.1). The first 3D US device for prostate biopsy to receive regulatory approval incorporated a 1D array with a mechanical fixed arm, receiving clearance in 2008 (Artemis, Table 9.1). Many other platforms were subsequently brought to market using a variety of approaches to track probe movement as described below.

INTEGRATED ENCODER

Integrated encoders (Figure 9.3a) are one of the simplest methods for monitoring US probe position. Encoders built into the casing of the probe monitor the roll, pitch, and yaw while scanning the prostate [25]. However, as the encoders only record probe angle, this system has no means to monitor translation. In addition, this approach is generally limited to side-fire probes that are able to scan the prostate by only rotating the probe. While user-friendly and able to provide general correlation with MRI planes, this tracking mechanism is not as rigorous as other approaches.

TABLE 9.1

List of FDA Approved Commercial 3D Ultrasound Machines for Prostate Biopsy and Their Respective Tracking Mechanism

Device Name	Manufacturer	Approval	Tracking Method
Artemis	Eigen	2008 - FDA	Mechanical arm
Trinity/Urostation	Koelis	2010 - FDA	Mechanical scanning
Navigo	UC-Care	2011 - FDA	Electromagnetic
iSR'obot Mona Lisa	BioBot	2011 - FDA	Mechanical arm
BioJet	Medical Targeting Technologies	2012 - FDA	Mechanical arm
UroNav	Philips	2012 - FDA	Electromagnetic
Virtual Navigator	Esaote	2014 - FDA	Electromagnetic
Bk3000	BK Ultrasound/MIM	2016 - FDA	Electromagnetic
Fusion Bx	Focal Healthcare	2016 - FDA	Mechanical arm
ExactVu	Exact Imaging	2016 - FDA	Integrated encoder
BiopSee	MedCom	2017 - FDA	Mechanical arm
SmartTarget	SmartTarget Ltd.	2017 - FDA	Mechanical arm
RS85 Prestige	Samsung	2018 - FDA	Electromagnetic

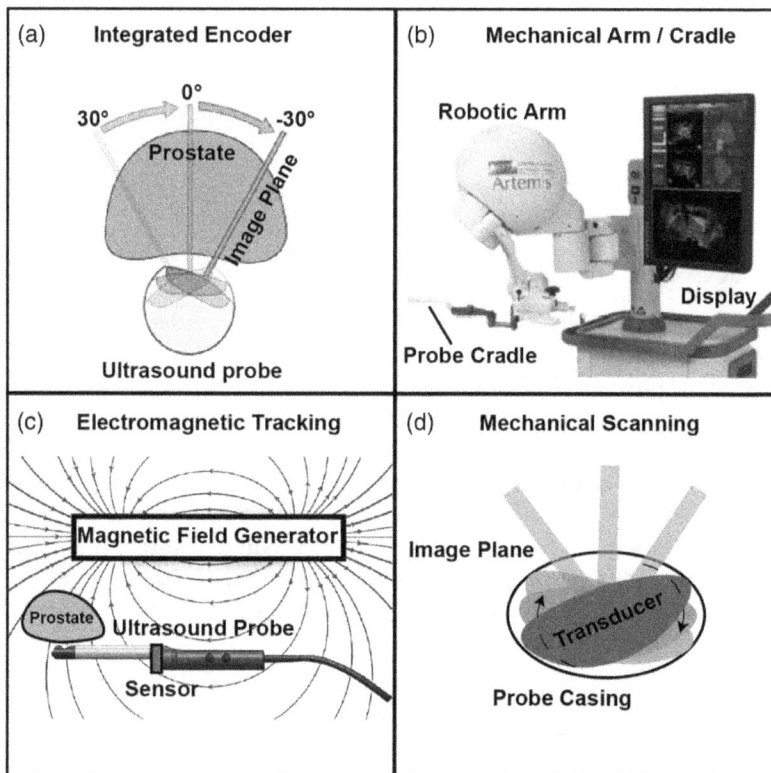

FIGURE 9.3 Tracking mechanisms for creating 3D US volumes. (a) Integrated encoders use built in encoders in the probe casing to track probe movement. (b) Mechanical arm or cradle systems stabilize and record the location of a US probe in a cradle using a jointed mechanical arm containing positional encoders. (c) Electromagnetic tracking records the position of a US probe in a time-varying magnetic field. (d) Mechanical scanning: The transducer array is automatically rotated inside the probe inside a casing to perform the sweep of the prostate while recording orientation.

MECHANICAL ARM OR CRADLE

In contrast to integrated encoders, mechanical arm or cradles offer improved tracking and while allowing for the full six degrees of freedom during scanning. Here, the US probe is mounted in a cradle connected to a jointed mechanical arm (Figure 9.3b) containing positional encoders which provide precise, real-time tracking in 3D space [23, 24]. This positional information along with the US image data is used to create highly accurate 3D US volumes. This ensures that existing MRI defined targets can be accurately fused with the real-time US scan. In addition, the exact biopsy site can be stored allowing re-biopsy in a subsequent session with an error of a few millimeters [26]. Although the mechanical arm removes movement error and improves tracking accuracy, the large form factor of these machines and unconventional fixed axis for scanning increase the learning curve and space required for TRUS prostate biopsies [23].

ELECTROMAGNETIC TRACKING

Electromagnetic tracking (Figure 9.3c), based on Faraday's law, generates current in a solenoid sensor based on its location in a time-varying magnetic field [23, 24]. Three sensors, oriented in three orthogonal directions, are mounted on an US probe and measure the strength of a time-varying magnetic field via the generated voltage in each sensor [24]. From the sensor values, it is possible to track the position and orientation of the probe relative to the magnetic field generator [23, 24]. With this tracked positional information, it is possible to generate 3D reconstructions of the prostate and track biopsy cores, while allowing the operator to maneuver the probe unrestricted [23]. However, this approach can be prone to distortion and reduced tracking accuracy due to electromagnetic interference and metallic objects [24]. In addition, without stabilization, errors may arise during biopsy needle deployment that reduce targeting accuracy relative to mechanical arm approaches [23]. Moreover, electromagnetic tracking has also been used to track biopsy needles independent of probe tracking, removing the need for a mounted guide for tracking core location [27].

MECHANICAL SCANNING

Mechanical scanning (Figure 9.3d) probes utilize a mechanized transducer housed within the probe casing [24]. The location of the transducer is electronically rotated inside the probe casing to interrogate the volume of the prostate [24]. Based on the angular position of the transducer, these systems may recreate a 3D volume, though these are influenced by precession during rotation. These systems do not require the operator to move the probe after positioning, as the mechanized transducer is able to rotate the image plane to the desired location within the prostate [28]. The accuracy of this approach can be degraded by the motion of the patient or the US probe; however, it is possible to update the 3D scan following deformation or substantial misalignment through image-based tracking [23, 24]. Furthermore, through sophisticated registration methods and organ tracking (image based tracking of capsule deformation during biopsy) it is possible to improve prediction accuracy of the location of a biopsy core prior to biopsy [29].

ALTERNATIVE TRACKING APPROACHES

Apart from the approaches in clinical use listed in Table 9.1, two other tracking techniques are in development: optical tracking and sensorless tracking. In optical tracking, a visual target is mounted to the handle of the US probe and tracked by a camera [30], which calculates the location of the US transducer. This technique is limited in clinical use due to difficulties maintaining line of sight [30]. In sensorless tracking, 3D US volumes are generated from array transducers using image data alone through software. Several groups have been able to achieve a drift error of ~5 mm using speckle decorrelation algorithms combined with Gaussian filtering to estimate the elevational distance between neighboring US frames [31–33], though in *ex vivo* setting. *In vivo*, drift errors of 10 mm were achieved using a type of deep learning network, a deep contextual-contrastive network (DC²-Net) with a 3D ResNeXt backbone structure that focused on areas of strong speckle patterns. This model also managed to achieve dice similarity scores of 0.89 [31]. With continued active research into deep learning, sensorless 3D US reconstruction is a promising approach to obviate errors associated with motion and deformation.

IMAGE VOLUME RECONSTRUCTION

Following the acquisition of a series of 2D US images using one of the various tracking techniques, the volume of the scan needs to be reconstructed before segmentation and fusion with MRI. There are a variety of techniques that can be employed to reconstruct the 3D volume, but all methods use the tracked location of the US plane for each image and the B-mode intensity data as inputs for reconstruction. Prostate volumetric correlation of MRI and US volumes, against excised whole gland measurements can vary (US: 0.70–0.90, MRI: 0.8–0.96) [18, 19], but in general it is possible to achieve accurate targeting with a ± 20% variation [26]. There are three general subcategories used for reconstruction of 3D US volumes: voxel-based algorithms, pixel-based algorithms, and function-based algorithms [24, 34].

VOXEL GRID SELECTION

In all three techniques, prior to reconstruction, a 3D grid is established across the data volume to determine sampling density, called the voxel grid [34]. An appropriate voxel grid has two main components: a sufficient amount of information for the defined volume to preserve information present in the original images; and a homogenous information density throughout the whole volume to reduce the amount of interpolated voxels [34, 35]. If voxel size and spacing in the 3D grid is too large, there will be a loss of information and resolution (aliasing) from the original 2D images, and if the voxel size and spacing is too small, there will be a large number of voxels that need to be interpolated in the 3D sweep inducing unnecessary error into the 3D reconstruction. One approach for selecting an appropriate voxel grid is to first perform principal component analysis (PCA) to select an appropriate coordinate system based on the pixel distribution in 3D space, and then calculating the appropriate voxel size based on probe resolution and the rotation of 2D image planes in 3D space as a trigonometric minimization function [35]. Another approach involves subdivision of a tetrahedral grid that is specifically adapted to match the complexity of the dataset, resulting in efficient and compressed volume grids [34, 36].

VOXEL-BASED ALGORITHMS

Voxel-based reconstruction algorithms assign every voxel in a predefined volume a value based on nearby pixel information (Figure 9.4). Depending on the specific technique a voxel can be assigned an intensity value based on one pixel or many [24]. A general approach for voxel-based reconstruction is as follows. First, a voxel grid is established for a desired volume within the US sweep. Next, for each voxel, relevant nearby US image frames are selected based on their proximity to the voxel in 3D space. Within those frames relevant pixel information is selected along with associated weights depending on distance from the voxel. Then, an intensity value is assigned to the voxel based on the selected surrounding pixels and their corresponding weights. For one-pixel approaches, each voxel is generally assigned the value of the nearest pixel, while for multi-pixel approaches, the value of a voxel is normally set to an interpolated value depending on the nearby pixels from multiple frames [37, 38]. For multi-pixel approaches, a weight factor is typically applied to the pixel values prior to interpolation that is inversely proportional to the pixels distance from the specific voxel [24, 38]. As a result, the closer pixel values have stronger influence on the voxel's intensity value. Since

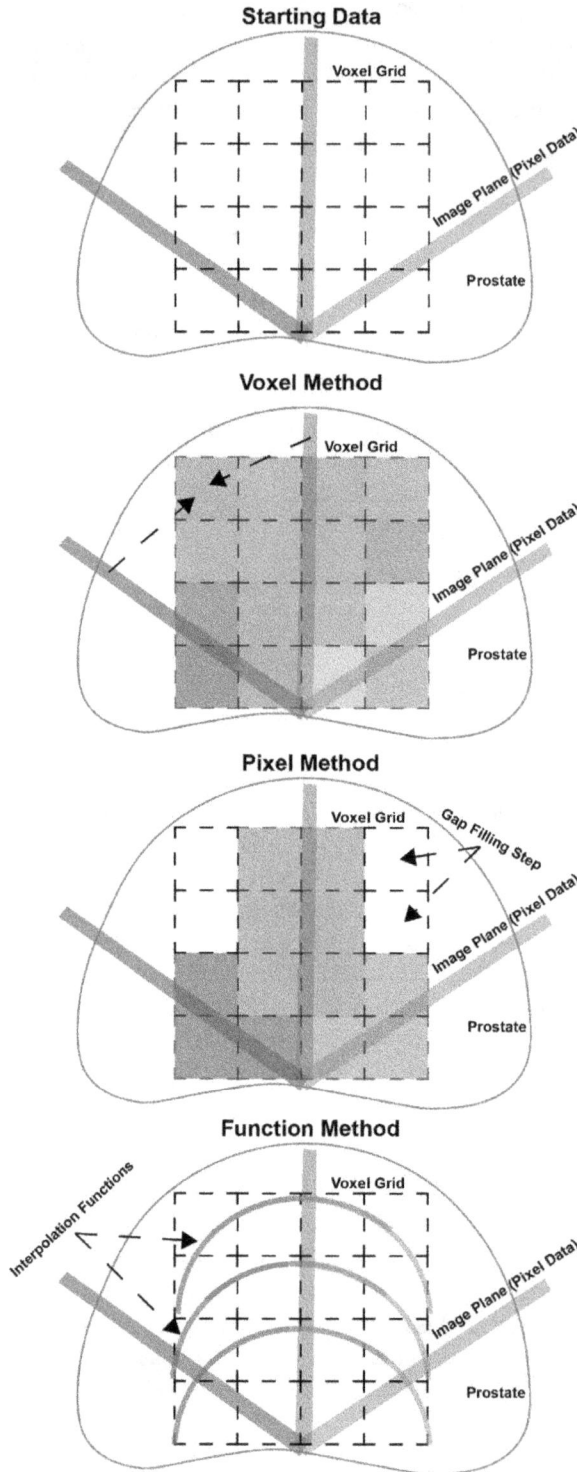

FIGURE 9.4 Example of various volume reconstruction techniques using three images at different orientations (red, green, blue). Voxel techniques assign a value to each voxel often based on weighted interpolation from nearby pixels. Pixel methods assign pixel values to nearby voxels and can have empty voxels based on scan distribution. Function-based methods fit a curve to the pixel data to populate voxels.

voxel-based algorithms loop through each voxel and assign an intensity value, there are generally no gaps in the resulting reconstruction.

PIXEL-BASED ALGORITHMS

Pixel-based reconstruction algorithms assign the value of each pixel in the series of images to one or more voxels [24]. A general pixel-based reconstruction approach would be as follows: a voxel grid is created around the imaging sweep. The algorithm then loops through each frame and assigns the pixel values from each frame to the voxels that intersect the corresponding pixel. Pixel distribution to various voxels can be based on a single pixel, an average of pixels, or an interpolated vale from a surrounding group of pixels. A common approach is to take a local kernel of pixels and apply various weights to each pixel value to populate the surrounding voxels [24, 39]. Depending on the scan distribution and voxel size, there may be unfilled voxels after pixel distribution due to voxels located away from any nearby frames. As such, a gap-filling or interpolation step is used to fill the remaining gaps in the data. Following pixel distribution and assignment to various voxels, a follow-up interpolation step based on nearby pixel values is applied to fill in the empty voxels; however, sometimes it is beneficial to leave empty voxels unfilled to let the reviewer know there is missing data [24].

FUNCTION-BASED ALGORITHMS

Function-based reconstruction algorithms utilize function-derived interpolation to populate voxel values. By taking small windows of nearby frames at known distances apart, a function can be fit to pixel values within these frames to interpolate to voxel values between frames. Commonly used functions are polynomials, splines, radial basis functions, or Bézier curves [24, 34]. By overlapping the series of neighboring scans for each section of interpolation, it is possible to further improve reconstruction accuracy through distance-weighted averaging of shared voxels [24].

COMPUTATION TIME VS. RECONSTRUCTION QUALITY

In order to choose an adequate reconstruction algorithm, balance between reconstruction speed and quality must be considered [24]. The more sophisticated algorithms, such as spline-based approaches or those that create local pixel kernels to approximate voxel values, generally allow for improved reconstruction accuracy at the cost of longer computation times. However, some approaches, such as Bézier curve interpolation, manage to achieve extremely fast reconstruction times (0.008 s/image) while maintaining a high level of reconstruction accuracy [24, 40]. In general, advances in computational power of computer and GPU performance have improved the overall speed of reconstruction, allowing for near real-time volume reconstruction, allowing for the use of 3D US in the clinic setting during biopsy [24].

SEGMENTATION TECHNIQUES

In addition to 3D reconstruction, a segmentation step is typically implemented before fusion with MRI can occur. The specific process varies from device to device, but in general the operator is asked to mark points around the prostate capsule and some fiducial points visible on both MR and US [4]. Examples of fiducials include: BPH nodules, calcifications, the urethra, seminal vesicles, or the boundary between the peripheral and transition zones. Typically, the user is asked to mark the prostate capsule on a subset of slides and an interpolative algorithm fills in the remaining slides to create the prostate capsule volume [41]. Prior to biopsy, the MRI is also segmented by a radiologist or experienced reviewer, these segmentations of the prostate capsule and any included fiducials are used as the basis for registration between MRI and US [4, 26]. There are a number of different prostate segmentation methods, as well as research into semi- and fully automated segmentation algorithms to speed up and improve registration accuracy. Manual segmentation tends to be time-consuming and prone to variability, while automated segmentation is difficult due to inherently low signal-to-noise and contrast-to-noise ratios of US imaging [41, 42]. A comprehensive review paper, published in 2022 by Jiang et al., describes the popular and developing techniques for prostate segmentation; a brief summary is included below [41].

EDGE-BASED ALGORITHMS

Edge-based algorithms employ an edge identification step followed by a point ordering step to create an outline of the prostate capsule [41]. For edge detection, a filtering process is applied to identify the boundaries of the prostate. This can be done via conventional filtering, probabilistic filtering, or through more complicated filtering processes such as superimposing the original image over an inverted multi-resolution filtered image, a method known as radial basis relief [43]. In general, edge-based filtering algorithms need to be finely tuned to allow for adequate edge detection, and tend to be very sensitive to image noise and artifacts [41]. As such they are commonly used alongside other filtering algorithms to achieve adequate segmentation, in both fully and semi-automated approaches.

REGION-BASED ALGORITHMS

Region-based algorithms utilize both regional and edge information to segment the prostate based on common areas (Figure 9.5). This can be achieved through regional clustering and statistical analysis [41]. Regional clustering invokes a parameter to divide an image into regions to find the boundaries of the prostate [41]. Common parameters include the greyscale value of pixels, or more robustly, a parameter known as the local binary pattern, a measure of a pixels value relative to its neighborhood and a metric for localized image contrast [41, 44]. By applying adaptive thresholding, optimizing for inter-class variance, and accounting for density and Euclidean distance of points (density-based spatial clustering of applications with noise, DBSCAN) it is possible to identify the boundaries of the prostate [41, 45]. Statistical analysis region identification uses region-based statistics, such as intensities, averages, and standard deviations, to identify similar areas within images through minimization techniques. This approach can be assisted through the use of *priori* knowledge of greyscale distributions and pixel intensity gradients of various regions [41, 46].

DEFORMABLE MODEL-BASED ALGORITHMS

Deformable model-based algorithms implement a template prostate model that is modified based on image data and energy minimization functions while balancing internal and external forces on the model that maintain contour smoothness [41, 47]. Three different frameworks are generally used: the active contour model framework (also known as snake or energy minimization curve framework), edge level set framework, and curve fitting framework [41]. The active contour model is a spline curve created from input points that is deformed based on image gradient data in a specified region of interest, characterized by an energy minimization function that balances preservation of the original shape's smooth contours with deformation to match the input image data [48]. This approach is dependent on proper assignment of internal and external forces to drive and control the deformation [41]. Edge level set framework utilizes level set implicit functions as desired boundaries, applying finite differences to approximate the solution of a partial differential equation [49]. The level set propagates from the seed point, perpendicular to the curve until it reaches the maximum of the intensity gradient. This approach is characterized by good signal to noise ratio and well defined edges, thus is commonly used [41, 46]. Curve fitting framework approximates the prostate boundary through fitting ellipsoids or splines. These ellipsoids or splines are initialized from a series of user

Initialization Points

Region Expansion

Final Segmentation

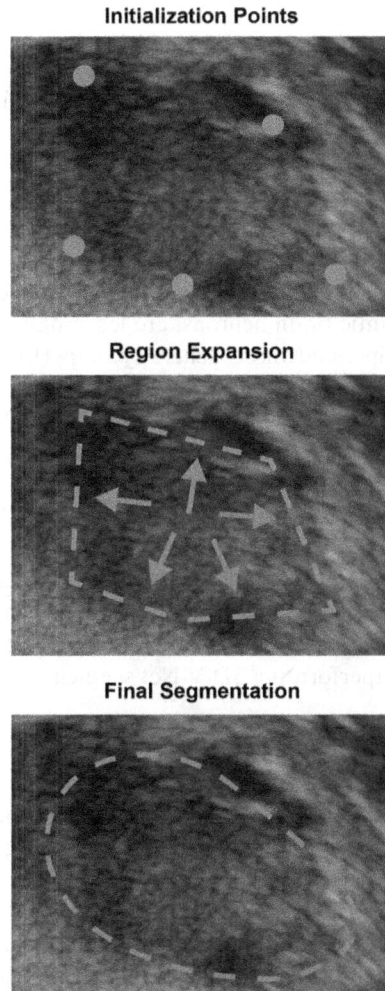

FIGURE 9.5   Region-based segmentation approach. After a urologist or an assistant places initialization points, a region is established that is then expanded to find the boundaries of the prostate.

selected points, and are then transformed using varying combinations of thin plate splines, Bayesian frameworks, *priori* knowledge, and a variety of other optimization techniques [41]. Without using *priori* knowledge, deformable model segmentations may differ significantly from the actual prostate boundaries reducing segmentation accuracy [50].

STATISTICAL MODEL-BASED ALGORITHMS

Statistical model algorithms utilize big data and computational power to delineate the prostate. Utilizing a priori database of prostate shape features and the associated B-mode intensity data an optimal segmentation is selected [41]. One such approach is a point distribution model (PDM) selection of an active shape model (ASM) framework, followed by a PCA of the PDM with a generalized Procrustes analysis to allow for model variation [41]. Building on this approach, the active appearance model (AAM) algorithm framework was developed, utilizing the shape information of the ASM framework in combination with greyscale information to generate texture information for improved segmentation results [41].

Learning Model-Based Algorithms

Learning model algorithms generally utilize either support vector machines (SVM) or neural networks to train a model to delineate the prostate boundary [41]. SVMs are supervised learning models that attempts to create a hyper-plane between different types of data, such as the prostate capsule and the surrounding tissue, with an optimal compromise between model complexity and training results to avoid overfitting and promote generalization. A general SVM approach utilizes Gabor filters applied to the image data to extract texture features which are then input into a SVM for classification. This classification allows for a statistical shape model which automatically delineates the prostate [51, 52]. SVM approaches tend to work well with limited nonlinear analytical data [41]. Neural networks, designed to mimic brain neurons, are learning algorithms capable of identifying complex multivariate relationships, and these generally outperform linear regression techniques when working with patient data. In particular, U-Net convolutional neural networks (CNN) perform particularly well at medical image segmentation tasks, and this holds true for prostate delineation [41, 53, 54]. Learning-based approaches often experience degraded performance in the far apex and base of the prostate due to poor contrast-to-noise ratios and increased variability in prostate shape. Recent research has managed to overcome this hurdle through radially re-slicing a diverse set of 3D prostate volumes about the center axis into 2D images. In doing so, each image slice passed through the approximate center of the prostate, avoiding images only containing the far apex or base. These images were then subsequently segmented by a modified 2D U-Net model and reconstructed back into a 3D US sweep with a dice similarity coefficient of around 0.95 and a processing speed of less than 0.7 seconds per prostate, outperforming 3D V-Net segmentation networks [55, 56].

VOLUME RENDERING

Following reconstruction and typically after segmentation, rendering is performed to relay 3D US information to the physician performing the biopsy. Given the large volume of data acquired in a 3D US scan and the time constraints associated with use in the clinic, it is important to consider the optimal way to render the data [24]. Displaying too much information can increase rendering times and consume resources, while on the other hand, rendering not enough information can diminish the benefit of 3D US leaving out information that could be potentially helpful to the physician. In general there are three main rendering approaches: slice projection, surface rendering, and volume rendering [24, 57].

Slice projection (Figure 9.6a) displays a 2D slice of the 3D US data, either in the native imaging plane or in a reconstructed plane. The operator can typically scroll through the rendered image plane to visualize various regions of the prostate. However, there can be an increased cognitive load for the physician in this approach as they are required to mentally correlate the location of the image plane in 3D space [24, 58].

Surface rendering (Figure 9.6b) is a fast-rendering approach that only displays a 3D model of the prostate capsule. This approach is dependent on a segmentation step prior to rendering in order to identify the capsule boundaries [24]. Visualizing the shape of the prostate during biopsy allows for simple real-time visualization of systematic and targeted biopsy locations as well an easy-to-read overlay model for any MRI targets.

Volume rendering (Figure 9.6c) provides all of the 3D US data together, typically utilizing translucent anatomy to allow for visualization of interior feature information [59]. While computationally intense, this approach provides physicians with the most comprehensive view of the prostate.

Primary roles of US in prostate biopsy are for gland visualization, biopsy needle targeting, and fusion with MRI, rather than for cancer identification. Consequently, it is often better to mitigate the processing times by hiding extraneous information that could potentially distract the user while providing no utility. As such, one common approach that 3D US fusion devices utilize is a hybrid surface rendering and slice projection approach. In this approach, a 3D model of the prostate is

Slice Projection

Surface Rendering

Volume Rendering

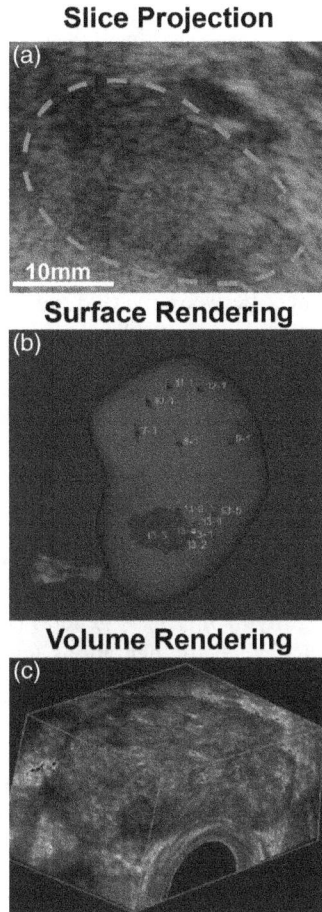

FIGURE 9.6 Various methods for rendering 3D US data of the prostate. Slice projection (a) displays a 2D planar image of the US data in native and nonnative orientations. Surface rendering (b) displays a model of the prostate capsule for localization, core placement tracking, and MRI target visualization. Volume rendering (c) presents the full 3D sweep of the prostate allowing for thorough review by the physician [59].

displayed with associated slice projections, while tracking probe locations on the model aid with probe navigation. As such, the surface model allows the physician to quickly reference MRI targets and identify biopsy sites; while slice projection allows the physician to review the 3D data as desired. The US probe itself can still be used to interrogate the prostate for biopsy core targeting, with some systems overlaying 3D targeting data onto the real-time US.

FUSION TECHNIQUES

Following image acquisition and segmentation of both 3D US and T2-weighted MRI scans, a fusion step is performed to co-register MRI image data and target information to the US scan [20, 23]. The goal of the fusion transformation is to map the MRI space and coordinates to the US space and coordinates to allow for real-time visualization of MRI information on the US scan, improving clinically significant prostate cancer detection rates [20]. This fusion step can also be performed multiple times throughout the biopsy to account for any changes to the prostate shape such as deformation and variation in US probe pressure on the prostate. One study found that the majority of missed lesions (1.9% of patients) during prostate biopsy were due to errors in targeting (51.2%

of misses), rather than the lesion being MRI invisible (40.5% of misses) or missed by the radiologist (7.1% of misses), highlighting the importance of accurate co-registration and targeting [60]. It is estimated in order to have a 90% biopsy hit rate on an MRI identified target, the target registration error (TRE) between MRI and US needs to be less than 2.87 mm not accounting for needle deflection, frequently substantial in clinical practice [61]. Existing MRI-US fusion systems have a validated TRE of approximately 3 mm [29, 62–65], which may be sufficient for diagnosis but may not be accurate for therapy guidance [66]. MRI-US fusion is typically performed using one of the following approaches: rigid registration, elastic (non-rigid) registration, or learning-based registration [20, 23, 67]. A 2018 meta-analysis found no significant difference in the odds ratio for identifying clinically significant prostate cancer between the two most common approaches: elastic (1.45) and rigid (1.40) registration [20]. However, some studies have found improved TRE for elastic registration (2–3 mm) over rigid registration alone (5–6 mm) [29, 62–65]. Regardless, software-based fusion outperforms cognitive based fusion for identifying clinically significant prostate cancer at time of biopsy [68].

RIGID REGISTRATION

Rigid or affine registration preserves the shape and relative positioning between pixels of the transformed data [20, 23]. Typical transformations encompassed by rigid registration algorithms include: translation, scaling, and rotating (Figure 9.7b) [23]. A general rigid registration approach for prostate biopsy starts with US and MRI surface segmentations (capsule-based registration) or image data (image-based registration). For capsule or surface-based registration, the MRI data is scaled to match the US data based on the capsule volume differences, and the MRI is translated to align the centroids of the prostate capsule on MRI and US. Following scaling and translation, the MRI is rotated to maximize alignment between the two gland segmentations. At this point the image data between MRI and US is relatively aligned, but there will still be some differences based on tissue deformation from the US probe, an issue that can be accounted for with subsequent elastic registration [23].

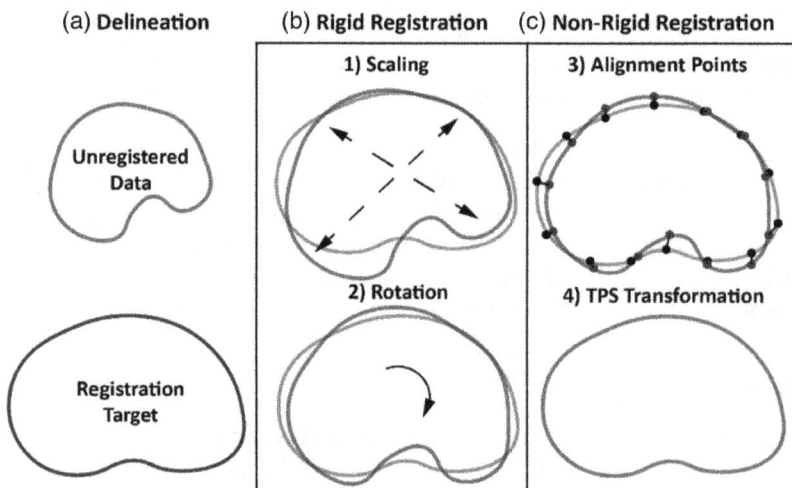

FIGURE 9.7 Simple example of a rigid and non-rigid capsule-based registration process in 2D. Using delineated capsules (a) of the unregistered MRI data and the US registration target, the MRI data is scaled (1) and rotated (2) to maximize alignment with the US capsule for rigid registration (B). The subsequent elastic registration step (c) identifies alignment points (3) between the two capsules and performs a thin-plate spline (TPS) transformation (4) to match the MRI data with the US data.

An alternative to gland-based registration uses anatomic landmarks (features) and other sources of mutual information in the image data to drive the registration. While image-based registration is more computationally intensive, as it compares raw image data, improved registration accuracy has been reported for image-based mutual information registration over surface-based registration with the added benefit of not requiring preliminary segmentation [66]. Image based registration often use a technique known as block matching consisting of several iterative steps of increasingly finer adjustments [69]. First, the source image data are divided into rectangular regions called blocks. These blocks are then matched to a similar block within the target image data based on image-to-image similarity metrics, in this case mutual information between image sets [70]. Next, a vector field is generated that maps each source image block to its best matching target image block. A regularization step is applied to the vector fields to generate the transformation map for the source image set to the target image set [66].

ELASTIC (NON-RIGID) REGISTRATION

Elastic or non-rigid registration is typically performed after rigid registration and attempts to account for tissue deformation through a transformation map based on alignment points (Figure 9.7c) [20, 23]. As the US probe pushes against the prostate gland, the shape can significantly deviate from the image of the gland on MRI. This deformation is also operator dependent and changes throughout the procedure as the probe is navigated to different locations of the prostate. Additionally, needle biopsy causes bleed and localized swelling, further affecting registration [20]. Alignment or control points may include a combination of points along prostate capsule or internal landmarks, (urethra, BPH nodules, zonal boundary points, calcifications, etc.) selected by the operator on both the MRI and US scan. The transformation algorithm, typically a spline based transform (b-spline or thin-plate-spline), will warp the MRI data to align the selected control points on MRI and US [23, 71, 72]. While warping the MRI alignment points to match the US control points, the transform will also warp the surrounding voxels proportionally depending on relative distance from the alignment points. For image based registration, instead of using the capsule segmentation, a moving image is compared to a fixed image and an optimization step compares image-to-image metrics (mutual information) through changing the parameters of a grid of b-splines to yield the transformation with the best match between the two images [66].

LEARNING-BASED REGISTRATION

Boundary- and feature-based registration generally rely on the operator to identify corresponding locations or gland boundaries on MRI and US scans, but are tedious for the operator, challenging to accurately find corresponding points, and are prone to error [67, 72]. To improve fusion quality, consistency, and ease there has been considerable research into automated learning-based MRI-US fusion approaches. Traditional machine learning fusion approaches utilize a deformable model and are generally employed as an elastic registration step following rigid registration, making them dependent on operator segmentation for initialization [67]. Utilizing patient-specific deformable models and a hybrid surface point matching method, it is possible to achieve a TRE of 1.44 mm [67, 73]. Recent research efforts have focused on developing deep learning models for performing MRI-US fusion, with the goal of predicting the necessary transformation matrix. Due to a lack of a ground truth co-registration transformation matrices for training, research primarily focuses on unsupervised or weakly supervised models [67]. Various approaches include combinations of fully convolutional networks (FCN), 2D and 3D CNNs, and multimodality generative adversarial networks (GAN) [67, 74–76]. With the addition of biomechanical constraints to deformation, one group was able to achieve a TRE of 1.57 mm and a dice similarity coefficient of 0.94. The framework is capable of first segmenting both MRI and US, then applying a point-cloud-based deformation for accurate registration in a manner that could potentially be translated to the clinic [67, 77].

VALIDATION APPROACHES

Validation of biopsy accuracy is a key step in demonstrating the clinical utility of a 3D US system for prostate biopsy. While systems may differ in image acquisition, tracking, reconstruction, and registration, the same fundamental sources of error can be independently characterized (Figure 9.8):

1. **Initial registration error**: Imperfect registration between the MRI defined lesion and the initial 3D US scan.
2. **Dynamic registration error**: Further error is introduced throughout the biopsy procedure due to tissue deformation from movement of the probe and tissue trauma induced by the needle.
3. **Targeting error**: Inaccuracy in positioning needle over the intended target which occurs for many reasons including user inexperience and the spacing of the holes in the biopsy template (if used).
4. **Needle deflection**: Caused by tissue mechanical properties and needle design
5. **Stored location**: In many systems, the coordinates of the biopsy needle are digitally recorded, to enable re-biopsy at a future date. However, the capture of this information may involve a user-confirmed action to mark the biopsy location or needle depth, which introduces perceptual error in the needle localization. This error is also introduced when registering this location with a previously obtained 3D US scan during a follow-up biopsy, typically 6 months after the initial biopsy.

It is important to note that these errors do not sum together linearly, and the overall error will be less than the sum. It should also be noted that many of the preclinical validation studies were performed on early system versions, sometimes before commercialization, thus, existing literature may not correctly reflect biopsy accuracy. In existing validation studies of commercially available systems, the metric for biopsy accuracy can be defined in numerous ways thus preventing direct comparison between studies. Moreover, due to the nature of regulation of 3D US systems for prostate biopsy, many commercially available systems report benchtop validation, with unknown accuracy in a real-world clinical setting. Moreover, clinical studies often report on cancer detection rate in comparison to conventional systematic or cognitive fusion biopsy. This further complicates the

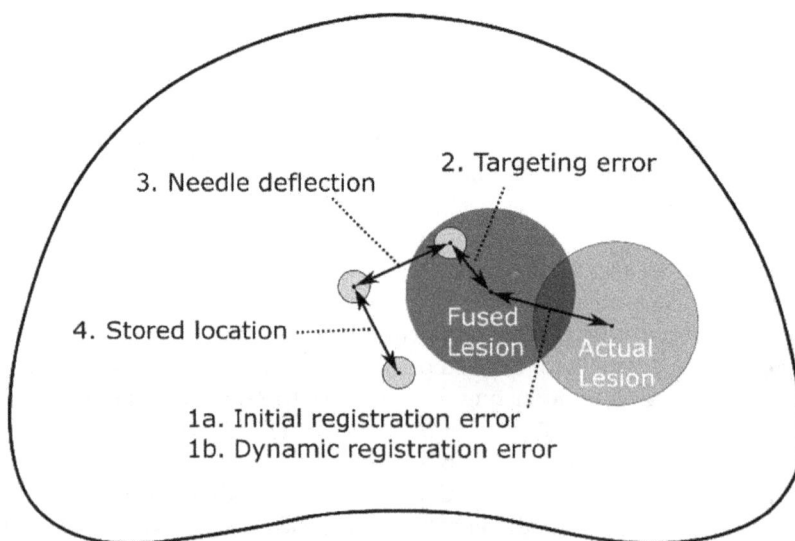

FIGURE 9.8 Sources of error when performing MRI-US fusion biopsy. Here, the goal is to biopsy the center of the true lesion.

analysis as it is dependent on the reliability of the MRI defined lesion which is essentially a function of radiologist skill and cancer grade. Consequently, when comparing clinical studies, it is important to account for the PI-RADS score and radiologist experience.

Preclinical

Preclinical studies typically employ prostate tissue mimicking phantoms composed of agar or hydrogels to assess the targeting error of fusion guidance platforms [65]. In order to determine TRE, biopsy samples are taken of MR- or CT-visible targets the distance between the needle tracks and lesions can be calculated. Using this approach, Uronav, an EM-tracking system, reported a TRE of 2.4 ± 1.2 mm [65].

The Trinity system, which uses an image-based tracking approach, was validated using a similar study with a hydrogel phantoms except the needle tracks were identified by injecting a mixture of dye and gadolinium though the outer sheath of the biopsy needle immediately after deployment [64]. The phantoms were then scanned on MRI to reconstruct the 3D volume and also sectioned and photographed to identify each needle track based on dye color coding. Here, procedural targeting error (PTE) is defined as the shortest distance between the needle track and lesion center, was found to be 2.09 ± 1.28 mm. In addition, the system registration error was 0.83 ± 0.54 mm. It was then assumed that these errors are independent and additive giving an overall error of 2.92 mm.

The BiopSee system was validated using the same approach and found a targeting error of 0.83 ± 0.48 mm [78] which was defined as the distance between virtually planned cores and the confirmed needle tracks. The needle track was also derived from the US images during biopsy. Importantly, PTE is defined differently than above, corresponding to the distance between this tracked core and the virtually planned core. This measure of PTE was found to be 0.26 ± 0.46 mm. There was no statistical difference between the two metrics which was used to support the use of this automatically calculated error in later *in vivo* studies [79]. While this study supports the use of PTE, it is not clear if it provides an accurate assessment of overall error (distance from lesion center to true needle track). Indeed in a similar a follow-up phantom study [80], the PTE (0.39 mm) was found to be only a modest contributor to overall error (2.33 mm). Moreover, errors are likely to be even greater *in vivo* due to needle deflection from inhomogeneous tissue resistance and reduced precision of contouring.

A similar study was also performed during the initial validation of the Artemis platform [81]. Here, the authors constructed an agar phantom using an MRI scan of a patient's prostate and previously developed phantom recipe [82]. The total needle guidance error was found to be 2.13 ± 1.28 mm which largely arises due to needle deflection which accounts for 2.08 ± 1.59 mm. In addition, a commercially available hydrogel phantom (Zerdine, CIRS) was used for assessing the accuracy of segmentation and volume calculation which exhibited a 5% error. Agar phantoms are advantageous as they are inexpensive, can be easily cast in patient-specific molds and also permit placement of fiducials of any desired shape and size [83, 84]. Similar phantoms can be produced for extending the study to include analysis of US-guided focal therapy modalities such as laser ablation [85, 86], high-intensity focused US [87] and radiofrequency ablation [88].

Clinical: TRE

Clinical studies typically focus on cancer detection rates rather than attempting to assess biopsy accuracy because of the inability to assess ground truth needle tracks. However, there are some notable exceptions such as a study by Moldovan et al. which rigorously quantifies the registration accuracy of the Trinity fusion platform [89]. Three intraprostatic fiducials were inserted into patients prior to radiotherapy. The fiducials were inserted through a needle under US guidance using the same procedure that would have been used for biopsy. The patient then received an MRI on the same day. Using the US images, a virtual needle was positioned with the tip defined by the fiducial and the angle defined by that of the US probe. Virtual lesions were defined as a sphere overlaying the fiducial on the MRI image set. The US and MR images were then fused together using the software on the Trinity system. Registration error was defined as follows: (1) TRE_{3D} is the distance between the tip of the virtual needle and the virtual lesion, and 2) TRE_{2D} is the orthogonal distance between the virtual needle tract and the virtual lesion. TRE_{2D} is the better metric of registration accuracy as small errors along the needle trajectory are less critical given that most biopsy cores are 15–20 mm. Here, TRE_{3D} and TRE_{2D} ranged from 3.8 to 5.6 mm and 2.5 to 3.6 mm, respectively with much of the variation attributed to location and user experience. Importantly, the design of the study removes contributions from needle deflection thus providing an accurate assessment of registration error alone.

Studies such as this suffer from three intrinsic sources of error: (1) fiducial localization error (FLE), (2) fiducial migration, and (3) segmentation variability. When determining registration accuracy with fiducials, it is necessary to define the center of the fiducial on each image set. This is generally a manual process and as such it is user dependent. In an earlier study, the same user identified fiducials on the same image set for five consecutive days resulting in an FLE of 0.21 ± 0.11 mm [66]. This suggests that FLE is a relatively minor contributor to overall error. In contrast, the same study found that segmentation variability was responsible for ~50% of the calculated TRE. Segmentation error arises when using semi-automated algorithms in which the user is required to manually outline the prostate on a subset of the images. Finally, implanted fiducials are susceptible to migration with one study showing mean migration of 1.2 ± 0.2 mm in patients receiving radiotherapy [90]. FLE and migration can be mitigated by using anatomical landmarks (e.g., calcifications, capsule, peripheral zone – transition zone boundary) instead of implants. Accurately identifying these landmarks on different imaging modalities is quite challenging but can be achieved as demonstrated in multiple studies including US, MRI and wholemount registration [74, 84, 91, 92].

In one clinical study on 106 patients, TRE was determined to 1.7 mm [79] using a stepper-stabilizer, while image-based TRE in another system was found to be 2.8 mm in a clinical study on 88 patients [29]. In the same study, the accuracy of cognitive fusion was 7.1 mm. It is important to note that the methodology and definitions of error between the two systems are very different, thus, they cannot be directly compared. Moreover, these automatically calculated errors may not accurately reflect the true TRE.

CLINICAL CANCER DETECTION RATES

In general, clinical studies have continuously demonstrated that targeted biopsy (TB) has a greater detection rate than conventional systematic biopsy (SB) or cognitive fusion biopsy, though taking both targeted and systematic biopsy cores yields the highest rate. In an early clinical trial on 101 patients, the per-core cancer detection rate was significantly higher for TB than SB at 20.6% vs. 11.7% [93]. The difference was even more pronounced for targets deemed to be highly suspicious for cancer on MRI at 53.8% vs. 29.9%. The same group subsequently reported on targeted biopsy performed on patients with a prior negative systematic biopsy and found cancer in 37.5% of cases and high-grade cancer in 11% cases [94]. In addition, TB was responsible for pathological upgrading in 38.9% cases compared to SB alone. Similar results were reported in a later trial where 23.5% of cancers diagnosed as clinically insignificant on SB were upgraded after TB [95]. In a study targeting high-grade ROIs, fusion biopsy exhibited a 33% positivity rate in comparison to the 7% achieved via SB [26]. In another clinical trial, 91% of men with PCA had significant cancer when diagnosed via TB vs 54% with SB [96]. Comparable results have been reported for numerous biopsy platform [29, 79, 97–101].

DISCUSSION

The ability to image the prostate with 3D US and MRI-US fusion has transformed the diagnosis of prostate cancer by detecting clinically significant disease earlier, while reducing unnecessary biopsies [9]. In particular, 3D tracking has enabled longitudinal monitoring of localized disease to better understand tumor progression [96]. Consequently, adoption has exploded [9] and clinical guidelines (National Cancer Comprehensive Network, NCCN) now endorse the use of targeted, image-guided biopsy for staging of intermediate-risk disease [2]. Despite the improved outcomes, 3D US is not universally used for prostate biopsy due to the higher cost and learning curve. One key area of improvement is in error reduction. Existing TRE of biopsy platforms and validation to date have been largely performed in ideal patient settings or with the aid of stabilizers. In practice, registration errors can exceed 7 mm [29], and are location- and physician-dependent. Similarly, patient motion can significantly contribute to error as patients are typically given local anesthesia or conscious sedation.

As new techniques for image registration are developed, ground truth validation studies are of vital importance in understanding its clinical utility. While many validations use phantoms because they are highly controllable and do not require human or animal subjects, it is debatable if they are adequate for assessing true targeting error. Phantoms are generally homogenous and designed to mimic shape and acoustic properties of human tissue. However, they do not necessarily share similar mechanical properties of human tissue, and sources of error such as tissue deformation and needle deflection may not be properly represented. In existing studies of biopsy systems in this chapter, targeting error increased when moving from phantoms to patients. At the same time, it can be difficult to assess the true accuracy with prostate tissue in general due to swelling and time-based changes. For even highly controlled post-prostatectomy correlations with imaging, the associated error of matching a pathology slide with an associated MR image can be on the order of 1–3 mm, which could potentially stack with any targeting errors potentially inflating registration errors [91]. These issues may be mitigated by performing multiple types of validation studies with detailed reporting and large sample sizes.

FUTURE DIRECTIONS

While 3D US in its current state appears sufficient for use with biopsy, targeted or focal therapy requires considerably improved accuracy to be widely adopted. Treatment success rates are variable and poor [85–88], likely due to insufficient registration accuracy [60]. Newer imaging techniques that incorporate 3D US may also provide more information to achieve improved registration. Micro-ultrasound is high frequency (29MHz) imaging that has demonstrated potential in early studies to identify clinically significant prostate cancer during biopsy with comparable sensitivity to mpMRI [25]. Structures that are not apparent at lower frequencies (i.e., ejaculatory ducts, BPH nodules, tissue texture in the peripheral zone), may provide for better fiducials for registration [25]. In the future, improved and automated segmentation and registration, and sensor-less tracking all promise to significantly reduce the sources of error for biopsy.

REFERENCES

1. R. L. Siegel, K. D. Miller, H. E. Fuchs, and A. Jemal, "Cancer statistics, 2022," *CA. Cancer J. Clin*, vol. 72, no. 1, pp. 7–33, 2022.
2. J. L. Mohler et al., "Prostate cancer, version 2.2019," *J. Natl. Compr. Cancer Netw.*, vol. 17, no. 5, pp. 479–505, May 2019.
3. M. E. Noureldin, M. J. Connor, N. Boxall, S. Miah, T. Shah, and J. Walz, "Current Techniques of Prostate Biopsy: An Update from Past to Present," *Translational Andrology and Urology*, vol. 9, no. 3. AME Publishing Company, pp. 1510–1517, 2020.
4. C. J. Das, A. Razik, A. Netaji, and S. Verma, "Prostate MRI–TRUS fusion biopsy: A review of the state of the art procedure," *Abdom. Radiol*, vol. 45, no. 7, pp. 2176–2183, 2020.
5. H. Hricak, P. L. Choyke, S. C. Eberhardt, S. A. Leibel, and P. T. Scardino, "Imaging prostate cancer: A multidisciplinary perspective," *Radiology*, vol. 243, no. 1, pp. 28–53, 01-Apr-2007.
6. M. A. Bjurlin et al., "Update of the standard operating procedure on the use of multiparametric magnetic resonance imaging for the diagnosis, staging and management of prostate cancer," *J. Urol.*, vol. 203, no. 4, pp. 706–712, Apr. 2020.
7. A. Stabile et al., "Multiparametric MRI for prostate cancer diagnosis: Current status and future directions," *Nat. Rev. Urol.*, vol. 17, no. 1, pp. 41–61, Jul. 2019.

8. S. Woo, C. H. Suh, S. Y. Kim, J. Y. Cho, and S. H. Kim, "diagnostic performance of prostate imaging reporting and data system version 2 for detection of prostate cancer: A systematic review and diagnostic meta-analysis," *Eur. Urol*, vol. 72, no. 2, pp. 177–188, 2017.

9. V. Kasivisvanathan et al., "MRI-targeted or standard biopsy for prostate-cancer diagnosis," *N. Engl. J. Med.* vol 85, no, 10, pp. 908–920, Mar. 2018.

10. L. Marks, S. Young, and S. Natarajan, "MRI-ultrasound fusion for guidance of targeted prostate biopsy," *Curr. Opin. Urol.*, vol. 23, no. 1, pp. 43–50, 2013.

11. C. Arsov et al., "Prospective randomized trial comparing magnetic resonance imaging (MRI)-guided in-bore biopsy to MRI-ultrasound fusion and transrectal ultrasound-guided prostate biopsy in patients with prior negative biopsies," *Eur. Urol.*, vol. 68, no. 4, pp. 713–720, Oct. 2015.

12. H. Watanabe, D. Igari, Y. Tanahasi, K. Harada, and M. Saitoh, "Development and application of new equipment for transrectal ultrasonography," *J. Clin. Ultrasound*, vol. 2, no. 2, pp. 91–98, 1974.

13. S. W. Hardeman, R. W. Wake, and M. S. Soloway, "Two new techniques for evaluating prostate cancer the role of prostate-specific antigen and transrectal ultrasound," *Postgrad. Med.*, vol. 86, no. 2, pp. 1970198, 1989.

14. R. Clements, G. J. Griffiths, W. B. Peeling, and P. G. Ryan, "Experience with ultrasound guided transperineal prostatic needle biopsy 1985–1988," *Br. J. Urol.*, vol. 65, no. 4, pp. 362–367, 1990.

15. K. Shinohara, T. M. Wheeler, and P. T. Scardino, "The appearance of prostate cancer on transrectal ultrasonography: Correlation of imagining and pathological examinations," *J. Urol*, vol. 142, no. 1, pp. 76–82, 1989.

16. F. A. Distler et al., "The value of PSA density in combination with PI-RADS™ for the accuracy of prostate cancer prediction," *J. Urol.*, vol. 198, no. 3, pp. 575–582, Sep. 2017.

17. A. Verma, J. St. Onge, K. Dhillon, and A. Chorneyko, "PSA density improves prediction of prostate cancer," *Can. J. Urol.*, vol. 21, no. 3, pp. 7312–7321, 2014.

18. D. R. H. Christie, and C. F. Sharpley, "How accurately can prostate gland imaging measure the prostate gland volume? Results of a systematic review," *Prostate Cancer*, vol. 2019, 2019.

19. N. R. Paterson et al., "Prostate volume estimations using magnetic resonance imaging and transrectal ultrasound compared to radical prostatectomy specimens," *Can. Urol. Assoc. J.*, vol. 10, no. 7–8, p. 264, Aug. 2016.

20. W. Venderink, M. de Rooij, J. P. M. Sedelaar, H. J. Huisman, and J. J. Fütterer, "Elastic versus rigid image registration in magnetic resonance Imaging–transrectal ultrasound fusion prostate biopsy: A systematic review and meta-analysis," *Eur. Urol. Focus*, vol. 4, no. 2, pp. 219–227, 2018.

21. D. H. Turnbull, and F. S. Foster, "Fabrication and characterization of transducer elements in two-dimensional arrays for medical ultrasound imaging," *IEEE Trans. Ultrason. Ferroelectr. Freq. Control*, vol. 39, no. 4, pp. 464–475, 1992.

22. J. K. Logan et al., "Current status of magnetic resonance imaging (MRI) and ultrasonography fusion software platforms for guidance of prostate biopsies," *BJU Int.*, vol. 114, no. 5, pp. 641–652, Nov. 2014.

23. M. Kongnyuy, A. K. George, A. R. Rastinehad, and P. A. Pinto, "Magnetic resonance imaging-ultrasound fusion-guided prostate biopsy: Review of technology, techniques, and outcomes," *Curr. Urol. Rep.*, vol. 17, no. 4, pp. 1–9, 2016.

24. Q. Huang, and Z. Zeng, "A review on real-time 3D ultrasound imaging technology," *Biomed Res. Int.*, vol. 2017, 2017.

25. L. Klotz et al., "Comparison of micro-ultrasound and multiparametric magnetic resonance imaging for prostate cancer: A multicenter, prospective analysis," *Can. Urol. Assoc. J.*, vol. 15, no. 1, p. E11, Jul. 2021.

26. S. Natarajan et al., "Clinical application of a 3D ultrasound-guided prostate biopsy system," *Urol. Oncol. Semin. Orig. Investig.*, vol. 29, no. 3, pp. 334–342, May 2011.

27. P. Fletcher et al., "Vector prostate biopsy: A novel magnetic resonance Imaging/Ultrasound image fusion transperineal biopsy technique using electromagnetic needle tracking under local anaesthesia," *Eur. Urol.*, vol. 83, no. 3, pp. 249–256, 2023.

28. *https://koelis.com/*.

29. F. Cornud et al., "Precision matters in MR imaging-targeted prostate biopsies: Evidence from a prospective study of cognitive and elastic fusion registration transrectal biopsies," *Radiology*, vol. 287, no. 2, pp. 534–542, May 2018.

30. A. Shah et al., "An open source multimodal image-guided prostate biopsy framework," *Lect. Notes Comput. Sci. (Including Subser. Lect. Notes Artif. Intell. Lect. Notes Bioinformatics)*, vol. 8680, pp. 1–8, 2014.

31. H. Guo, H. Chao, S. Xu, B. J. Wood, J. Wang, and P. Yan, "Ultrasound volume reconstruction from freehand scans without tracking," *IEEE Trans. Biomed. Eng.*, vol. 70, no. 3, pp. 970–979, 2022.

32. L. Tetrel, H. Chebrek, and C. Laporte, "Learning for graph-based sensorless freehand 3D ultrasound," *Lect. Notes Comput. Sci. (Including Subser. Lect. Notes Artif. Intell. Lect. Notes Bioinformatics*, vol. 10019, pp. 205–212, 2016.

33. T. A. Tuthill, J. F. Krücker, J. B. Fowlkes, and P. L. Carson, "Automated three-dimensional US frame positioning computed from elevational speckle decorrelation," *Radiology*, vol. 209, no. 2, pp. 575–582, 1998.

34. M. H. Mozaffari, and W. S. Lee, "Freehand 3-D ultrasound imaging: A systematic review," *Ultrasound Med. Biol.*, vol. 43, no. 10, pp. 2099–2124, 2017.

35. R. San José-Estépar, M. Martín-Fernández, P. P. Caballero-Martínez, C. Alberola-López, and J. Ruiz-Alzola, "A theoretical framework to three-dimensional ultrasound reconstruction from irregularly sampled data," *Ultrasound Med. Biol.*, vol. 29, no. 2, pp. 255–269, 2003.

36. T. Roxborough, and G. M. Nielson, "Tetrahedron based, least squares, progressive volume models with application to freehand ultrasound data," In *Proceedings. IEEE Visualization Conference*, pp. 93–100, 2000.

37. S. Sherebrin, A. Fenster, R. N. R. M.D., and D. Spence, "Freehand three-dimensional ultrasound: Implementation and applications," https://doi.org/10.1117/12.237790, vol. 2708, pp. 296–303, 1996.

38. J. W. Trobaugh, D. J. Trobaugh, and W. D. Richard, "Three-dimensional imaging with stereotactic ultrasonography," *Comput. Med. Imaging Graph.*, vol. 18, no. 5, pp. 315–323, 1994.

39. C. D. Barry et al., "Three-dimensional freehand ultrasound: Image reconstruction and volume analysis," *Ultrasound Med. Biol.*, vol. 23, no. 8, pp. 1209–1224, 1997.

40. Q. Huang, Y. Huang, W. Hu, and X. Li, "Bezier interpolation for 3-D freehand ultrasound," *IEEE Trans. Human-Machine Syst.*, vol. 45, no. 3, pp. 385–392, 2015.

41. J. Jiang, Y. Guo, Z. Bi, Z. Huang, G. Yu, and J. Wang, "Segmentation of prostate ultrasound images: The state of the art and the future directions of segmentation algorithms," *Artif. Intell. Rev.*, vol. 56, pp. 615–651, 2022.

42. P. Wu, Y. Liu, Y. Li, and Y. Shi, "TRUS image segmentation with non-parametric kernel density estimation shape prior," *Biomed. Signal Process. Control*, vol. 8, no. 6, pp. 764–771, 2013.

43. Y. J. Liu, W. S. Ng, M. Y. Teo, and H. C. Lim, "Computerised prostate boundary estimation of ultrasound images using radial bas-relief method," *Med. Biol. Eng. Comput.*, vol. 35, no. 5, pp. 445–454, 1997.

44. T. Ojala, M. Pietikäinen, and T. Mäenpää, "Multiresolution gray-scale and rotation invariant texture classification with local binary patterns," *IEEE Trans. Pattern Anal. Mach. Intell.*, vol. 24, no. 7, pp. 971–987, 2002.

45. R. Manavalan, and K. Thangavel, "TRUS image segmentation using morphological operators and DBSCAN clustering," In *Proceedings 2011 World Congress On Information and Communication Technology WICT 2011*, pp. 898–903, 2011.

46. A. Sarti, C. Corsi, E. Mazzini, and C. Lamberti, "Maximum likelihood segmentation with Rayleigh distribution of ultrasound images," *Comput. Cardiol.*, vol. 31, pp. 329–332, 2004.

47. F. Shao, K. V. Ling, W. S. Ng, and R. Y. Wu, "Prostate boundary detection from ultrasonographic images," *J. Ultrasound Med.*, vol. 22, no. 6, pp. 605–623, 2003.

48. M. Kass, A. Witkin, and D. Terzopoulos, "Snakes: Active contour models," *Int. J. Comput. Vis.*, vol. 1, no. 4, pp. 321–331, 1988.

49. S. Osher, and J. A. Sethian, "Fronts propagating with curvature-dependent speed: Algorithms based on Hamilton-Jacobi formulations," *J. Comput. Phys.*, vol. 79, no. 1, pp. 12–49, 1988.

50. S. Ghose et al., "A survey of prostate segmentation methodologies in ultrasound, magnetic resonance and computed tomography images," *Comput. Methods Programs Biomed.*, vol. 108, no. 1, pp. 262–287, Oct. 2012.

51. Y. Zhan, and D. Shen, "Deformable segmentation of 3-D ultrasound prostate images using statistical texture matching method," *IEEE Trans. Med. Imaging*, vol. 25, no. 3, pp. 256–272, 2006.

52. X. Yang, and B. Fei, "3D prostate segmentation of ultrasound images combining longitudinal image registration and machine learning," In *Medical Imaging 2012: Image-Guided Procedures, Robotic Interventions, and Modeling*, vol. 8316, pp. 803–811, Feb. 2012, https://doi.org/10.1117/12.912188.

53. W. Weng, and X. Zhu, "U-net: Convolutional networks for biomedical image segmentation," *IEEE Access*, vol. 9, pp. 16591–16603, 2015.

54. R. J. G. Van Sloun et al., "Zonal Segmentation in Transrectal Ultrasound Images of the Prostate Through Deep Learning." In *IEEE Int. Ultrason. Symp. (IUS)*, vol. 2018-October, Dec. 2018.

55. N. Orlando, D. J. Gillies, I. Gyacskov, C. Romagnoli, D. D'Souza, and A. Fenster, "Automatic prostate segmentation using deep learning on clinically diverse 3D transrectal ultrasound images," *Med. Phys.*, vol. 47, no. 6, pp. 2413–2426, 2020.

56. N. Orlando et al., "Effect of dataset size, image quality, and image type on deep learning-based automatic prostate segmentation in 3D ultrasound," *Phys. Med. Biol.*, vol. 67, no. 7, p. 074002, Mar. 2022.

57. T. R. Nelson, and T. T. Elvins, "Visualization of 3D ultrasound data," *IEEE Comput. Graph. Appl.*, vol. 13, no. 6, pp. 50–57, 1993.

58. J. N. Welch, J. A. Johnson, M. R. Bax, R. Badr, and R. Shahidi, "A real-time freehand 3D ultrasound system for image-guided surgery," *Proc. IEEE Ultrason. Symp.*, vol. 2, pp. 1601–1604, 2000.

59. A. Fenster, G. Parraga, and J. Bax, "Three-dimensional ultrasound scanning," *Interface Focus*, vol. 1, no. 4, p. 503, 2011.

60. C. Williams et al., "Why does magnetic resonance imaging-targeted biopsy miss clinically significant cancer?," *J. Urol.*, vol. 207, no. 1, pp. 95–107, Jan. 2022.

61. W. J. M. Van De Ven, S. Litjens, J. O. Barentsz, T. Hambrock, and H. J. Huisman, "Required accuracy of MR-US registration for prostate biopsies," *Lect. Notes Comput. Sci. (Including Subser. Lect. Notes Artif. Intell. Lect. Notes Bioinformatics)*, vol. 6963, pp. 92–99, 2011.

62. Y. Guo et al., "Image registration accuracy of a 3-dimensional transrectal Ultrasound–Guided prostate biopsy system," *J. Ultrasound Med.*, vol. 28, no. 11, pp. 1561–1568, Nov. 2009.

63. S. Il Hwang et al., "Comparison of accuracies between real-time nonrigid and rigid registration in the MRI–US fusion biopsy of the prostate," *Diagnostics*, vol. 11, no. 8, p. 1481, Aug. 2021.

64. O. Ukimura et al., "3-dimensional elastic registration system of prostate biopsy location by real-time 3-dimensional transrectal ultrasound guidance with magnetic resonance/transrectal ultrasound image fusion," *J. Urol.*, vol. 187, no. 3, pp. 1080–1086, 2012.

65. S. Xu et al., "Real-time MRI-TRUS fusion for guidance of targeted prostate biopsies," *Comput. Aided Surg.*, vol. 13, no. 5, pp. 255–264, 2008.

66. V. V. Karnik et al., "Assessment of image registration accuracy in three-dimensional transrectal ultrasound guided prostate biopsy," *Med. Phys.*, vol. 37, no. 2, pp. 802–813, 2010.

67. H. Li et al., "Machine learning in prostate MRI for prostate cancer: Current status and future opportunities," *Diagnostics*, vol. 12, no. 2, p. 289, Jan. 2022.

68. D. W. Cool, X. Zhang, C. Romagnoli, J. I. Izawa, W. M. Romano, and A. Fenster, "Evaluation of MRI-TRUS fusion versus cognitive registration accuracy for MRI-targeted, TRUS-guided prostate biopsy," *Am. J. Roentgenol.*, vol. 204, no. 1, pp. 83–91, 2015.

69. S. Ourselin, A. Roche, S. Prima, and N. Ayache, "Block matching: A general framework to improve robustness of rigid registration of medical images," *Lect. Notes Comput. Sci. (Including Subser. Lect. Notes Artif. Intell. Lect. Notes Bioinformatics)*, vol. 1935, pp. 557–566, 2000.

70. W. M. Wells, P. Viola, H. Atsumi, S. Nakajima, and R. Kikinis, "Multi-modal volume registration by maximization of mutual information," *Med. Image Anal.*, vol. 1, no. 1, pp. 35–51, 1996.

71. B. Fei, C. Kemper, and D. L. Wilson, "A comparative study of warping and rigid body registration for the prostate and pelvic MR volumes," *Comput. Med. Imaging Graph*, vol. 27, no. 4, pp. 267–281, 2003.

72. W. Alyami, A. Kyme, and R. Bourne, "Histological validation of MRI: A review of challenges in registration of imaging and whole-mount histopathology," *J. Magn. Reson. Imaging*, vol. 55, no. 1, pp. 11–22, 2022.

73. Y. Wang et al., "Towards personalized statistical deformable model and hybrid point matching for robust MR-TRUS registration," *IEEE Trans. Med. Imaging*, vol. 35, no. 2, pp. 589–604, Feb. 2016.

74. Y. Hu et al., "Weakly-supervised convolutional neural networks for multimodal image registration," *Med. Image Anal.*, vol. 49, pp. 1–13, Oct. 2018.

75. P. Yan, S. Xu, A. R. Rastinehad, and B. J. Wood, "Adversarial image registration with application for MR and TRUS image fusion," *Lect. Notes Comput. Sci. (Including Subser. Lect. Notes Artif. Intell. Lect. Notes Bioinformatics)*, vol. 11046, pp. 197–204, 2018.

76. Q. Zeng et al., "Label-driven magnetic resonance imaging (MRI)-transrectal ultrasound (TRUS) registration using weakly supervised learning for MRI-guided prostate radiotherapy," *Phys. Med. Biol.*, vol. 65, no. 13, p. 135002, Jun. 2020.

77. X. Yang et al., "Deformable MRI-TRUS registration using biomechanically constrained deep learning model for tumor-targeted prostate brachytherapy," *Int. J. Radiat. Oncol.*, vol. 108, no. 3, p. e339, Nov. 2020.

78. T. Kuru et al., "Phantom study of a novel stereotactic prostate biopsy system integrating preinterventional magnetic resonance imaging and live ultrasonography fusion," *J. Endourol.*, vol. 26, no. 7, pp. 807–813, 2012.

79. B. A. Hadaschik et al., "A novel stereotactic prostate biopsy system integrating pre-interventional magnetic resonance imaging and live ultrasound fusion," *J. Urol.*, vol. 186, no. 6, pp. 2214–2220, 2011.

80. O. Wegelin et al., "An ex vivo phantom validation study of an MRI-transrectal ultrasound fusion device for targeted prostate biopsy," *J. Endourol.*, vol. 30, no. 6, pp. 685–691, 2016.

81. J. Bax et al., "Mechanically assisted 3D ultrasound guided prostate biopsy system," *Med. Phys.*, vol. 35, no. 12, pp. 5397–5410, 2008.

82. D. W. Rickey, P. A. Picot, D. A. Christopher, and A. Fenster, "A wall-less vessel phantom for Doppler ultrasound studies," *Ultrasound Med. Biol.*, vol. 21, no. 9, pp. 1163–1176, 1995.

83. J. Pensa, W. Brisbane, A. Priester, A. Sisk, L. Marks, and R. Geoghegan, "A System for Co-Registration of High-Resolution Ultrasound, Magnetic Resonance Imaging, and Whole-Mount Pathology for Prostate Cancer," *2021 43rd Annu. Int. Conf. IEEE Eng. Med. Biol. Soc.*, pp. 3890–3893, Nov. 2021.

84. A. Priester et al., "Registration accuracy of patient-specific, three-dimensional-printed prostate molds for correlating pathology with magnetic resonance imaging," *IEEE Trans. Biomed. Eng.*, vol. 66, no. 1, pp. 14–22, Jan. 2019.

85. R. Geoghegan et al., "A tissue-mimicking prostate phantom for 980 nm laser interstitial thermal therapy," *Int. J. Hyperth.*, vol. 36, no. 1, pp. 993–1002, 2019.

86. R. Geoghegan, L. Zhang, A. M. Priester, H. Wu, L. Marks, and S. Natarajan, "Interstitial optical monitoring of focal laser ablation," *IEEE Trans. Biomed. Eng.*, vol. 69, no. 8, pp. 2545–2556, 2022.

87. A. Eranki, A. S. Mikhail, A. H. Negussie, P. S. Katti, B. J. Wood, and A. Partanen, "Tissue-mimicking thermochromic phantom for characterization of HIFU devices and applications," *Int. J. Hyperth.*, vol. 36, no. 1, pp. 518–529, 2019.

88. Z. Bu-Lin, H. Bing, K. Sheng-Li, Y. Huang, W. Rong, and L. Jia, "A polyacrylamide gel phantom for radiofrequency ablation," *Int. J. Hyperthermia*, vol. 24, no. 7, pp. 568–576, 2008.

89. P. Moldovan et al., "Accuracy of elastic fusion of prostate magnetic resonance and transrectal ultrasound images under routine conditions: A prospective multi-operator study," *PLoS One*, vol. 11, no. 12, pp. 1–11, 2016.

90. M. M. Poggi, D. A. Gant, W. Sewchand, and W. B. Warlick, "Marker seed migration in prostate localization," *Int. J. Radiat. Oncol. Biol. Phys.*, vol. 56, no. 5, pp. 1248–1251, 2003.

91. H. H. Wu et al., "A system using patient-specific 3D-printed molds to spatially align in vivo MRI with ex vivo MRI and whole-mount histopathology for prostate cancer research," *J. Magn. Reson. Imaging*, vol. 49, no. 1, pp. 270–279, Jan. 2019.

92. Y. Hu et al., "MR to ultrasound registration for image-guided prostate interventions," *Med. Image Anal.*, vol. 16, no. 3, pp. 687–703, Apr. 2012.

93. P. A. Pinto et al., "Magnetic resonance imaging/ultrasound fusion guided prostate biopsy improves cancer detection following transrectal ultrasound biopsy and correlates with multiparametric magnetic resonance imaging," *J. Urol.*, vol. 186, no. 4, pp. 1281–1285, 2011.

94. S. Vourganti et al., "Multiparametric magnetic resonance imaging and ultrasound fusion biopsy detect prostate cancer in patients with prior negative transrectal ultrasound biopsies," *J. Urol.*, vol. 188, no. 6, pp. 2152–2157, 2012.

95. A. R. Rastinehad et al., "Improving detection of clinically significant prostate cancer: Magnetic resonance imaging/transrectal ultrasound fusion guided prostate biopsy," *J. Urol.*, vol. 191, no. 6, pp. 1749–1754, 2014.

96. G. A. Sonn et al., "Value of targeted prostate biopsy using magnetic resonance-ultrasound fusion in men with prior negative biopsy and elevated prostate-specific antigen," *Eur. Urol.*, vol. 65, no. 4, pp. 809–815, 2014.

97. S. Shoji et al., "Original article : Clinical investigation accuracy of real-time magnetic resonance imaging-transrectal ultrasound fusion image-guided transperineal target biopsy with needle tracking with a mechanical position-encoded stepper in detecting significant prostate cancer in biopsy-naïve men," *Int. J. Urol.*, vol. 24, no. 4, pp. 288–294, 2017.

98. R. Sun, A. Fast, I. Kirkpatrick, P. Cho, and J. Saranchuk, "Assessment of magnetic resonance imaging (MRI)-fusion prostate biopsy with concurrent standard systematic ultrasound-guided biopsy among men requiring repeat biopsy," *Can. Urol. Assoc. J*, vol. 15, no. 9, pp. 495–500, 2021.

99. A. Magnier, C. Nedelcu, S. Chelly, M. C. Rousselet-Chapeau, A. R. Azzouzi, and S. Lebdai, "Prostate cancer detection by targeted prostate biopsy using the 3D navigo system: A prospective study," *Abdom. Radiol.*, vol. 46, no. 9, pp. 4381–4387, 2021.

100. S. Miah et al., "A prospective analysis of robotic targeted MRI-US fusion prostate biopsy using the centroid targeting approach," *J. Robot. Surg.*, vol. 14, no. 1, pp. 69–74, 2020.

101. S. Hamid et al., "The SmartTarget Biopsy Trial: A prospective, within-person randomised, blinded trial comparing the accuracy of visual-registration and magnetic resonance imaging/ultrasound image-fusion targeted biopsies for prostate cancer risk stratification," *Eur. Urol.*, vol. 75, no. 5, pp. 733–740, 2019.

10 Carotid Atherosclerosis Segmentation from 3D Ultrasound Images

Ran Zhou

INTRODUCTION

Stroke as a neurological disease is one of the most serious causes of death and disability worldwide, which has been increasingly threatening human health.[1,2] Furthermore, one-third of stroke survivors have sequelae, such as hemiplegia and aphasia, and are unable to carry out normal life activities. Stroke, including hemorrhagic and ischemic, is a sudden acute cerebrovascular disease with obvious clinical symptoms. Ischemic stroke accounts for 87% of cerebrovascular accidents[3] and is caused mainly due to the rupture of carotid atherosclerotic plaques.[4] Once the carotid plaque ruptures, the substances in the plaque interact with the thrombolytic factors in the blood and cause coagulation. These clots (emboli) can then travel into the brain and block the cerebral arterial blood vessels, resulting in insufficient oxygen supply to parts of the brain, leading to ischemic stroke and even death.[5] The population with carotid atherosclerosis disease tends to be younger[6] and even young people might have early atherosclerosis because their unhealthy dietary habits lead to obesity, metabolic syndrome, impaired glucose tolerance, etc.[7]

Figure 10.1a shows the structure of the common carotid artery (CCA), the internal carotid artery (ICA), and the external carotid artery (ECA). The ICA mainly supplies blood to the brain, while ECA mainly supplies blood to the facial organs, scalp, and skull. The carotid bifurcation (BF) is the enlarged part at the beginning of the ICA, which is also called the carotid bulb. Figure 10.1b shows the longitudinal view of the carotid artery and Figure 10.1c is a schematic diagram of the cross-sectional structure of blood vessels, including the intima, media, and adventitia. The intima is located in the innermost layer of the vessel wall, which is composed of smooth endothelium and elastic membrane. Media is located between the intima and adventitia of vessels, which is composed of smooth muscle cells and elastic fibers. The adventitia is the outermost layer of the vessel wall, including connective tissue, collagen, and elastic fibers.

Medical imaging is widely used for the diagnosis of carotid atherosclerosis, which can be used to directly measure the characteristics of carotid plaques.[9,10] Monitoring the progression and regression of plaques is important for managing patients at risk for cerebrovascular events. Clinically, contrasted computed tomography (CT),[11,12] magnetic resonance imaging (MRI),[13,14] and ultrasound imaging (US)[15,16] can be used to diagnose carotid atherosclerosis. Although CT and MRI provide high-resolution images, the radiation of CT imaging and the high cost of MRI imaging limit their applications for carotid plaque imaging and especially for monitoring the progression and regression of the disease. Carotid US imaging can show the intima-media thickness clearly, plaque location, size, and vessel stenosis,[17] and it can also be used to visualize the internal characteristics of plaques[18,19] through the echogenic characteristics of plaque constituents.[20] Thus, US has become the most common clinical method used for carotid plaque diagnosis and monitoring changes in the plaques.

Although two-dimensional (2D) US is regarded as a clinically convenient way of carotid plaque examination, the quality of two-dimensional carotid US imaging is affected by many factors, such as the operator's skill and the angle of the US probe relative to the vessel resulting in inconsistent evaluation results. Three-dimensional ultrasound (3DUS) provides a more efficient, convenient,

DOI: 10.1201/9781003299462-12

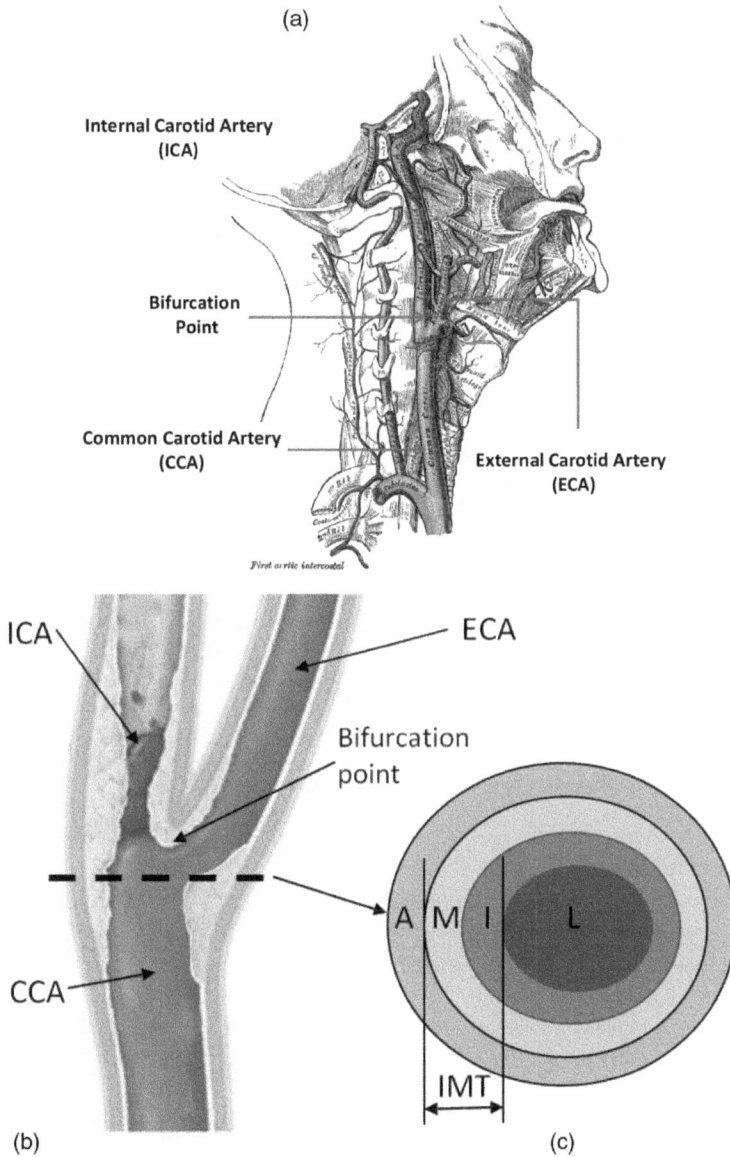

FIGURE 10.1 The anatomy and the structure of a carotid artery. (a) The anatomy of the carotid artery. (b) Schematic sagittal cross-section of a carotid artery showing the bifurcation point (BF), common carotid artery (CCA), internal carotid artery (ICA), and external carotid artery (ECA). (c) Transverse cross-section of a carotid artery showing the adventitia (A), media (M), intima (I), And lumen (L), as well as the IMT measurement. (Image (a): Courtesy of the 20th U.S. Edition of Gray's Anatomy page #549 and https://en.Wikipedia.org/wiki/Common_carotid_artery); (b) and (c) are from the Electronic Thesis and Dissertation Repository and available at https://ir.lib.uwo.ca/etd/1568.[8])

repeatable, and reliable method for analyzing carotid plaques. Furthermore, the imaging quality of 3DUS is less dependent on the operator's skill. Therefore, carotid 3DUS can more reliably be used to analyze the structure, morphology, and composition of plaques, and any changes in these biomarkers.[21,22] Recently, many research results have shown that carotid 3DUS measurements (i.e., total plaque volume (TPV), [23,24] carotid wall volume (VWV),[25,26] vessel-wall-plus-plaque thickness (VWT)[27]) are stronger biomarkers than 2DUS measurement (i.e., carotid intima-media thickness (IMT)).[28]

These 3DUS measurements require the segmentation of the media-adventitia (MAB), lumen-intima boundaries (LIB), and plaque boundaries from the carotid 3DUS image. Although manual segmentation of MAB, LIB, and plaques in 3DUS images can be performed effectively, they require experienced observers and a significant investment of time. To alleviate the manual segmentation burden, computer-assisted methods are important and necessary for segmenting the carotid arteries and plaques from 3DUS images and some of these methods have demonstrated high accuracy and low variability.

TRADITIONAL METHODS FOR CAROTID 3DUS IMAGE SEGMENTATION

Traditional carotid 3DUS segmentation methods were mainly focused on geometric prior knowledge-based segmentation approaches, such as active contour models and level-set methods.

GEOMETRIC PRIOR KNOWLEDGE-BASED SEGMENTATION METHODS

Gill et al. proposed a dynamic spherical method to directly segment the vessel wall from carotid 3DUS images and used the edge-based energy function to refine the LIB contour.[29] Solovey et al. used weak geometric prior knowledge to segment the LIB contour from 3D carotid US images.[30] However, these two methods are based on geometric prior information making them useful for segmentation of the LIB contour of the carotid artery. However, the shape of LIB is irregular, especially at the carotid artery BF for cases in which plaques are present, making these methods suboptimal. Destrempes et al. used an ultrasonic echo signal (RF) to estimate the motion field and input of an estimated prior into the Bayesian model to segment the plaque in a longitudinal carotid section B-mode ultrasonic image.[31] However, this method is sensitive to image quality and noise.

ACTIVE CONTOUR MODEL-BASED METHODS

Yang et al. used the active contour model to directly segment the CCA wall from 3DUS images and used it in the evaluation of carotid plaques.[32] Loizou et al. proposed a semi-automatic plaque segmentation method based on the snake model. They used two-dimensional Doppler US images of a longitudinal section of the carotid artery.[33,34] Delsanto further used a gradient map to initialize the active contour model to segment the inner and outer membrane of the carotid artery wall in the longitudinal section to obtain the plaque.[35]

LEVEL-SET METHODS

Ukwatta et al. proposed a method based on the 2D level-set method to segment the MAB and LIB of the CCA.[36] They first sliced the 3DUS carotid image, but manually marked several anchor points on the MAB and LIB on each 2D image, and finally used the level-set method to complete the segmentation. However, this method is very time-consuming and requires 2.8 ± 0.4 min to segment a blood vessel with a length of about 1.5 cm. Ukwatta et al. improved the 2D level-set method and proposed a carotid artery MAB and LIB segmentation method based on 3D sparse field level sets (3D SFLS).[37] However, this method still relies heavily on the manual marking of anchor points. The observer needs to mark multiple points ($n \geq 4$) on the MAB and LIB contours visualized in the cross-sectional images of the carotid artery at certain distance intervals, and the location of the carotid BF and the axis of the carotid artery needed to be marked. Cheng et al. used the level-set method to segment the plaque in the 2D cross-sections of the 3DUS carotid US images.[38] However, this method needed to know the boundary of the carotid artery MAB and LIB in advance, which affected the accuracy of the MAB and LIB contour segmentation. Cheng et al. further improved the level-set method, extended it to 3D, and directly established a 3D level-set model to segment the plaque in carotid 3DUS images.[39] However, this method still required the use of the carotid artery wall intima and adventitia segmentation as prior knowledge.

DEEP LEARNING METHODS FOR CAROTID 3DUS IMAGE SEGMENTATION

Most of the traditional methods, which relied on user interaction to initialize the segmentation or place segmentation anchor points lengthen the segmentation time, require a trained user, and introduce user variability. These methods are sensitive to contour initialization and image quality and generally provide suboptimal performance, limiting the clinical applications of these methods. Thus, investigators have sought to develop fully automated segmentation methods, such as deep learning,[40] which has achieved success in many medical imaging tasks, including classification, segmentation, registration, and computer-aided diagnosis. Most of the deep-learning-based methods for carotid US image segmentation fall into the following three categories:

1. Methods that focus on network architecture selection or design. Fundamentally, these methods aim at selecting or devising network architectures that are more accurate and robust to segment carotid US images.
2. Methods that focus on loss function design. A popular approach is to develop a novel loss function that is more suitable for plaque segmentation, such as the Dice similarity coefficient loss (DSC).
3. Methods that focus on post-processing design. Methods in this class can improve and refine the coarse segmentation provided by the deep-learning models.

NETWORK ARCHITECTURE

U-Net

U-Net and its variants are popular deep-learning networks for carotid 3DUS image segmentation, which consist of an encoder to extract image features and a decoder to upsample feature maps to their original size.[41] Figure 10.2 shows a modified network architecture of U-Net, in which the left branch is the encoder with five stages, where each stage contains two convolutional layers followed by a max-pooling operation.[42] The kernel size of the convolutions is typically 3×3 pixels for 2D images or $3 \times 3 \times 3$ voxels for 3D images. A parametric rectified linear unit is used as activation neurons in each convolutional layer.[43] The right branch of the network is the decoder with a structure similar to that of the encoder branch. However, it uses an up-convolution operation instead of the max-pooling operation at each stage to recover the size of the feature map as the same size as the input image. The last layer is a 1×1 convolution (or $1 \times 1 \times 1$ convolution) with a pixel-wise softmax layer, which is used to classify each pixel into different categories (i.e., within the lumen, between MAB and LIB, or outside the MAB). The shortcut connections can be added to the original U-Net

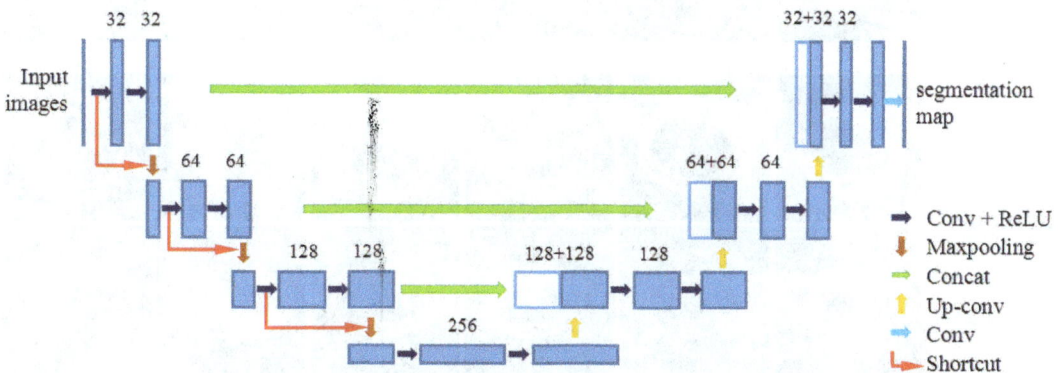

FIGURE 10.2 The modified architecture of U-Net. The U-Net consists of an encoder to extract image features and a decoder to up-sample feature maps to their original size. (This figure is from Fig. 6 in Zhou et al.[44])

architecture to avoid overfitting, which connects the input to the output of each convolutional block, skipping two convolutional layers.

Zhou et al. used the U-Net to design a semi-automatic algorithm by manually setting several anchor points on the MAB contours.[44] A total of 144 3DUS images were used in their experiments. The comparison of the algorithm segmentation results to manual segmentations by an expert showed that the modified U-Net can achieve an average DSCs of 96.46 ± 2.22% and 92.84 ± 4.46% for the MAB and LIB, respectively. Although this method achieved high segmentation accuracy, it required manual interactions resulting in a long segmentation time, which was about 34.4 ± 9.8 s per 3DUS image. We refer the reader to Zhou et al.[44] for more results.

Zhou et al. also implemented the U-Net to carotid 3DUS image segmentation without user interactions. Trained on 150 subjects and tested on 100 subjects, U-Net achieved average DSCs of 91.24 ± 3.86% and 88.41 ± 5.65% for MAB and LIB, respectively.[45] Although the results were lower than that in their previous work,[44] this method didn't have any user interactions during testing, and the computation time was shorter than 1 s per 3DUS image.

Figure 10.3 shows two examples of the MAB and LIB segmentation generated by 3D U-Net. The segmentations were very close to the manual segmentation; however, the contours were not smooth with some burrs.

UNet++

UNet++ is an improvement of the U-Net network and outperforms variants of U-Net networks.[46] As shown in Figure 10.4a, UNet++ consists of several U-Nets of varying depths with decoders that are densely connected at the same resolution via redesigned skip connections. The loss function is generated by combining losses from these U-Nets with different depths. Figure 10.4b shows the nested dense convolution blocks in the first level of the UNet++.

The UNet++ algorithm was applied to 3DUS carotid image segmentation using 699 3DUS images from 250 subjects and achieved average DSCs of 91.39 ± 3.60% and 89.03 ± 5.51% for MAB

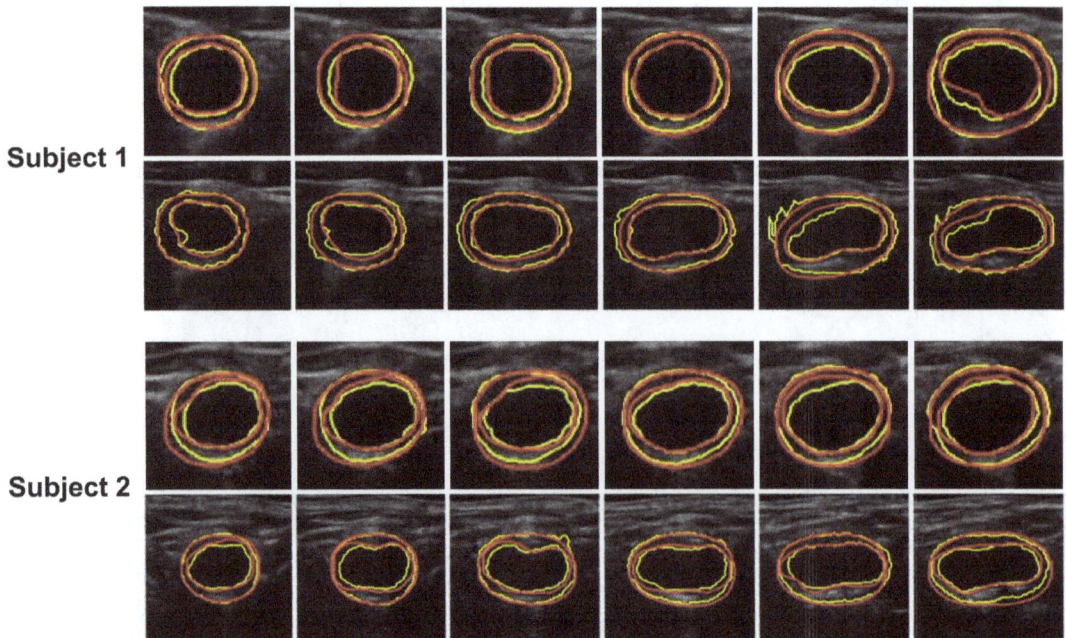

FIGURE 10.3 Examples of MAB and LIB segmentation generated by 3D U-Net. Yellow contours are algorithm segmentations and red contours are manual segmentations.

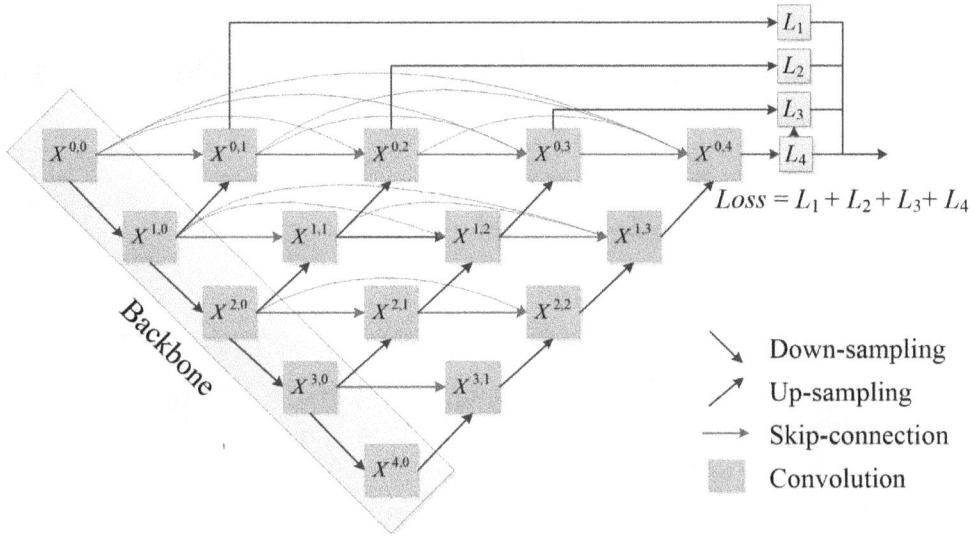

(a)

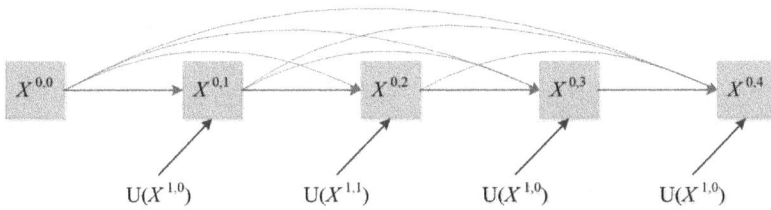

$$X^{0,1} = H[X^{0,0}, U(X^{1,0})]$$

$$X^{0,2} = H[X^{0,0}, X^{0,1}, U(X^{1,1})]$$

$$X^{0,3} = H[X^{0,0}, X^{0,1}, X^{0,2}, U(X^{1,2})]$$

$$X^{0,4} = H[X^{0,0}, X^{0,1}, X^{0,2}, X^{0,3}, U(X^{1,3})]$$

(b)

FIGURE 10.4   The network architecture of UNet++. (a) The UNet++ consists of an encoder and decoder, which are connected by a series of nested dense convolution blocks. (b) An example of a detailed analysis of the nested dense convolution blocks of the UNet++. (Figure a is adapted from Fig. 2 in Zhou et al.[47].)

and LIB, respectively, which were higher than that of U-Net. Figure 10.5 shows two examples of the MAB and LIB segmentation generated by UNet++. The segmentations were smoother than that generated by U-Net (shown in Figure 10.3).

Voxel-FCN

The complex structure of the decoder path in U-Net and UNet++ requires many parameters and a long time for network training. In addition, the global prior that the LIB is nested within the MAB may facilitate the segmentation task but this prior is difficult to implement using the existing U-Net. A voxel-based FCN, named voxel-FCN, for the carotid 3DUS image segmentation task, was proposed by Zhou et al.[48] We note that the proposed voxel-FCN differs from previous work in that it requires fewer parameters and can learn both the images' spatial and contextual information.

Figure 10.6 shows the architecture of the proposed voxel-FCN. The encoder is composed of a general convolutional neural network (CNN) to extract spatial image information and a

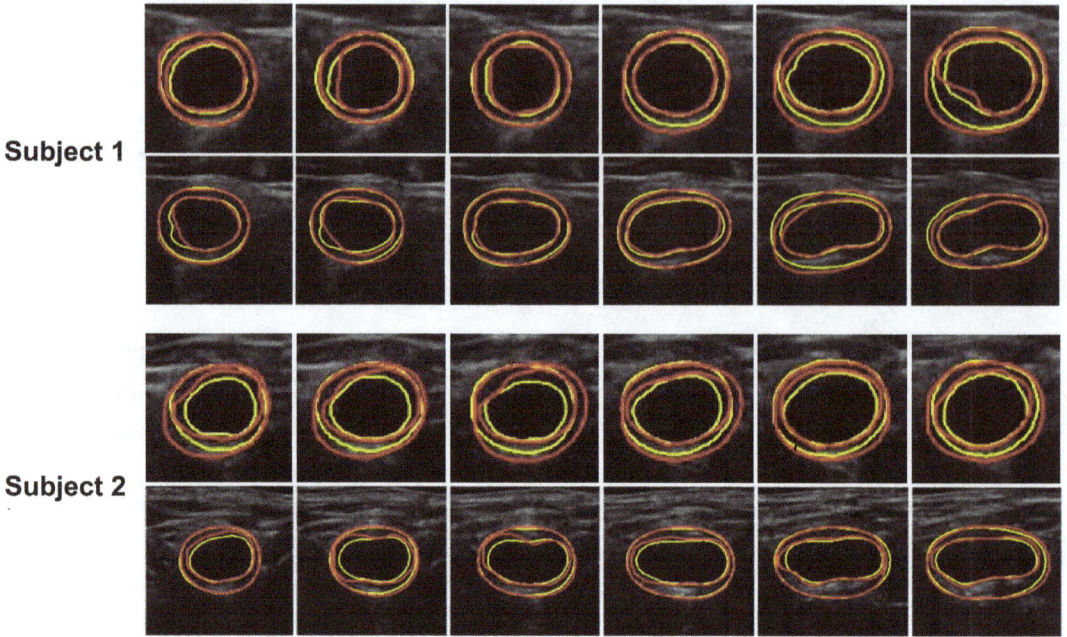

FIGURE 10.5 Examples of MAB and LIB segmentation generated by UNet++ with VGG19 as the backbone. Yellow contours are algorithm segmentations and red contours are manual segmentations.

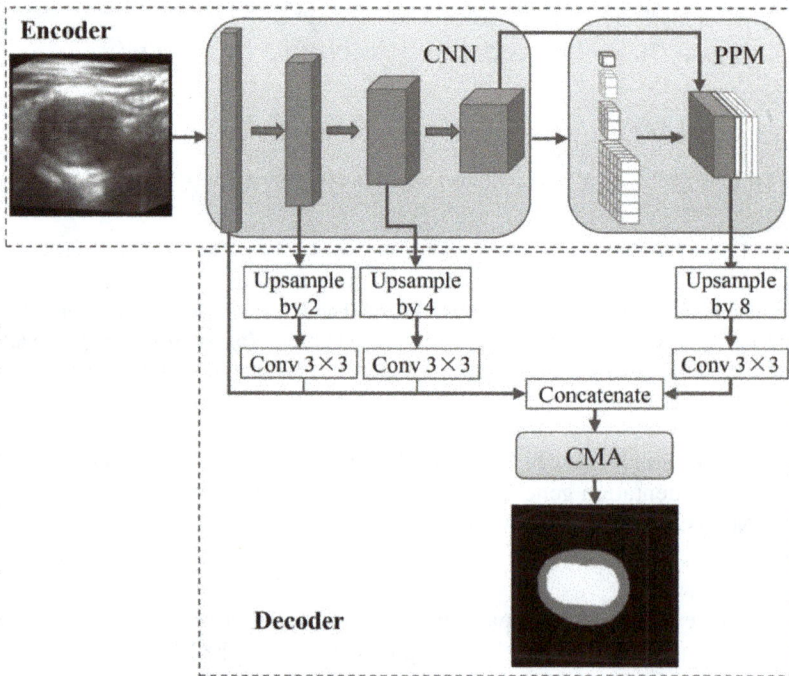

FIGURE 10.6 The architecture of the voxel-FCN. The encoder consists of a general CNN and a pyramid-pooling module (PPM) to extract the spatial and contextual image features. The decoder employs a concatenation module with an attention mechanism (CMA) to combine multi-level features extracted in the encoder. The convolution operations after up-sampling are used to reduce the dimensions of the feature maps. All the convolution and max-pooling operations are implemented in 3D. (This figure is adapted from Fig. 1 in Zhou et al.[48].)

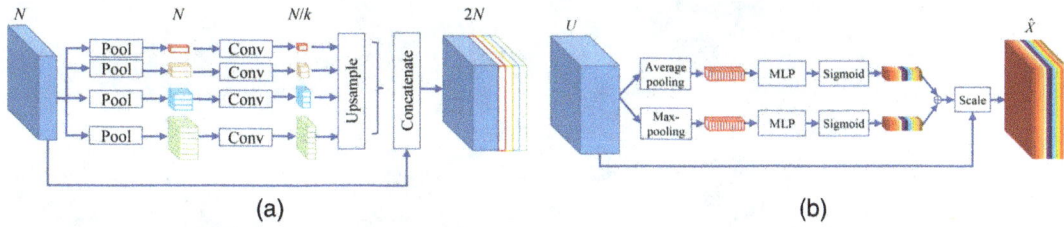

(a) **(b)**

FIGURE 10.7 Architectures of PPM and CMA modules. (a) Pyramid-pooling module (PPM): Each of the four pooling levels ($k = 4$) consists of a pooling layer, a convolution layer, and an up-sampling layer. The feature maps in the four levels are convolved (feature channels = N/k), up-sampled, and concatenated to generate the PPM output. (b) Concatenate module with an attention mechanism (CMA): Global average pooling and max-pooling operations are used to extract global features of each channel and a fully connected layer (MLP) with a sigmoid activation is applied to encode the correlation between channels. The outputs from the average pooling and the max-pooling paths are merged ("+") and scaled as the outputs of the CMA module. (These figures are adapted from Figs. 3 and 4 in Zhou et al.[48].)

pyramid-pooling module (PPM) to learn the contextual information of MAB and LIB. The decoder concatenates multi-level features with an attention mechanism (CMA), which assigns weights to different feature channels to make the CNN focus on salient features dynamically. The decoding part of the voxel-FCN allows the combination of both spatial and contextual information learned by the encoder. VGG16 network was extended to 3D to build the encoder of the Voxel-FCN. The PPM is commonly used to reduce the context information loss between different feature sub-regions caused by the fixed-size constraint of the CNN (as shown in Figure 10.7a). The concatenation module with an attention mechanism (CMA) fused the multi-level feature maps extracted in the encoder by adaptively assigning channel-wise feature relationships at different levels of the encoder (as shown in Figure 10.7b).

Using 1007 3DUS images, voxel-FCN yielded a DSC of $93.4 \pm 3.1\%$ for the MAB in the CCA, and $91.6 \pm 5.1\%$ in the BF by comparing the algorithm and expert manual segmentations and achieved a DSC of $89.8 \pm 6.2\%$ and $89.5 \pm 6.8\%$ for the LIB in the CCA and the bifurcation respectively. Figure 10.8 shows two examples of the MAB and LIB segmentation generated by voxel-FCN, which were very close to those generated by U-Net (shown in Figure 10.3). Although the segmentation accuracy of voxel-FCN was similar to the U-Net and V-Net, the architecture of the decoder in voxel-FCN is simpler than the U-Net and V-Net and required fewer parameters, greatly reducing the computational demands. The number of network parameters of voxel-FCN was about 1.6 M; however, the number of network parameters was 16 M for the original U-Net and 73M for the original V-Net.

Loss Function

Many studies focused on the model architecture and training procedures; however, the loss function is also important as it affects the robustness of deep learning models and the accuracy of the predictions.

Weighted Binary-Entropy Loss

Cross-entropy is the most popular loss function used in carotid 3DUS image segmentation. To better understand the position of the arterial lumen, Zhou et al. proposed a weighted binary-entropy loss by using a probability map to weight the binary-entropy loss value for each pixel.[49]

In this application, it is a binary classification problem, where the foreground was the lumen and the region outside the lumen was the background. In this method, it was assumed that the lumen region was labeled by 1 and the background was labeled by 0. Then, the weighted cross entropy penalizes the deviation of the output of the softmax layer from 1 using:

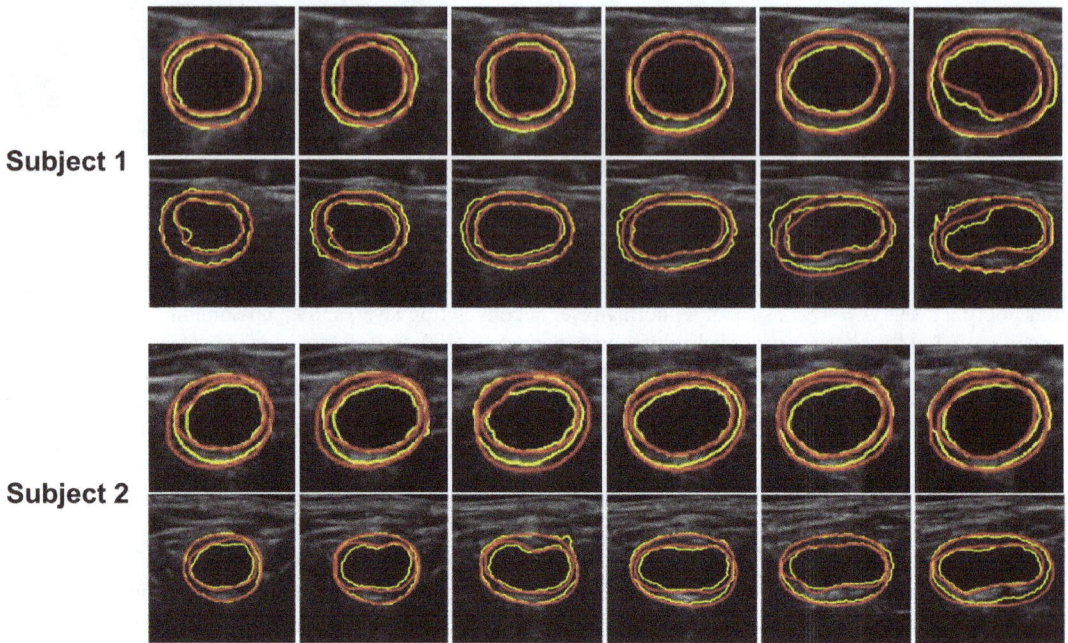

FIGURE 10.8 Examples of MAB and LIB segmentation generated by Voxel-FCN. Yellow contours are algorithm segmentations and red contours are manual segmentations.

$$E = \sum_{(x,y)\in Z^2} w(x,y)\log\left(p_{l(x,y)}(x,y)\right) \tag{10.1}$$

where l is the true label of each pixel, $p_{l(x,y)}(x,y)$ is the probability output of the softmax layer for pixels classified to 1, and w is a weight map.

The weight map was pre-computed using masks of images used for training and was then used to calculate the probability of a pixel belonging to a certain class at every position in the training images. The following equation gave the weight map belonging to the lumen:

$$w(x,y) = \frac{1}{N}\sum_{i}^{N} mask_i(x,y) \tag{10.2}$$

where N is the number of training images, i is the index of images, and mask (x, y) is the label in a training image at position (x, y) where the lumen pixels in a mask image are labeled as 1 and background pixels are 0.

Dice similarity coefficient Loss

The dice similarity coefficient (DSC) is another popular metric for segmentation performance evaluation and has been used for a loss function in medical image segmentation methods, as follows:

$$Loss_{dsc}(y,\hat{y}) = 1 - DSC(y,\hat{y}) \tag{10.3}$$

$$DSC(y,\hat{y}) = \frac{2|y\cap\hat{y}|}{|y\cup\hat{y}|}\times 100\% \tag{10.4}$$

where y is the binary image denoted by experts, and $\hat{y}$ is the network output.

To segment the CCA and ICA from 3DUS images, Jiang et al. proposed an adaptive triple loss function by assigning weights of the three DSC losses adaptively,[50] which was designed to minimize the total DSC loss associated with the MAB, LIB, and the carotid vessel wall (CVW).

$$L = \alpha\ Loss_{dsc}\left(y_{MAB}, \hat{y}_{MAB}\right) + \beta Loss_{dsc}\left(y_{LIB}, \hat{y}_{LIB}\right) + \gamma Loss_{dsc}\left(y_{CVW}, \hat{y}_{CVW}\right) \tag{10.5}$$

where $y_{MAB}, y_{LIB}, y_{CVW}$ are binary images denoting whether each pixel is enclosed by the manually segmented MAB, LIB, and CVW, respectively; $\hat{y}_{MAB}, \hat{y}_{LIB}, \hat{y}_{CVW}$ are the network outputs.

Jiang et al. reported that the adaptive triple loss could improve the original U-Net. They used 224 3DUS images for experiments and achieved a DSC of 95.1% ± 4.1% and 91.6% ± 6.6% for the MAB and LIB, in the CCA, respectively, and 94.2% ± 3.3% and 89.0% ± 8.1% for the MAB and LIB, in the ICA, respectively.[50]

POST-PROCESSING OF DEEP-LEARNING SEGMENTATION

CNN may provide inaccurate segmentation results due to the low echogenicity of some plaques, resulting in non-smooth boundaries, small spurious blobs, and labeling inconsistency, which require further post-processing to correct. Zhou et al. proposed a multi-class continuous max-flow (CMF) algorithm and an ensemble algorithm by fusing the output of multiple CNNs to improve and refine the deep-learning segmentation.[48,51]

Continuous Max-Flow (CMF) Algorithm

In a method reported by Zhou et al.,[48] they denoted $p_l(x) \in [0, 1]$, $l \in L$ as the probability maps associated with the initial segmentation provided by the CNNs. These probability maps can be interpreted as the cost of assigning a label l to each pixel $x \in \Omega$. For example, the greater probability $p_l(x)$ suggests a higher likelihood and lower cost of assigning a label l, $l \in L$, to x, and *vice versa*. Let $u_l(x) = \{0,1\}, l \in L$ be the labeling function for the segmentation region R_l such that $u_l(x) = 1$ for $x \in R_l$ and $u_l(x) = 0$ otherwise. In addition, the MRF constraint was incorporated to enforce the smoothness of the final segmentation u. To this end, the initial CNN segmentation of a 3D image $V(x)$ was refined by minimizing the total labeling cost and the surface area of u as follows:

$$\min_{u_l(x)=\{0,1\}} \sum_{l \in L} \left\{ \langle u_l, 1 - \rho_l \rangle + \alpha \int_\Omega |\nabla u_l| dx \right\} \tag{10.6}$$

Subject to $\sum_{l \in L} u_l(x) = 1$, $\alpha > 0$ weighs the surface area measurement of u_l in the form of total-variation. Direct optimization of Eq. 10.6 is challenging because of the non-convex labeling function $u_l(x) = \{0,1\}$ and the nonsmooth total variation term. Following the previous work,[52] this challenging optimization problem was investigated through convex relaxation, i.e., by relaxing $u_l(x)$ from $\{0, 1\}$ to continuous convex sets $[0,1]$ as follows:

$$\min_{u_l(x)=[0,1]} \sum_{l \in L} \left\{ \langle u_l, 1 - \rho_l \rangle + \alpha \int_\Omega |\nabla u_l| dx \right\} \tag{10.7}$$

subject to $\sum_{l \in L} u_l(x) = 1$. In the convex domain, robust and efficient global optimization algorithms exist and can be employed to solve the convex optimization problem Eq. 10.7. The max-flow theory[43] shows that this convex relaxed segmentation model can be solved globally and exactly through flow maximization.

Zhou et al. used 1007 3DUS images for the experiment and found that the CMF post-processing could greatly improve the baseline voxel-FCN vessel-wall-volume (VWV) and VWV% measurements. For example, the baseline voxel-FCN VWV error was improved from 6.1 ± 50.1 mm³ to 0.04 ± 51.2 mm³. Similarly, the baseline voxel-FCN VWV% was also higher and statistically significantly different from U-Net and V-Net with an improvement of VWV% from $7.6 \pm 35.5\%$ to $2.0 \pm 26.4\%$. In addition, the voxel-FCN + CMF post-processing reduced the standard deviation (SD) of VWV% from 35.3 mm³ to 26.4 mm³.

Ensemble Algorithm

Because the performances of deep-learning models vary on different network architectures, Zhou et al. reported an ensemble algorithm by fusing the segmentations of multiple CNNs for carotid US image segmentation, aiming at combining the advantages of different CNN models to achieve higher accuracy and better generalization performance.[45,51]

The ensemble algorithm was reported that it employed eight individual UNet++ networks with different backbones and slightly different architectures in the encoder for carotid artery segmentation in 3D/2D US images, as shown in Figure 10.9.

Each trained network in the ensemble was applied to a test US image and the eight network predictions $p_i(x) \in \{0,1\}, x \in \Omega, i = 1,2,\ldots,8$ were fused to generate a single consensus segmentation $\hat{p}(x) \in \{0,1\}, x \in \Omega$, using an in-house developed image fusion algorithm.[53] Briefly, the image fusion algorithm enforced a pixel-wise similarity between each of the eight initial predictions $p_i(x)$ and the final consensus segmentation $\hat{p}(x)$ as $\int_\Omega |p_i(x) - \hat{p}(x)| dx$, dubbed as the similarity term. In addition, the final consensus segmentation $\hat{p}(x)$ was regularized for spatial smoothness in terms of total variation: $\int_\Omega |\nabla \hat{p}(x)| dx$, coined as the regularization term. The similarity and regularization terms were combined into a single objective function, which was solved through primal-dual analysis and convex optimization techniques on a GPU. The image fusion algorithm aims to generate a single consensus segmentation, which is similar to each of the eight segmentations and spatially smooth (to mimic manual segmentation).

Figure 10.10 shows examples of MAB and LIB segmentations generated by the ensemble of eight UNet++ networks with different backbones. The ensemble algorithm generated MAB and LIB contours were very close to manual segmentations, which were also smoother and better than a single model generated contours (shown in Figures 10.3, 10.5, and 10.8). In the study by Zhou et al., they reported that the UNet++ ensemble could improve the performance when using a small training dataset. Using 669 3DUS images from 250 subjects for the experiment, the UNet++ ensemble yielded better performance than the popular U-Net, voxel-FCN, and single UNet++ when small training datasets were available for training. For example, trained on 30 subjects (R1~R5), the UNet++ ensemble achieved a DSC of {91.01~91.56%, 87.53~89.44%} and ASSD of {0.10~0.11 mm, 0.32~0.39 mm} for {MAB, LIB}, which were similar to the results produced by U-Net and Voxel-FCN trained on 150 subjects.

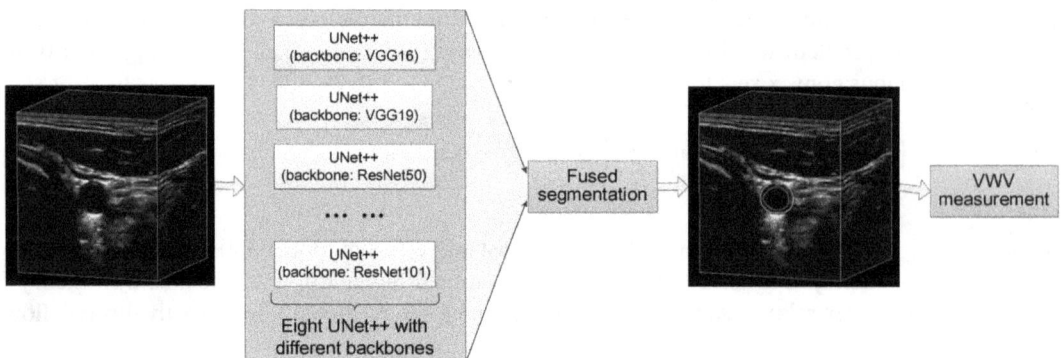

FIGURE 10.9 Eight segmentation maps generated by the trained ensemble UNet++ models were fused to output a single final segmentation. The vessel-wall-volume (VWV) was calculated from the final segmentation. (This figure is adapted from Fig. 1 in Zhou et al.[45])

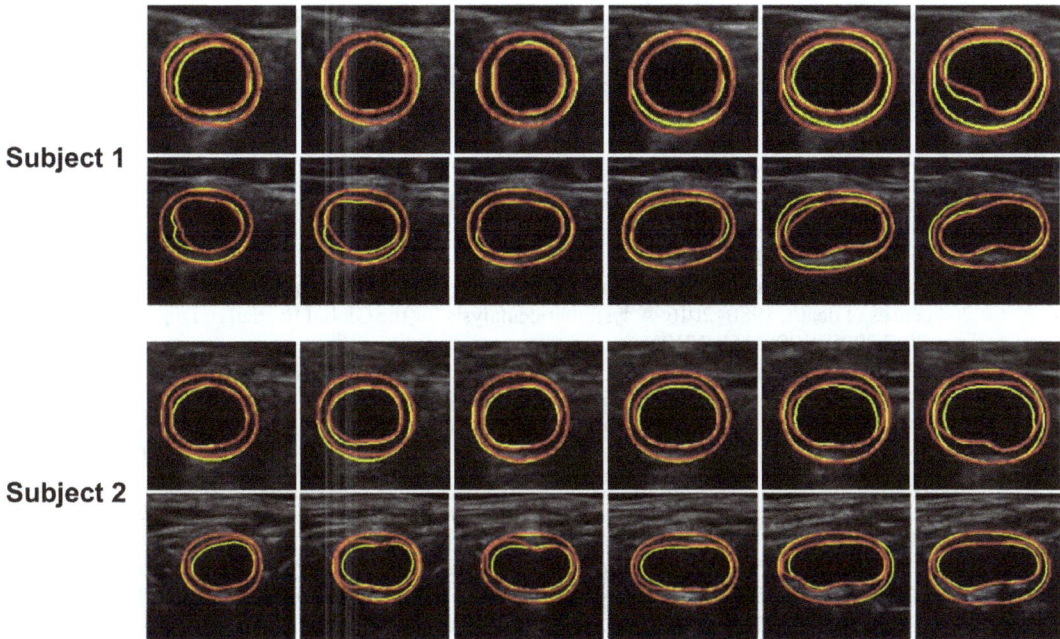

FIGURE 10.10 Examples of MAB and LIB segmentation generated by the ensemble of eight UNet++ networks with different backbones. Yellow contours are algorithm segmentations and red contours are manual segmentations.

DISCUSSION

3DUS provides an efficient, convenient, repeatable, and reliable method for analyzing carotid plaques. In this chapter, we described the applications of traditional methods and deep-learning methods for carotid 3DUS image segmentation. The traditional methods provide possible ways for semi/auto-segmentation of carotid plaques in US images. However, these types of methods are sensitive to initialization and image quality, and generally provide suboptimal performance, limiting their clinical applications. Recently, deep-learning-based methods have shown utility in automatic carotid US image segmentation. We summarized the deep-learning methods into three categories: network architecture design, loss function modification, and post-processing. Methods of network architecture design aim at devising network architectures that are more accurate and robust to carotid US images. Methods that focus on loss function design are used to develop novel loss functions that are more suitable for plaque segmentation. Methods that focus on post-processing designs can refine the coarse segmentations provided by the deep-learning models. Although deep learning has achieved some successes in carotid 3DUS image segmentation, there are still many challenges in clinical practice:

1. High performance of deep learning requires datasets with many labeled images for training, which is very labor-intensive in clinical practice;
2. The imbalance distribution of VWVs or TPAs will cause an inferior performance of the models on under-represented groups, thereby reducing the generalizability and fairness of deep-learning models; and
3. Manual segmentation might have incorrect annotations, which will cause a decrease in segmentation accuracy.

Thus, future work should focus on these practical problems and develop more robust deep-learning algorithms for carotid 3DUS image segmentation. For example, improvements can be achieved through the application of semi-supervised and self-supervised methods to reduce the requirements of annotated ultrasound images in training; the generation of synthetic images to balance the dataset; and the use of data selection methods to clean the dataset.

REFERENCES

1. GBD Diseases and Injuries Collaborators. Global burden of 369 diseases and injuries in 204 countries and territories, 1990–2019: A systematic analysis for the Global Burden of Disease Study 2019. *Lancet.* 2020;396(10258):1204–1222.

2. Naghavi M, Abajobir AA, Abbafati C, et al. Global, regional, and national age-sex specific mortality for 264 causes of death, 1980–2016: A systematic analysis for the Global Burden of Disease Study 2016. *Lancet.* 2017;390(10100):1151–1210.

3. Benjamin EJ, Muntner P, Alonso A, et al. Heart disease and stroke statistics-2019 update: A report from the American Heart Association. *Circulation.* 2019;139(10):e56–e528.

4. Howard DPJ, Gaziano L, Rothwell PM, Oxford VS. Risk of stroke in relation to degree of asymptomatic carotid stenosis: A population-based cohort study, systematic review, and meta-analysis. *Lancet Neurol.* 2021;20(3):193–202.

5. Ait-Oufella H, Tedgui A, Mallat Z. Atherosclerosis: An inflammatory disease. *Sang Thromb Vaiss.* 2008;20(1):25–33.

6. Niu LL, Zhang YL, Qian M, et al. Standard deviation of carotid young's modulus and presence or absence of plaque improves prediction of coronary heart disease risk. *Clin Physiol Funct I.* 2017;37(6):682–687.

7. Song P, Fang Z, Wang H, et al. Global and regional prevalence, burden, and risk factors for carotid atherosclerosis: A systematic review, meta-analysis, and modelling study. *Lancet Global Health.* 2020;8(5):e721–e729.

8. Ukwatta E. *Vascular Segmentation Algorithms for Generating 3D Atherosclerotic Measurements.* London, Ontario, Canada: School of Graduate and Postdoctoral Studies, The University of Western Ontario, 2013.

9. Yuan C, Kerwin WS, Ferguson MS, et al. Contrast-enhanced high resolution MRI for atherosclerotic carotid artery tissue characterization. *J Magn Reson Imaging.* 2002;15(1):62–67.

10. Steinl DC, Kaufmann BA. Ultrasound imaging for risk assessment in atherosclerosis. *Int J Mol Sci.* 2015;16(5):9749–9769.

11. Herzig R, Burval S, Krupka B, Vlachova I. Comparison of ultrasonography, CT angiography, and digital subtraction angiography in severe carotid stenoses. *Eur J Neurol.* 2004;11(11):774–781.

12. Wintermarka M, Jawadif SS, Rappe JH, et al. High-Resolution CT imaging of carotid artery atherosclerotic plaques. *Am J Neuroradiol.* 2008;29(5):875–882.

13. Chu B, Ferguson MS, Chen H, et al. Magnetic resonance imaging features of the disruption-prone and the disrupted carotid plaque. *JACC Cardiovasc Imaging.* 2009;2(7):883–896.

14. Lopez Gonzalez MR, Foo SY, Holmes WM, et al. Atherosclerotic carotid plaque composition: A 3T and 7T MRI-histology correlation study. *J Neuroimaging: Off J Am Soc Neuroimaging.* 2016;26(4):406–413.

15. van Engelen A, Wannarong T, Parraga G, et al. Three-dimensional carotid ultrasound plaque texture predicts vascular events. *Stroke.* 2014;45(9):2695–2701.

16. Saba L, Sanagala SS, Gupta SK, et al. A multicenter study on carotid ultrasound plaque tissue characterization and classification using six deep artificial intelligence models: A stroke application. *IEEE Trans Instrum Meas.* 2021;70.

17. Acharya UR, Krishnan MMR, Sree SV, et al. Plaque tissue characterization and classification in ultrasound carotid scans: A paradigm for vascular feature amalgamation. *IEEE Trans Instrum Meas.* 2013;62(2):392–400.

18. Zhou R, Luo Y, Fenster A, Spence JD, Ding M. Fractal dimension based carotid plaque characterization from three-dimensional ultrasound images. *Med Biol Eng Comput.* 2019;57(1):135–146.

19. Awad J, Krasinski A, Parraga G, Fenster A. Texture analysis of carotid artery atherosclerosis from three-dimensional ultrasound images. *Med Phys.* 2010;37(4):1382–1391.

20. Johri AM, Herr JE, Li TY, Yau O, Nambi V. Novel ultrasound methods to investigate carotid artery plaque vulnerability. *J Am Soc Echocardiog.* 2017;30(2).

21. Chiu B, Egger M, Spence JD, Parraga G, Fenster A. Quantification of carotid vessel wall and plaque thickness change using 3D ultrasound images. *Med Phys.* 2008;35:3691–3710.

22. Nanayakkara ND, Chiu B, Samani A, Spence JD, Samarabandu J, Fenster A. Non-rigid registration of 3D ultrasound images to monitor carotid plaque changes. *Radiother Oncol.* 2007;84:S77–S77.

23. Buchanan D, Gyacskov I, Ukwatta E, Lindenmaier T, Fenster A, Parraga G. Semi-automated segmentation of carotid artery total plaque volume from three dimensional ultrasound carotid imaging. In Medical Imaging 2012: Biomedical Applications in Molecular, Structural, and Functional Imaging. 2012;8317.

24. Pollex RL, Spence JD, House AA, et al. Comparison of intima media thickness and total plaque volume of the carotid arteries: Relationship to the metabolic syndrome. *Arterioscl Throm Vas.* 2006;26(5):E85.

25. Buchanan DN, Lindenmaier T, Mckay S, et al. The relationship of carotid three-dimensional ultrasound vessel wall volume with age and sex: Comparison to carotid intima-media thickness. *Ultrasound Med Biol.* 2012;38(7):1145–1153.

26. Egger M, Spence JD, Fenster A, Parraga G. Validation of 3D ultrasound vessel wall volume: An imaging phenotype of carotid atherosclerosis. *Ultrasound Med Biol.* 2007;33(6):905–914.

27. Chiu B, Egger M, Spence DJ, Parraga G, Fenster A. Area-preserving flattening maps of 3D ultrasound carotid arteries images. *Med Image Anal.* 2008;12(6):676–688.

28. Raggi P, Stein JH. Carotid intima-media thickness should not be referred to as subclinical atherosclerosis: A recommended update to the editorial policy at atherosclerosis. *Atherosclerosis.* 2020;312:119–120.

29. Gill JD, Ladak HM, Steinman DA, Fenster A. Accuracy and variability assessment of a semiautomatic technique for segmentation of the carotid arteries from three-dimensional ultrasound images. *Med Phys.* 2000;27(6):1333–1342.

30. Solovey I. *Segmentation of 3D carotid ultrasound images using weak geometric priors* [M.S. Thesis]. Canada, The University of Waterloo; 2010.

31. Destrempes F, Meunier J, Giroux MF, Soulez G, Cloutier G. Segmentation of plaques in sequences of ultrasonic B-mode images of carotid arteries based on motion estimation and a Bayesian model. *IEEE Trans Bio-Med Eng.* 2011;58(8):2202–2211.

32. Yang X, Jin J, He W, Yuchi M, Ding M. Segmentation of the Common Carotid Artery with Active Shape Models from 3D Ultrasound Images. SPIE Medical Imaging: Computer-Aided Diagnosis; 2012; San Diego, CA.

33. Loizou CP, Pattichis CS, Pantziaris M, Nicolaides A. An integrated system for the segmentation of atherosclerotic carotid plaque. *IEEE Trans Inf Technol B.* 2007;11(6):661–667.

34. Petroudi S, Loizou C, Pantziaris M, Pattichis C. Segmentation of the common carotid intima-media complex in ultrasound image using active contours. *IEEE Trans Bio-Med Eng.* 2012;59(11):3060–3070.

35. Delsanto S, Molinari F, Liboni W, Giustetto P, Badalamenti S, Suri JS. User-independent plaque characterization and accurate IMT measurement of carotid artery wall using ultrasound. *Conf Proc IEEE Eng Med Biol Soc.* 2006;2006:2404–2407.

36. Ukwatta E, Awad J, Ward AD, et al. Three-dimensional ultrasound of carotid atherosclerosis: Semiautomated segmentation using a level set-based method. *Med Phys.* 2011;38(5):2479–2493.

37. Ukwatta E, Yuan J, Buchanan D, et al. Three-dimensional segmentation of three-dimensional ultrasound carotid atherosclerosis using sparse field level sets. *Med Phys.* 2013;40(5).

38. Cheng JY, Li H, Xiao F, et al. Fully automatic plaque segmentation in 3-D carotid ultrasound images. *Ultrasound Med Biol.* 2013;39(12):2431–2446.

39. Cheng JY, Chen YM, Yu YY, Chiu B. Carotid plaque segmentation from three-dimensional ultrasound images by direct three-dimensional sparse field level-set optimization. *Comput Biol Med.* 2018;94:27–40.

40. Litjens G, Kooi T, Bejnordi BE, et al. A survey on deep learning in medical image analysis. *Med Image Anal.* 2017;42:60–88.

41. Ronneberger O, Fischer P, Brox T. U-net: Convolutional networks for biomedical image segmentation. Paper presented at: International Conference on Medical image computing and computer-assisted intervention 2015.

42. Zhou R, Ma W, Fenster A, Ding M. U-Net based automatic carotid plaque segmentation from 3D ultrasound images. Paper presented at: Medical Imaging 2019: Computer-Aided Diagnosis 2019.

43. Simonyan K, Zisserman A. Very Deep Convolutional Networks for Large-Scale Image Recognition. 3rd International Conference for Learning Representations; 2015.

44. Zhou R, Ma W, Fenster A, Ding M. U-Net based automatic carotid plaque segmentation from 3D ultrasound images. SPIE Medical Imaging - Computer-Aided Diagnosis; 2019.

45. Zhou R, Guo F, Azarpazhooh MR, et al. Carotid vessel-wall-volume ultrasound measurement via a UNet++ ensemble algorithm trained on small datasets. *Ultrasound Med Biol.* 2023;49(4):1031–1036.

46. Zhou Z, Siddiquee MMR, Tajbakhsh N, Liang J. Unet++: Redesigning skip connections to exploit multiscale features in image segmentation. *IEEE Trans Med Imaging.* 2019;39(6):1856–1867.

47. Zhou R, Ou Y, Fang X, et al. Ultrasound carotid plaque segmentation via image reconstruction-based self-supervised learning with limited training labels. *Mathemat Biosci Eng.* 2023;20(2):1617–1636.

48. Zhou R, Guo F, Azarpazhooh MR, et al. A voxel-based fully convolution network and continuous max-flow for carotid vessel-wall-volume segmentation from 3D ultrasound images. *IEEE Trans Med Imaging*. 2020;39(9):2844–2855.
49. Zhou R, Fenster A, Xia Y, Spence JD, Ding M. Deep learning-based carotid media-adventitia and lumen-intima boundary segmentation from three-dimensional ultrasound images. *Med Phys*. 2019;46(7):3180–3193.
50. Jiang M, Zhao Y, Chiu B. Segmentation of common and internal carotid arteries from 3D ultrasound images based on adaptive triple loss. *Med Phys*. 2021;48(9):5096–5114.
51. Zhou R, Guo F, Azarpazhooh MR, et al. Deep learning-based measurement of total plaque area in B-mode ultrasound images. *IEEE J Biomed Health*. 2021;25(8):2967–2977.
52. Chan TF, Esedoglu S, Nikolova M. Algorithms for finding global minimizers of image segmentation and denoising models. *SIAM J Appl Math*. 2006;66(5):1632–1648.
53. Yuan J, Shi J, Tai X-C. A convex and exact approach to discrete constrained tv-l1 image approximation. *East Asian J Appl Math*. 2011;1(2):172–186.

Section III

3D Ultrasound Algorithms

11 Segmentation of Neonatal Cerebral Lateral Ventricles from 3D Ultrasound Images

Zachary Szentimrey and Eranga Ukwatta

INTRODUCTION

Intraventricular hemorrhaging (IVH) within cerebral lateral ventricles affects 20–30% of very-low-birth-weight (VLBW) infants (<1500 g).[1–3] The blood may build up and lead to post-hemorrhagic ventricle dilatation (PHVD), which is an enlargement of the cerebral ventricles.[3] Not all patients with IVH will have PHVD; however, because of the potential for neurological degradation, it is important to diagnose and treat patients at the onset.[1] Medical imaging methods are important for the diagnosis and prognosis of such patients.

The most widely used imaging tool for measuring IVH is cranial two-dimensional (2D) ultrasound (US), which can be performed at the bedside.[3,4] Although IVH is relatively easy to diagnose with 2D US and is cost-effective, estimating volumetric changes over time is unreliable due to a high user dependency on 2D image slice acquisition.[1,5,6] Magnetic resonance imaging (MRI) is another commonly used modality for measuring cerebral ventricle volumes. It provides more detailed images than US and segmentation methods for MRI have already been developed for measuring ventricular volume.[1,7] An issue with MRI is the requirement for special equipment to maintain and move the neonate to the MR imager whereas US can be performed while the neonate is in an incubator.[1,8]

To address the limitations of 2D US and MRI, some studies[1] have investigated the use of 3D US for measuring ventricular volumes in preterm babies. Compared to 2D US, 3D US has shown to be more sensitive to ventricular size and shape changes and has reduced user dependency when acquiring images.[1] Quantification of volumes requires segmentation of the cerebral lateral ventricles from 3D US images which is where deep-learning segmentation methods can assist.

BACKGROUND ON NEONATAL CEREBRAL LATERAL VENTRICLE

CEREBRAL VENTRICLE ANATOMY AND PATHOPHYSIOLOGY

VLBW infants (infants less than 1500 g) represent only ~1.2% of births every year in the United States.[9] However, 20–30% of this VLBW population develop IVH, which is bleeding located inside or around the cerebral ventricles.[10] The cerebral ventricles consist of four chambers that house the choroid plexus which generates cerebral spinal fluid (CSF).[11] The chambers are the right and left lateral ventricles, the third ventricle and the fourth ventricle, shown in Figure 11.1. IVH incidence and severity are inversely proportional to gestational age so those born the earliest are at higher risk of IVH.[10,12] Severe hemorrhages cause clots and blood breakdown products inside the ventricles, which can block the flow of CSF. When CSF is blocked, it can cause abnormal enlargement of the ventricles known as PHVD. The enlarged ventricles compress the surrounding brain tissue leading to acute problems, such as vomiting or chronic neurological problems lasting throughout the subject's life. In fact, 45–85% of neonates with moderate-to-severe IVH develop cognitive deficits later in life and 75% require learning disability education in school.[10] Therefore, it is important to monitor

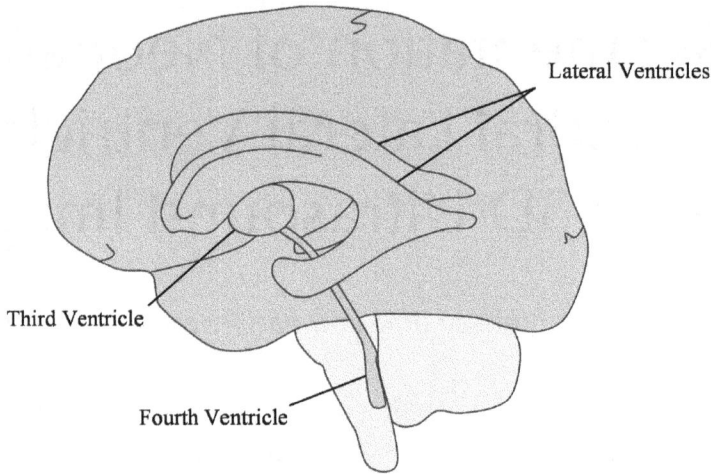

FIGURE 11.1 Diagram of the ventricular cavities, viewed at an oblique sagittal vantage point.

the progression of IVH and PHVD and apply the appropriate treatment measure at the optimal time. Typically, cranial 2D US with manual measurements is used to measure and monitor the ventricles but there is no widespread consensus on when to intervene and perform treatments.[13,14] Using 3D US could provide quantitative volumetric values and reduce variability between users and institutions.

Current Monitoring Techniques

Each institution has their own screening standards and time frames when diagnosing IVH and PHVD. Of all, the most common screening tool is transfontanelle 2D cranial US.[13–15] Typically, neonates born less than 1500 g are screened at regular intervals during the first month of life but IVH mainly occurs within the first three days of life.[16,17] Scans within the first week often detect the worst signs of IVH and most screening occurs while trying to plan a treatment strategy.[17] The physical size and shape of the head are measured and clinical signs and symptoms, such as brachycardia, are monitored.[18] However, imaging tools are very valuable resources for diagnosing IVH especially at the onset.

Magnetic Resonance Imaging

MRI is another commonly used modality for measuring cerebral ventricle volumes. It provides clearer images than US, which suffers from speckle, low soft-tissue contrast and few details on anatomical structure of the brain.[19] MRI has especially high tissue contrast between white and grey matter as well as with hemorrhages. MRI has many issues that can make it difficult for use in day-to-day monitoring. One such issue is the requirement for special equipment to maintain and move the neonate to the MR imager whereas US can be performed while the neonate is in an incubator.[19,20] MRI-compatible incubators are very expensive and minimally used. Even still, they are only imaged if absolutely necessary and take patient health into consideration before making the decision.[20] Even after getting the patient to the imager, they must remain stationary with their breath held. Immobilizing blankets have worked on some patients but not on every neonate.[20] With all the risks and problems outlined with MRI, the largest problem is the use of IVs, ventilation machines, and infusion pumps which may be unfit for MRI use and work improperly.[20] All these issues can make MRI a difficult screening tool to use for measuring IVH and PHVD.

2D Ultrasound

The most widely used imaging tool for measuring IVH is cranial 2D US due to the ability to rapidly acquire images at bedside without radiation. IVH is relatively easy to diagnose with 2D US and is cost effective compared to MRI and most other images techniques.[19] For incubator-bound neonates, 2D US has been the imaging choice for diagnosis historically. 2D US is performed using a mid-frequency transducer (5–8 MHz) with a wide view curvilinear array.[5,19] This frequency allows for proper depth of penetration in tissues in the skull (7–14 cm).[5] To perform a 2D US scan, the technician acquires images in the coronal and sagittal planes through the anterior fontanelle on the patient.[5] They take screen captures and manually choose the captures that best show the relevant landmarks of the cerebral ventricles.[5] CSF in the brain typically has lower scatter and attenuation with US waves and thus, appears darker (black) in images. On the other hand, locations of hemorrhaging and blood clots appear lighter (white) where surrounding brain tissue appear grey. PHVD would appear as a large black area surrounded by lighter tissue. Measuring through the anterior fontanelle was initially performed to measure changes in the anterior horn width (AHW) but it was later determined that different areas of the 3D ventricles undergo dilatation at different rates for each patient and measuring the AHW may not always correlate to patients with PHVD.[21,22] In addition to AHW, the ventricle index (VI) and third ventricle width can be measured in from the anterior fontanelle. To better measure dilatation in the posterior horns of the ventricle, the US probe can be used to image along the sagittal plane to capture the thalamo-occipital distance (TOD).[21,22] While these measurements provide quantitative values for monitoring IVH and PHVD, they can suffer from user variability. Separate technicians could acquire different 2D slices when making multiple measurements over time or even the same technician could acquire different slices.[23] There is high user variability and limitations using linear measurement for estimating an irregular 3D shape which makes monitoring IVH difficult.

3D Ultrasound

In the late 1990s and early 2000s, clinical research began using 3D US as it became available.[24] 3D US systems first used 2D transducers that were mechanically tilted, translated, and rotated about the patient for image acquisition. These images were then stitched together since the exact transducer movement was precisely recorded, allowing for relative positions to be captured between each image.[1,25] A 3D volume is the result when reconstructed using geometric algorithms, shown in Figure 11.2. Recently, matrix array transducers have been developed which can capture entire 3D volumes; however, these systems are very expensive compared to 2D transducers.[25] Instead of purchasing an expensive and new 3D US system, research centers are using 2D US transducers in conjunction with fast reconstruction algorithms, allowing the clinician to view the images as soon they are acquired. These 2D US systems with reconstruction algorithms have been shown to be reliable for creating 3D US images quickly and are used in practice.[1] Such systems have been used to observe and analyze a variety of clinical applications including carotid arteries, prostate, and liver.[26-28] Important landmarks are labeled in Figure 11.2, including the inferior, posterior, and anterior horns of one ventricle, the septum between lateral ventricles, the left and right Foramen of Monro and the midpoint of the thalamus and choroid plexus intersection boundary.

Neonatal ventricle measurements such as AHW and VI can be performed accurately using 3D images and scanning time can be reduced to 2 min using 3D US compared to 10 min with 2D US.[1,29] Measurements, such as AHW and VI, have been used historically, but they are poor indicators of ventricular volumes changes compared to actually measuring the ventricles volume directly.[5] By measuring volumes, clinicians can track volumetric changes over time to determine when and if to intervene and the treatment plan going forward. Estimating ventricle volume and areas of enlargement can be performed using 3D US better than 2D US. Studies have shown that neonatal ventricle volumes from 3D US images correlate strongly with volumes removed when performing a cerebral tap.[8] In comparison to MRI, volumes measured with 3D US has been shown to be biased lower than

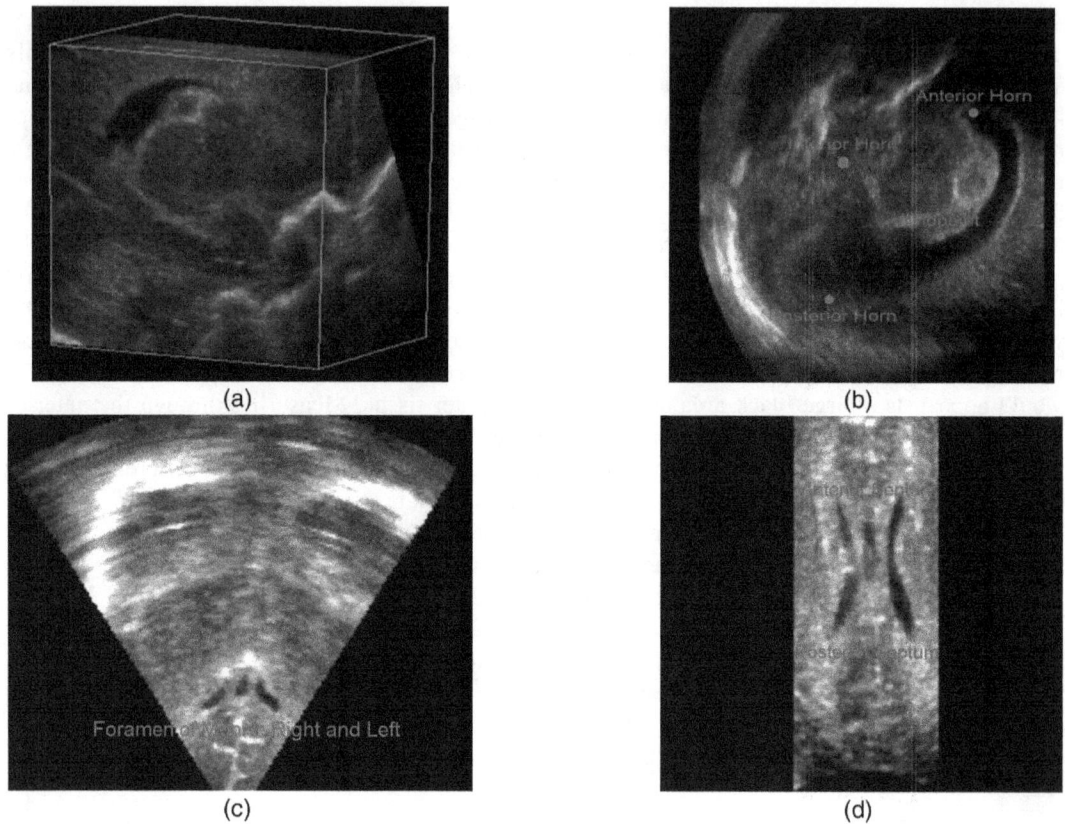

FIGURE 11.2 Example of a 3D US image showing the (a) reconstructed volume, (b) a slice of the sagittal plane of the right ventricle, (c) a slice of the coronal plane, and (d) a slice of the transverse plane. All images show important landmarks that can be used to track ventricle changes over time.

that of MRI but those differences are considered systemic and can be adjusted through calibration.[8] Also, measuring volume is not as clinically important as measuring volumetric changes, which can be measured well with 3D US. A drawback with using 3D US is when patients get older, the opening at the anterior fontanelle shrinks causing reduced image quality and shadows to appear at the edges of image.[4,30] The reduced image quality and shadowing can make measuring the posterior horns difficult. Even still, 3D US can be performed fast, as inexpensive as 2D US and safer than MRI while providing volumetric measurements.

NEED FOR CEREBRAL LATERAL VENTRICLE SEGMENTATION FROM 3D US IMAGES

3D US is more sensitive to ventricular size changes compared to 2D US, but to determine the ventricle size and localized regions of enlargement, ventricles need to be segmented. Segmenting the cerebral lateral ventricle is tedious and time-consuming and is currently performed by clinicians manually, taking between 20 and 45 minutes to segment one image.[3,8] There can be millions of voxels in a 3D US image, so clinicians segment the few sagittal slices that are 1 mm apart. The spacing between is then interpolated to create a 3D surface. An example of a manually segmented slice in the sagittal plane is shown in Figure 11.3. Because of the high user dependence in manual segmentation of the ventricles, a fast and automated method is desirable.

In the section below, semi- and fully automated methods for lateral ventricles are described.

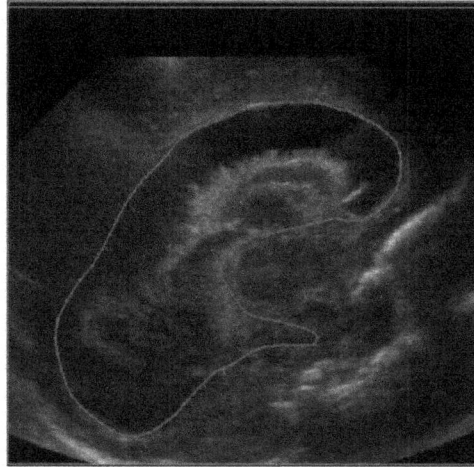

FIGURE 11.3 Manual segmentation of a 3D US image, showing the right lateral ventricle, in one slice along the sagittal plane.

NEONATAL CEREBRAL LATERAL 3D US SEGMENTATION WORK

CONVENTIONAL (NON-DEEP LEARNING) SEGMENTATION OF CEREBRAL LATERAL VENTRICLES FROM 3D US

An overview of conventional methods for cerebral ventricle segmentation are shown in Table 11.1 with Dice similarity coefficient (DSC) and mean absolute surface distance (MAD) as metrics.

Several techniques for segmenting the lateral ventricles in 3D US images were based on conventional techniques, such as convex optimization techniques and required some manual initializations.[31-33] The first semi-automated method by Qiu et al. in 2013 used subject-specific shape priors models that were created from expert annotated segmentations.[31] The shape prior models are rigidly registered to images captured in later time points using six manually labeled landmarks (at the

---

TABLE 11.1

Studies Using Conventional (Non-Deep Learning) Segmentation Techniques for Cerebral Ventricle Segmentation from 3D US Images

Automation Type	Reference	Type of Method	Number of Images	Time per Image	DSC ± STD	MAD ± STD (mm)
Semi-automated	(Qiu, Yuan, Kishimoto, et al. 2013)[31]	Convex optimization with shape prior	20	1.5 min	0.724 ± 0.025	0.7 ± 0.1
	(Qiu et al. 2014)[32]	Convex optimization with voxel pre-labels	25	2.5 min	0.789 ± 0.034	0.7 ± 0.2
	(Qiu, Yuan, Kishimoto, McLeod, et al. 2015)[33]	Convex optimization with voxel pre-labels and phase asymmetry maps	50	3 min	0.782 ± 0.044	0.65 ± 0.3
Automated	(Qiu, Yuan, Kishimoto, Chen, et al. 2015)[34]	Multi-phase Geodesic level-sets	15	54 min	0.732 ± 0.030	0.64 ± 0.3
	(Qiu et al. 2017)[23]	Multi-phase Geodesic level-sets	30	54 min	0.767 ± 0.062	1.0 ± 0.3

anterior, posterior and inferior horns). The rigid registration models provide an initial guess for a convex optimization-based surface evolution segmentation method. The study used 20 images from four patients (five images per patient) with an average DSC and MAD of 0.724 ± 0.025 and 0.7 ± 0.1 mm. The total segmentation time including manual initialization was 1.5 minutes for a single 3D US image.

A similar semi-automated convex optimization-based segmentation approach was devised by Qiu et al. in 2014.[32] This method requires a different user initialization compared to the landmark registration approach mentioned earlier. For initialization, the user labels some voxel inside and outside the ventricles as foreground and background voxels respectively on a few sagittal views. These initializations are used to estimate prior intensity probability density functions for both foreground and background. These are used as constraints and costs in the optimization procedure. The energy function is then set-up into a min-cut problem while also maintaining smooth boundaries of the object. The max-flow graph algorithm is used to solve the problem. The experiments performed by Qiu et al. yielded a mean DSC, MAD, and absolute volumetric difference (VD) of 0.789 ± 0.034, 0.7 ± 0.2 mm, and 2.5 ± 2.9cm^3, respectively with a computational time of 2.5 minutes per 3D US image.[32]

A semi-automated approach by Qiu et al. in 2015 tried to improve upon the work from Qiu et al. in 2014 by including phase asymmetry maps of the 3D US images to regularize the energy function.[32,33] The study used 50 3D US images with DSC, MAD and VD values of 0.782 ± 0.044, 0.65 ± 0.26 mm and 1.9 ± 3.2 cm^3 respectively and a total run time of 3 minutes max per 3D US image on a GPU.

Qiu et al. in 2017 implemented a method for fully automated segmentation using multi-phase geodesic level-sets (MGLS) algorithms in combination with phase congruency maps, a shape prior and atlas selection methods.[23] This study is the first to report a method for the automated segmentation of neonatal cerebral ventricles from 3D US images. The algorithm MGLS technique starts by performing a region growing method to extract a ROI and the 3D US images are converted into phase congruency maps, which works well for detecting step edges in US images. Next, the phase congruency maps are inputs for an atlas registration framework, which compares the current image to a subject-specific atlas. An atlas selection strategy was applied using to find the optimal atlas subset based on characteristics of the images. The selected atlases were next used as inputs for STAPLE to construct a spatial shape prior for the segmentation task. The segmentation task is formulated to evolve multiple contours by means of a level set approach to minimize the energy function with respect to changes in multiple segmentation regions. These regions/level sets are the left ventricle, right ventricle and background. The problem is formulated as a min-cut problem, solved using convex relaxation and the max-flow algorithm. The MGLS method was used on 30 3D US images with mean DSC and MAD of 0.767 ± 0.062 and 1.0 ± 0.3 mm respectively with a total run time of 54 min per image. The study by Qiu et al. has multiple limitations including the long run time and computational inefficiency.[23] The same MGLS method was also used by Qiu et al. in 2015 on 15 patients only with DSC and MAD of 0.732 ± 0.030 and 0.64 ± 0.3 mm, respectively.[34]

FULLY AUTOMATED DEEP LEARNING-BASED SEGMENTATION OF CEREBRAL LATERAL VENTRICLES FROM 3D US

There have been recent studies using deep learning for cerebral lateral ventricle segmentation of neonates in US images. Work has been done on both 2D US and 3D US images of neonatal cerebral ventricles but more emphasis will be placed on 3D US images. A table summarizing the automated methods using deep learning is shown in Table 11.2.

Many recent deep-learning-based applications use 2D models for segmentation of 3D US images. Gontard et al. worked with 3D US scans and deep learning, developing a 2D SegNet CNN with 152 images of 10 patients and 230 ventricles providing a mean intersection over union (IoU) of 0.54 on

TABLE 11.2

Fully Automated Work for Cerebral Ventricle Segmentation from 3D US Images Using Deep Learning-Based Algorithms and Techniques

Automation Type	Reference	Type of Method	Number of Images	Time per Image	DSC ± STD
Automated	(Gontard et al. 2021)[35]	2D SegNet	152 total images	1.5 min	0.70 ± N/A
	(M. Martin et al. 2018)[36]	2D U-Net	15 total images	5 s	0.816 ± 0.04
	(Szentimrey et al. 2021)[37]	2D Multiplane U-Net	193 total images	<60 s	0.76 ± 0.09
	(Szentimrey et al. 2022)[38]	3D U-Net Ensemble	190 total images	5 se	0.74 ± 0.06 (Two ventricles) 0.85 ± 0.05 (One ventricle)

the training data.[35] However, their validation results used only 83 3D US images from 6 patients (122 ventricles).[35] Their SegNet method required 60 s to segment one image and they did not report VD or MAD metrics.[35] The method by Gontard et al. used manually segmented boundaries by an expert.[35] The images were inputted into the SegNet model three at a time and resized to 200×200 pixels followed by a median filter of size [1,3,3]. The group used three methods for correcting class imbalance. These methods included a reweighing scheme in the loss function, using three classes instead of two (these classes are ventricle, black background and non-black background) and they only used sagittal 2D slices which contain the ventricles. The work by Gontard et al. also used a pretrained SegNet model, trained on ImageNet data and did not actually train a model using only 3D US data.[35] Image augmentation techniques including rotation, translation, scaling and shear were also implemented to prevent overfitting.

Martin et al. used the 2D U-Net deep learning model for cerebral ventricle segmentation in 3D US images.[36] Their work included 15 3D volumes and reported results for four images only. They used a 2D US transducer and developed a 3D reconstruction algorithm for generating their images. For post-processing, a morphological binary operation was implemented to extract the five largest connected regions with a volume greater than 1% of the mean ventricle volume followed by a Gaussian filter.

Szentimrey et al. in 2021 developed a 2D multiplane U-Net that can segment 3D US images from multiple orientations.[37] The 2D model extracts slices from orthogonal views during training and fuses the results from each plane using a mean vote when making predictions. Their model was developed on 193 images total, of which images that contain both one and two ventricles are present, and requires up to 60 seconds to segment one image which is much longer than the work by Martin et al.[36]

In 2022, Szentimrey et al. extended the 2D multiplane model by developing a 3D U-Net ensemble that considers all image voxels at once during training and is not slice-based.[38] This means more 3D features can be learned and areas more difficult to identify can be segmented more easily. The 3D U-Net ensemble is comprised of three individual models, these include a 3D Attention U-Net, 3D U-Net++ and 3D U-Net with shape prior. The results from each individual model are combined using a mean voting strategy and the final segmentation are post-processed by keeping the three largest segmented objects and removing the rest. A total of 190 images were used and they were separated by number of ventricles present (one or two ventricles). Their work is the first to use cross-validation, presenting results on the largest test set of its kind and only takes five seconds to segment one image. They prove that 3D deep learning models are faster and more accurate at segmenting 3D US images when compared to 2D deep-learning models.

3D US NEONATAL SEGMENTATION USING A 3D U-NET ENSEMBLE METHOD

The most recent and novel deep learning segmentation method was developed by Szentimrey et al. which improves upon the other works by being more accurate, faster, trained on the largest dataset of its kind and having the ability to work on images that only contains one ventricle.[38] The 3D US dataset and methods developed by Szentimrey et al. are discussed below as well as a discussion of the results.[38]

DESCRIPTION OF THE 3D US DATASET

The dataset consisted of 190 3D US images, of which 103 images contained only one lateral ventricle and 87 included both ventricles (a total of 277 ventricles).[38] The images were collected by a motorized 3D US system developed specifically for cranial imaging of neonates.[1] The transducers used include the HDI 5000 (Philips, Bothel WA) and C8-5 (Philips, Bothel WA) curved array 5–8 MHz transducers.[1] The scan angle was in the range of 30°–72° at 25 frames s^{-1} and angular spacing between 0.2° and 0.3°.[1] The 3D US scanner used a housing to hold the transducers with a Faulhaber IE2400 1409 DC motor with 2100:1 gear ratio (MicroMo Electronics, Clearwater FL) and a MCDC motor controller.[1] The Epiphan VGA2USB LR 2D US frame grabber (RB Computing, Nepean ON) was used.[1] In order to perform a 3D US scan, the motor controller tilts the transducer in a sweeping motion while the frame grabber acquires 2D images of the brain while in motion. Software developed at the University of Western Ontario reconstructs the 3D US images from 2D scans.[39] Bilinear interpolation is used to fill in the gaps between successive 2D US images to form the final 3D US image.[1] The most common voxel size in the images was $0.22 \times 0.22 \times 0.22 \text{mm}^3$ but images with varying voxel spacing were also used.[1] For all images, each voxel had the same volume per image but the voxel size may be different between images. During image acquisition, the technician or clinician locates the center of the target of interest while the device tilts the transducer on the axis at the tip of the probe.[1] Due to the limited FOV in the 3D US system, some images contained only one ventricle.[40] The dataset consisted of images from 30 patients with varying degrees of IVH, patients with PHVD and over multiple acquisition days. Each image had a manual segmentation of the lateral ventricles, where the images were manually segmented on the sagittal view at 1mm intervals using multi-planar reformatting software. These manual segmentations were used as ground truth when comparing segmentation methods.

The deep learning models tested by Szentimrey et al. used the same data split and the same images in each split to ensure a proper comparison can be made.[38] Separate models were made for images that contained one ventricle versus two ventricles. The first part of the study separated the images into train/validation/test sets then further testing was conducted using 5-fold cross-validation where the data was split into five folds with data left over for validation.

OVERVIEW OF SEGMENTATION PIPELINE

The 3D U-Net ensemble segmentation pipeline is shown in Figure 11.4. The data were pre-processed using normalization and down-sampling to create images with consistent size and image intensities between 0 and 1. Data augmentation techniques were applied to artificially increase the dataset size and counteract data imbalance. The ensemble was made from three U-Net-based models: a 3D U-Net with attention gates, a 3D U-Net with additional skip connections, and a 3D U-Net with an encoder-decoder shape prior embedded into the loss function. A mean voting strategy calculated the voxel-wise mean on the predicted segmentation masks from the three individual models. A threshold was applied and the largest objects in the mask were retained to remove potential outliers.

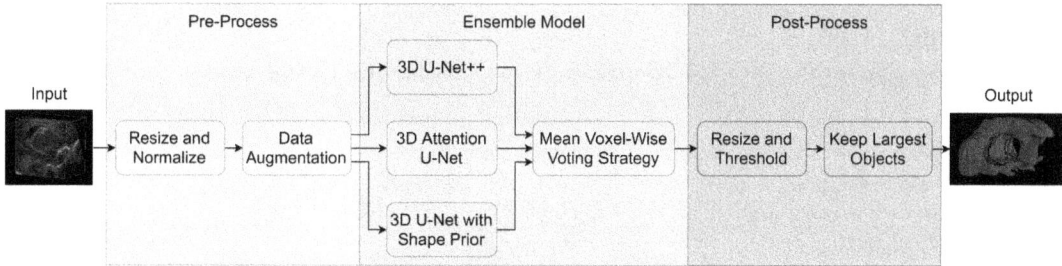

FIGURE 11.4   Deep-learning segmentation pipeline. The ensemble consists of a 3D U-Net++, 3D attention U-Net, and 3D U-Net with shape prior performing a mean voting strategy. During test time, data augmentation is not applied. (From Szentimrey et al. in *Medical Physics, vol. 49, issue 2.*[38])

3D U-Net Ensemble Model

The segmentation model is an ensemble of three 3D U-Net-based models combined using a voxel-wise mean voting strategy. The three models used in the ensemble are a 3D U-Net++, 3D Attention U-Net, and a 3D U-Net with a shape prior autoencoder embedded in the loss function. The ensemble selection was based on the desire to combine models that each add their own inductive biases. Ensemble selection is a difficult task as there are many possible combinations but only one may outperform the rest. Previous studies using ensembles have shown that ensembles between more diverse models tend to provide more unique information per model as compared to less diverse models.[41,42] The problem with solely choosing models based on diversity is that models which are less accurate can be very diverse so a trade-off between accuracy and diversity is necessary.[42]

Mean voting was used instead of majority voting because the output from each model was a probability map created from the sigmoid activation function on the last convolutional layer of each model. Much of the boundary surrounding the ventricles in US images is not clearly defined so the calculated segmentation masks had a value between 0 and 1 at the boundaries and not exactly a 0 or 1. By using mean voting, models with higher confidence will have a greater impact in determining voxel-wise classification.[41]

Hyperparameters

An outline of the hyperparameters used by Szentimrey et al. can be seen in Table 11.3.[38]

All the U-Net-based models used $3 \times 3 \times 3$ convolutional blocks with the number of filters being [32, 64, 128, 256, and 512] at each corresponding step in the model. In the contracting path, the $2 \times 2 \times 2$ maximum pooling operation with stride two was used. In the expanding path, $2 \times 2 \times 2$ transposed convolution operations were used. The convolution blocks were initialized with the He normal distribution and used the ReLU activation function.[43] Batch normalization and dropout layers were applied after each step to mitigate overfitting.[44] The final layer used a sigmoid activation function to return values between 0 and 1 and a global adaptive threshold was applied. All models were trained until the training and validation loss plateaued with the trained weights selected that produced the best validation performance. Each was trained with the Adam optimizer (learning rate 1e-4) and a loss function merging binary cross-entropy (BCE) and DSC between ground truth and predicted segmentations. The loss function was chosen as it decreased training time while maintaining performance compared to only using BCE or DSC. The exact loss function can be seen in Eq. (11.1)

$$\mathcal{L}(v) = 0.5 \left(\frac{-1}{N} \sum_{i=1}^{N} \left[y_i \log(\hat{y}_i) + (1 - y_i) \log(1 - \hat{y}_i) \right] \right) - \frac{2 \sum_{i=1}^{N} [y_i \hat{y}_i]}{\sum_{i=1}^{N} [y_i] + \sum_{i=1}^{N} [\hat{y}_i]} \qquad (11.1)$$

TABLE 11.3

Overview of the Final Hyperparameters Used for the U-Net Models

Type of Hyperparameter	Final Hyperparameter Value
Model depth	5 Steps Deep
Number of filter at each step	[32, 64, 128, 256, 512]
Convolutional kernel used	$3 \times 3 \times 3$ for all 3D models
Pooling operation	Maximum pooling
Up-sampling operation	Transposed convolution
Weight initialization	He normal distribution[43]
Convolutional layer activation function	Rectified linear unit (ReLU)
Final layer activation function	Sigmoid
Batch size	1 for all 3D models
Optimizer	Adam
Learning rate	1e-4
Loss Function	Combination of binary cross-entropy and Dice similarity coefficient (Eq. (11.1)) for all models except 3D U-Net with Shape Prior that used Eq. (11.2).

where N is the number of voxels in the current volume (v), y_i is the ground truth voxel value of voxel i, and $\hat{y}_i$ is the predicted voxel value of voxel i. The 3D U-Net with a shape prior did not use this loss function.

The autoencoder incorporated a similar model architecture to the U-Net, except it did not make use of skip connections and was only four steps deep. The autoencoder decoder and encoder used $3 \times 3 \times 3$ convolutional blocks with batch normalization and dropout layers. The number of filters per step was [8, 16, 32, and 64]. The contracting and expanding paths used $2 \times 2 \times 2$ maximum pooling with a stride of two and $2 \times 2 \times 2$ transposed convolution operations, respectively. The auto-encoder used the Adam optimizer (learning rate 1e-4) with the same BCE and DSC loss function, from Eq. (11.1), for training.

The 3D U-Net with a shape prior loss function combined the original DSC and BCE function with mean squared error (MSE) losses comparing the ground truth mask, model output mask, shape prior output mask, and encoded mask. The loss function was initially developed by Ravishankar et al. but adapted by Szentimrey et al. for 3D images.[38,45] The shape prior was introduced to the 3D U-Net model training after 10 epochs to ensure the model established a shape in its predicted mask before regularization occurred. The loss function is given by Eq. (11.2) where y_i and $\hat{y}_i$ are the ground truth and predicted value of voxel i respectively, $\hat{p}_i$ is the shape prior output and e_i and $\hat{e}_i$ are the encoded values of the ground truth and predicted masks respectively. λ_1, λ_2 and λ_3 are hyperparameters, which can be tuned, in the range of 0 to 1, to provide the optimal balance of regularization between the shape prior loss and original loss. The λ_1, λ_2, and λ_3 values for all experiments were given to be 0.5, 1, and 0.5 respectively for Eq. (11.2).

$$\mathcal{L}(v) = \lambda_1 \left(0.5 \ BCE\left(y_i, \hat{y}_i\right) - DSC\left(y_i, \hat{y}_i\right) \right) + \lambda_2 \left(\frac{1}{N} \sum_{i=1}^{N} \left| \hat{y}_i - \hat{p}_i \right| \right) + \lambda_3 \left(\frac{1}{N} \sum_{i=1}^{N} \left| e_i - \hat{e}_i \right| \right) \quad (11.2)$$

The masks generated from the models were automatically upscaled to their original image size (each image had a different size) using linear interpolation and a global adaptive threshold was applied. Upscaling allowed for volumetric and surface-based differences to be calculated between ground truth and predicted masks. In addition, a morphological operation was implemented to remove the smallest segmentation objects in the mask to improve distance-based metrics. The operation identifies the three largest objects in the image and removes all other objects.

SEGMENTATION RESULTS

Visualizations of example segmented surfaces are shown in Figure 11.5. The surface generated by the 3D U-Net ensemble method closely matched the shape of the manually generated surface, although this surface is not smooth due to the challenges in manual segmentation of 3D US images. The summary of quantitative results for 26 test images containing both ventricles are shown in Table 11.4. Table 11.5 shows the summary of quantitative results for 28 3D US images containing only a single ventricle. The results for fivefold cross-validation using 75 3D US test images containing both ventricles are shown in Table 11.6 and 90 3D US images from one ventricle are shown in Table 11.7. DSC, MAD, and VD were used for comparison between the ground truth segmentations and predicted segmentations. DSC is shown in Eq. (11.3) where N is the number of voxels in the current volume (v), y_i is the ground truth voxel value of voxel i, and $\hat{y}_i$ is the predicted voxel value of voxel i. DSC describes the degree of spatial overlap between two binary segmentation masks and is a common metric in image segmentation.[46] The MAD and VD metrics are useful measurements clinically, especially VD, which has been used for patients with PHVD.[3] The calculation for MAD is shown in Eq. (11.4) where n is the number of minimum distances between each voxel $\hat{s}$ on the predicted surface $\hat{S}$ and surface ground truth segmentation surface S. The distance function is based on the Euclidian distance $d(\hat{s}, S) = \min_{s \in S} \|\hat{s} - s\|_2$ where s is a point on ground truth surface S. For calculating VD, the size of each voxel is multiplied by the number of TP elements to yield the ground truth segmentation area (GT_{vol}) and the predicted segmentation are (P_{vol}) for the ground truth

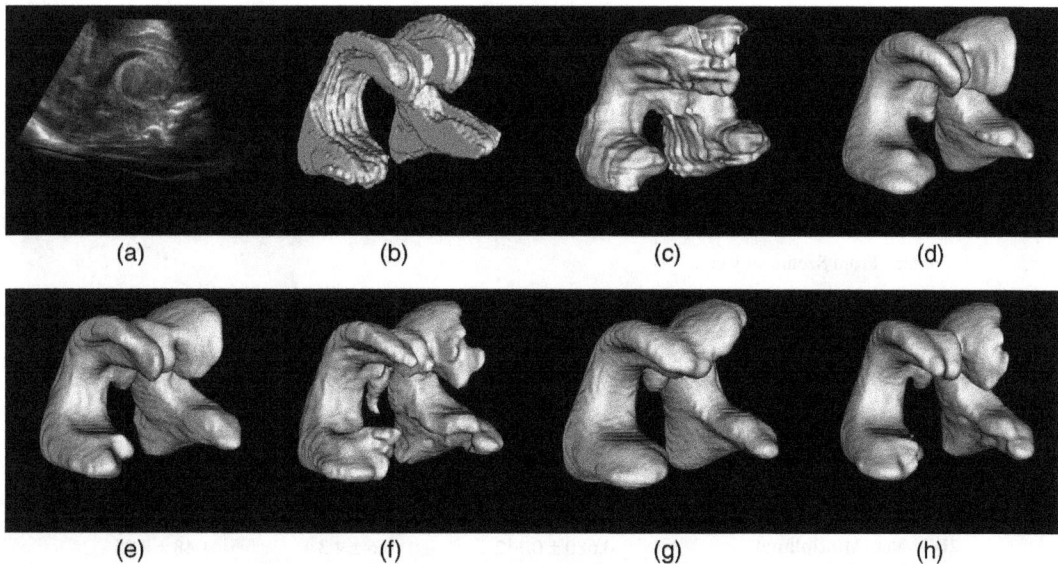

FIGURE 11.5   Examples of segmented surfaces from the two-ventricle cross-validation results. (a) The original 3D ultrasound image. (b) The manually segmented surfaces. (c) The surfaces using 2D multiplane U-Net with DSC of 0.719 and VD of 2.3 cm³. (d) The surfaces using 3D U-Net with DSC of 0.804 and VD of 1.9 cm³. (e) The surfaces using 3D attention U-Net with DSC of 0.838 and VD of 1.8 cm³. (f) The surfaces using 3D U-Net++ with DSC of 0.822 and VD of 1.2 cm³. (g) The surfaces using 3D U-Net with shape prior with DSC of 0.819 and VD of 1.0 cm³. (h) The surfaces using the 3D U-Net ensemble with DSC of 0.840 and VD of 1.0 cm³. (Figure 11.5a: From Szentimrey et al. In *Medical Physics, vol. 49, issue 2.*[38])

TABLE 11.4

Summary of the Results for the 26 Two-Ventricle Test Images. The Results Are Reported as DSC, VD, and MAD. The Values Reported Are the Mean and Standard Deviation from the Test Volumes

Model	DSC	VD (cm³)	MAD (mm)
2D SegNet	0.688 ± 0.084	2.9 ± 1.9	1.32 ± 0.46
2D U-Net (Multiplane)	0.708 ± 0.081	2.3 ± 2.4	1.26 ± 0.32
3D U-Net	0.704 ± 0.086	3.1 ± 3.1	1.25 ± 0.44
3D Attention U-Net	0.718 ± 0.063	2.6 ± 2.3	1.16 ± 0.31
3D U-Net++	0.725 ± 0.071	2.5 ± 2.0	1.15 ± 0.36
3D U-Net with Shape Prior	0.730 ± 0.060	2.4 ± 2.4	1.07 ± 0.30
3D Attention U-Net with Shape Prior	0.706 ± 0.071	2.8 ± 2.5	1.12 ± 0.30
3D U-Net++ with Shape Prior	0.704 ± 0.088	2.9 ± 2.5	1.13 ± 0.33
Ensemble (U-Net++, Attention U-Net, U-Net with Shape Prior)	**0.738 ± 0.062**	**2.3 ± 2.1**	**1.03 ± 0.29**

Source: From Szentimrey et al.[38]

TABLE 11.5

Summary of the Results for the 28 One-Ventricle Test Images. The Results Are Reported as DSC, VD, and MAD. The Values Reported Are the Mean and Standard Deviation from the Test Volumes

Model	DSC	VD (cm³)	MAD (mm)
2D U-Net (Multiplane)	0.787 ± 0.105	2.6 ± 2.2	2.16 ± 1.60
3D U-Net	0.816 ± 0.090	2.6 ± 2.2	1.94 ± 2.34
3D Attention U-Net	0.827 ± 0.113	2.6 ± 1.6	**1.12 ± 0.73**
3D U-Net++	0.810 ± 0.075	2.3 ± 1.5	1.65 ± 1.25
3D U-Net with Shape Prior	0.823 ± 0.088	2.7 ± 2.5	1.26 ± 0.70
Ensemble	**0.845 ± 0.051**	**2.1 ± 1.5**	1.19 ± 0.88

Source: From Szentimrey et al.[38]

TABLE 11.6

Summary of the Results for 75 3D US Test Images with Two-Ventricle Images Using Fivefold Cross-Validation. The Values Reported Are the Mean and Standard Deviation

Model	DSC	VD (cm³)	MAD (mm)
2D U-Net (Multiplane)	0.680 ± 0.112	4.6 ± 4.3	1.48 ± 0.65
3D U-Net	0.674 ± 0.123	4.1 ± 4.0	1.17 ± 0.44
3D Attention U-Net	0.681 ± 0.100	4.4 ± 5.1	1.26 ± 0.59
3D U-Net++	0.711 ± 0.075	3.9 ± 3.8	1.22 ± 0.42
3D U-Net with Shape Prior	0.676 ± 0.103	4.5 ± 4.8	1.15 ± 0.39
Ensemble	**0.720 ± 0.074**	**3.7 ± 4.1**	**1.14 ± 0.41**

Source: From Szentimrey et al.[38]

TABLE 11.7

Summary of the Results for 90 3D US Test Images with One-Ventricle Images Using a Fivefold Cross-Validation. The Values Reported Are the Mean and Standard Deviation

Model	DSC	VD (cm³)	MAD (mm)
2D U-Net (Multiplane)	0.776 ± 0.123	**3.3 ± 2.9**	1.73 ± 1.53
3D U-Net	0.790 ± 0.142	4.1 ± 3.7	1.75 ± 3.29
3D Attention U-Net	0.796 ± 0.127	3.9 ± 3.1	**1.36 ± 1.31**
3D U-Net++	0.793 ± 0.118	3.4 ± 3.1	1.57 ± 1.68
3D U-Net with Shape Prior	0.790 ± 0.135	4.5 ± 4.5	1.69 ± 3.51
Ensemble	**0.806 ± 0.111**	3.5 ± 2.9	**1.37 ± 1.70**

Source: From Szentimrey et al.[38]

and predicted masks, respectively. The absolute difference between these volumes is then calculated as shown in Eq. (11.5).

$$DSC(v) = \frac{2\sum_{i=1}^{N}[y_i \hat{y}_i]}{\sum_{i=1}^{N}[y_i] + \sum_{i=1}^{N}[\hat{y}_i]} \tag{11.3}$$

$$MAD(v) = \frac{1}{n}\sum_{i=1}^{n}|d(\hat{s}, S)| \tag{11.4}$$

$$VD = |GT_{vol} - P_{vol}| \tag{11.5}$$

In addition to these metrics, two statistical tests were conducted by Szentimrey et al.[38] on the cross-validation results to compare performance statistically. These tests included the paired t-test used to compare VD measurement and the non-parametric Wilcoxon signed-rank test, used for DSC comparison.[47] To determine if the difference in models' performance is statistically significant, $p < 0.05$ was used to determine significance.

A fast, accurate, and fully automated method for the segmentation of neonatal cerebral lateral ventricles was developed by Szentimrey et al. using a deep-learning-based ensemble of three 3D U-Net variant models.[38] Developing a method that can provide clinicians with automated and fast quantitative information is useful for the prognosis of neonates with IVH and PHVD. Compared to other studies, a large number of images were used as well as images with one ventricle, representing the adaptability of the developed method. It was determined that each variant 3D model outperformed the standard 3D U-Net model and the 3D U-Net ensemble had the best overall performance.

The performance of all the models during cross-validation decreased when compared to the single run tests. The 3D U-Net ensemble model performed best in DSC, MAD, and VD on the two-ventricle images and the best in DSC and MAD for the one-ventricle images. The ensemble was only bested by the 2D multiplane U-Net with regards to VD on the one-ventricle images but it was found that there was not a significant difference between the VD for the ensemble and 2D multiplane U-Net ($p < 0.20$).

DISCUSSION AND CONCLUSION

The first segmentation methods developed for neonatal cerebral lateral ventricles focused on semi-automated and fully automated non-deep-learning methods. The semi-automated methods were based on a convex optimization method using pre-labeled voxels from a clinician as a starting point.[31–33] Next, a fully automated method was developed using a multi-phase geodesic level-sets method which removed the requirement for a user to pre-label voxels; however, the accuracy yield from this method was lower than semi-automated ones.[23,34] The fully automated method also required more time at 54 minutes to produce one segmentation output.

More recently fully automated deep-learning methods were developed to improve segmentation time and accuracy of neonatal cerebral ventricles when compared to the non-deep-learning methods. The first such work used a 2D U-Net model on only 15 images.[36] This improved the segmentation time compared to previous works but introduced the need for more training images when developing deep-learning models. Next a 2D SegNet and 2D multiplane U-Net models were developed by Gontard et al. and Szentimrey et al., respectively.[35,37] These methods used over 100 images to train their models, achieved reasonable performance compared to non-deep-learning methods and required less than 2 min to segment one image. The problem with these methods is that their models are only 2D and interpret images slice-wise and not on the entire volumetric image. The most recent fully automated segmentation work used a 3D U-Net ensemble method that required only 5 s to segment one image and used the largest dataset of its kind.[38] The reported accuracy was the highest among recent works and was shown to outperform 2D models.

Deep learning has greatly improved the segmentation of neonatal cerebral lateral ventricles from 3D US images but more work is needed. The 3D U-Net ensemble method required much computation power and trainable parameters which can limit its deployment in a clinical setting. In addition, deep-learning models require much labeled data to be effective. For future improvement, semi-supervised areas could be explored that can incorporate unlabeled images into the training of deep-learning models. Semi-supervised learning means simpler models can be used while also requiring less labeled images, increasing the viability of using deep-learning models in a clinical setting. The use of these models in a clinical setting would allow for quantitative and visual support for physicians and clinicians resulting in improved treatment planning and monitoring of patient conditions.

REFERENCES

1. Kishimoto J, de Ribaupierre S, Lee DSC, Mehta R, St Lawrence K, Fenster A. 3D ultrasound system to investigate intraventricular hemorrhage in preterm neonates. *Phys Med Biol.* 2013;58(21):7513–7526. doi:10.1088/0031-9155/58/21/7513
2. Synnes AR, Chien LY, Peliowski A, Baboolal R, Lee SK. Variations in intraventricular hemorrhage incidence rates among Canadian neonatal intensive care units. *J Pediatr.* 2001;138(4):525–531. doi:10.1067/mpd.2001.111822
3. Kishimoto J, Fenster A, Lee DSC, de Ribaupierre S. Quantitative 3-D head ultrasound measurements of ventricle volume to determine thresholds for preterm neonates requiring interventional therapies following posthemorrhagic ventricle dilatation. *J Med Imaging.* 2018;5(02):1. doi:10.1117/1.jmi.5.2.026001
4. Haiden N, Klebermass K, Rücklinger E, et al. 3-D ultrasonographic imaging of the cerebral ventricular system in very low birth weight infants. *Ultrasound Med Biol.* 2005;31(1):7–14. doi:10.1016/j.ultrasmedbio.2004.07.017
5. Kishimoto J, de Ribaupierre S, Salehi F, Romano W, Lee DSC, Fenster A. Preterm neonatal lateral ventricle volume from three-dimensional ultrasound is not strongly correlated to two-dimensional ultrasound measurements. *J Med Imaging.* 2016;3(4). doi:10.1117/1.jmi.3.4.046003
6. Brouwer MJ, De Vries LS, Pistorius L, Rademaker KJ, Groenendaal F, Benders MJ. Ultrasound measurements of the lateral ventricles in neonates: Why, how and when? A systematic review. *Acta Paediatr Int J Paediatr.* 2010;99(9):1298–1306. doi:10.1111/j.1651-2227.2010.01830.x
7. Qiu W, Yuan J, Rajchl M, et al. 3D MR ventricle segmentation in pre-term infants with post-hemorrhagic ventricle dilatation (PHVD) using multi-phase geodesic level-sets. *Neuroimage.* 2015;118:13–25. doi:10.1016/j.neuroimage.2015.05.099

8. Kishimoto J, Fenster A, Lee DSC, de Ribaupierre S. In vivo validation of a 3-d ultrasound system for imaging the lateral ventricles of neonates. *Ultrasound Med Biol.* 2016;42(4):971–979. doi:10.1016/j.ultrasmedbio.2015.11.010

9. Martin JA, Hamilton BE, Osterman MJK, Driscoll AK. *Births: Final Data for 2019.* Vol 70.; 2019.

10. Ballabh P. Intraventricular hemorrhage in premature infants: Mechanism of disease. *Pediatr Res.* 2010;67(1):1–8. doi:10.1203/PDR.0b013e3181c1b176

11. Stratchko L, Filatova I, Agarwal A, Kanekar S. The ventricular system of the brain: Anatomy and normal variations. *Semin Ultrasound, CT MRI.* 2016;37(2):72–83. doi:10.1053/j.sult.2016.01.004

12. Ahn SY, Shim S-Y, Sung IK. Intraventricular hemorrhage and post hemorrhagic hydrocephalus among very-low-birth-weight infants in Korea. *J Korean Med Sci.* 2015;30:S52–S58. doi:10.3346/jkms.2015.30.S1.S52

13. Mohammad K, Scott JN, Leijser LM, et al. Consensus approach for standardizing the screening and classification of preterm brain injury diagnosed with cranial ultrasound: A Canadian perspective. *Front Pediatr.* 2021;9:618236:1–19. doi:10.3389/fped.2021.618236

14. Klebermass-Schrehof K, Rona Z, Waldhör T, et al. Can neurophysiological assessment improve timing of intervention in posthaemorrhagic ventricular dilatation? *Arch Dis Child Fetal Neonatal Ed.* 2013;98(4):291–297. doi:10.1136/archdischild-2012-302323

15. El-Atawi K. Risk factors, diagnosis, and current practices in the management of intraventricular hemorrhage in preterm infants: A review. *Acad J Pediatr Neonatol.* 2016;1(3):1–7. doi:10.19080/ajpn.2016.01.555561

16. Whitelaw A. Intraventricular haemorrhage and posthaemorrhagic hydrocephalus: Pathogenesis, prevention and future interventions. *Semin Neonatol.* 2001;6(2):135–146. doi:10.1053/siny.2001.0047

17. Paneth N, Pinto-Martin J, Gardiner J, et al. Incidence and timing of germinal Matrix/Intraventricular hemorrhage in low birth weight infants. *Am J Epidemiol.* 1993;137(11):1167–1176. doi:10.1176/appi.psychotherapy.1970.24.4.545

18. Dorner RA, Burton VJ, Allen MC, Robinson S, Soares BP. Preterm neuroimaging and neurodevelopmental outcome : A focus on intraventricular hemorrhage, post-hemorrhagic hydrocephalus, and associated brain injury. *J Perinatol.* 2018;38:1431–1443. doi:10.1038/s41372-018-0209-5

19. Beijst C, Dudink J, Wientjes R, et al. Two-dimensional ultrasound measurements vs. magnetic resonance imaging-derived ventricular volume of preterm infants with germinal matrix intraventricular haemorrhage. *Pediatr Radiol.* 2020;50(2):234–241. doi:10.1007/s00247-019-04542-x

20. Arthurs OJ, Edwards A, Austin T, Graves MJ, Lomas DJ. The challenges of neonatal magnetic resonance imaging. *Pediatr Radiol.* 2012;42:1183–1194. doi:10.1007/s00247-012-2430-2

21. Davies MW, Swaminathan M, Chuang SL, Betheras FR. Reference ranges for the linear dimensions of the intracranial ventricles in preterm neonates. *Arch Dis Child Fetal Neonatal Ed.* 2000;82(3):218–223. doi:10.1136/fn.82.3.f218

22. Brouwer MJ, De Vries LS, Groenendaal F, et al. New reference values for the neonatal cerebral ventricles. *Radiology.* 2012;262(1):224–233. doi:10.1148/radiol.11110334

23. Qiu W, Chen Y, Kishimoto J, et al. Automatic segmentation approach to extracting neonatal cerebral ventricles from 3D ultrasound images. *Med Image Anal.* 2017;35:181–191. doi:10.1016/j.media.2016.06.038

24. Campbell S. A short history of sonography in obstetrics and gynaecology. *Facts, Views Vis ObGyn.* 2013;5(3):213–229.

25. Huang Q, Zeng Z. A review on real-time 3D ultrasound imaging technology. *Biomed Res Int.* 2017;2017:1–20. doi:10.1155/2017/6027029

26. Ukwatta E, Awad J, Ward AD, et al. Three-dimensional ultrasound of carotid atherosclerosis: Semiautomated segmentation using a level set-based method. *Med Phys.* 2011;38(5):2479–2493. doi:10.1118/1.3574887

27. Qiu W, Yuan J, Ukwatta E, Tessier D, Fenster A. Three-dimensional prostate segmentation using level set with shape constraint based on rotational slices for 3D end-firing TRUS guided biopsy. *Med Phys.* 2013;40(7):1–12. doi:10.1118/1.4810968

28. Neshat H, Cool DW, Barker K, Gardi L, Kakani N, Fenster A. A 3D ultrasound scanning system for image guided liver interventions. *Med Phys.* 2013;40(11):1–13. doi:10.1118/1.4824326

29. Romero JM, Madan N, Betancur I, et al. Time efficiency and diagnostic agreement of 2-D versus 3-D ultrasound acquisition of the neonatal brain. *Ultrasound Med Biol.* 2014;40(8):1804–1809. doi:10.1016/j.ultrasmedbio.2014.03.013

30. Abdul-Khaliq H, Lange PE, Vogel M. Feasibility of brain volumetric analysis and reconstruction of images by transfontanel three-dimensional ultrasound. *J Neuroimaging.* 2000;10(3):147–150. doi:10.1111/jon2000103147

31. Qiu W, Yuan J, Kishimoto J, Ukwatta E, Fenster A. Lateral ventricle segmentation of 3D pre-term neo-nates US using convex optimization. *Med Image Comput Comput Interv – MICCAI 2013*. 2013;8151:559–566. doi:10.1007/978-3-642-40760-4_70

32. Qiu W, Yuan J, Kishimoto J, Ribaupierre S, De, Ukwatta E, Fenster A. Semi-automatic segmentation of preterm neonate ventricle system from 3D ultrasound images. 2014 IEEE 11th Int Symp Biomed Imaging, ISBI 2014. 2014:1222–1225.

33. Qiu W, Yuan J, Kishimoto J, et al. User-guided segmentation of preterm neonate ventricular system from 3-d ultrasound images using convex optimization. *Ultrasound Med Biol*. 2015;41(2):542–556. doi:10.1016/j.ultrasmedbio.2014.09.019

34. Qiu W, Yuan J, Kishimoto J, et al. Automatic 3D US brain ventricle segmentation in pre-term neonates using multi-phase geodesic level-sets with shape prior. *Med Image Comput Comput Interv – MICCAI 2015*. 2015;9351:89–96. doi:10.1007/978-3-319-24574-4_11

35. Gontard LC, Pizarro J, Peña BS, López SPL. Automatic segmentation of ventricular volume by 3D ultrasonography in post haemorrhagic ventricular dilatation among preterm infants. *Sci Rep*. 2021;11(567):1–13. doi:10.1038/s41598-020-80783-3

36. Martin M, Sciolla B, Sdika M, Wang X, Quetin P, Delachartre P. Automatic Segmentation of the Cerebral Ventricle in Neonates Using Deep Learning with 3D Reconstructed Freehand Ultrasound Imaging. *2018 IEEE Int Ultrason Symp*. 2018:1–4. doi:10.1109/ULTSYM.2018.8580214

37. Szentimrey Z, de Ribaupierre S, Fenster A, Ukwatta E. Automatic deep learning-based segmentation of neonatal cerebral ventricles from 3D ultrasound images. *Proc SPIE 11600, Med Imaging 2021 Biomed Appl Mol Struct Funct Imaging*. 2021:1–7. doi:10.1117/12.2581749

38. Szentimrey Z, de Ribaupierre S, Fenster A, Ukwatta E. Automated 3D U-net based segmentation of neo-natal cerebral ventricles from 3D ultrasound images. *Med Phys*. 2022;49(2):1034–1046. doi: 10.1002/mp.15432

39. Fenster A, Downey DB, Cardinal HN. Three-dimensional ultrasound imaging. *Phys Med Biol*. 2001;46:R67–99.

40. Harris A, de Ribaupierre S, Gardi L, Fenster A, Kishimoto J. Automated registration and stitching of multiple 3D ultrasound images for monitoring neonatal intraventricular hemorrhage. SPIE 10580, Med Imaging 2018 Ultrason Imaging Tomogr. 2018:1–7. doi:10.1117/12.2292925

41. Sagi O, Rokach L. Ensemble learning: A survey. *Wiley Interdiscip Rev Data Min Knowl Discov*. 2018;8:1–18. doi:10.1002/widm.1249

42. Gong Z, Zhong P, Hu W. Diversity in machine learning. *IEEE Access*. 2019;7:64323–64350. doi:10.1109/ACCESS.2019.2917620

43. He K, Zhang X, Ren S, Sun J. Delving Deep into Rectifiers: Surpassing Human-Level Performance on ImageNet Classification. *2015 IEEE Int Conf Comput Vis*. 2015:1026–1034.

44. Ioffe S, Szegedy C. Batch Normalization: Accelerating Deep Network Training by Reducing Internal Covariate Shift. *Proc 32nd Int Conf Mach Learn PMLR*. 2015;37:448–456. doi:10.1080/17512786.2015.1058180

45. Ravishankar H, Venkataramani RTS, Sudhakar P, Vaidya V. Learning and incorporating shape models for semantic segmentation. *Med Image Comput Comput Assist Interv – MICCAI 2017 MICCAI 2017 Lect Notes Comput Sci*. 2017;10433:203–211. doi: 10.1007/978-3-319-66182-7_24

46. Zou KH, Warfield SK, Bharatha A, et al. Statistical validation of image segmentation quality based on a spatial overlap index. *Acad Radiol*. 2004;11(2):178–189. doi:10.1016/S1076-6332(03)00671-8

47. Wilcoxon F. Individual comparisons by ranking methods. *Biometrics Bull*. 1945;1(6):80–83. doi:10.1093/jee/39.2.269

12 A Review on the Applications of Convolutional Neural Networks in Ultrasound-Based Prostate Brachytherapy

Jing Wang, Tonghe Wang, Tian Liu, and Xiaofeng Yang

INTRODUCTION

Prostate cancer was the most commonly diagnosed cancer for males in 12 regions of the world and caused more than 375,000 deaths worldwide in 2020 [1]. The treatment of prostate cancer involves radical prostatectomy and radiotherapy, e.g., external beam radiotherapy (EBRT), brachytherapy, or a combination of them [2]. Brachytherapy delivers doses to a highly localized region, taking the forms of a permanent low-dose-rate (LDR) seed implant or a high-dose-rate (HDR) afterloading. HDR prostate brachytherapy has been used as a monotherapy to treat low-risk or low-to-intermediate-risk prostate cancer [3–6], or it serves as a boost treatment in combination with EBRT for intermediate-to-high-risk patients, though some other studies revealed that HDR could be used as monotherapy even for intermediate- and high-risk patients with good biomedical control rates [7–11]. As a critical treatment option for prostate cancer, brachytherapy is a minimally invasive surgery that works by placing radioactive seeds/source in the prostate through catheters, sometimes called needles. Taking HDR prostate brachytherapy as an example, the treatment process involves: (1) dose pre-planning and predicting the number of needles/catheters to be implanted; (2) ultrasound (US)-guided needle insertion into prostate under anesthesia; (3) CT and/or MRI scanning for patient anatomy and needle locations; (4) treatment planning for radioactive source dwell positions and durations within needles; and (5) treatment delivery via a remote afterloader. In HDR treatment, the radioactive sources are of high doses and they are withdrawn after treatment; while for an LDR therapy, the radioactive seeds will be inserted into prostate and permanently left inside patient. Brachytherapy is expected to provide highly localized doses that is conformal to target lesion but spares organs-at-risk (OARs) within and adjacent prostate.

Transrectal ultrasound (TRUS) is widely adopted as an intraoperative imaging tool in both prostate cancer diagnosis [12] and prostate brachytherapy [13] stages. A probe is placed inside the rectum of patient to obtain real-time images of prostate and adjacent organs such as bladder, urethra, and rectum. A 3D TRUS image is usually obtained by reconstruction from 2D transverse-plane TRUS images by moving probe from prostate apex to base using a mechanic stepper [14], though several studies also reported the usage of sagittal plane images by rotation [15, 16] or a combination of both planes [17]. 3D TRUS plays an important role in prostate cancer lesion detection [18, 19], biopsy needle guidance [20, 21], pre-implant planning, and needle guidance in the operating room (OR) for brachytherapy [22, 23]. Besides the spatial 3D TRUS, another application called temporal-enhanced ultrasound (TeUS) has emerged as a new paradigm for prostate tissue characterization, which consists of a sequence of TRUS images taken within a continuous time period at a fixed position [18]. TeUS has demonstrated the feasibility in detecting dominant intraprostatic lesions (DILs), so it could potentially be integrated into the intraoperative planning workflow in OR to deliver a focal boost on DILs [24].

DOI: 10.1201/9781003299462-15

Though TRUS shows unique advantages of nonionizing and real-time imaging, it suffers from low image quality (e.g., image speckles, artifacts, and poor contrast) compared to other imaging modalities like CT or MRI, which poses great difficulty to segment prostate, OARs, and needles or seeds in 3D TRUS images. Those contours are of critical importance in generating treatment plan. The low image quality makes manual contouring more laborious and more reliant on the expertise level and experience of physicians. Image fusion is proposed to register the preprocedural multi-parametric MR images (or other high-quality imaging modalities containing anatomy and physiology information) to intraoperative 3D TRUS images for prostate patients to leverage the strength of image-guided needle insertion [25–28]. However, it requires a high-performance deformable registration method that is challenging between multiple modalities.

With the development of computer technology, more sophisticated and powerful artificial intelligence (AI) tools have been developed in medicine, including the popular machine learning (ML) algorithms [29]. Deep learning (DL) is a branch of ML developed by adding more hidden layers to the original artificial neural networks (ANNs). Compared to handcraft feature selection methods, DL adaptively extracts deep features from medical images to learn representations with multiple levels of abstraction and demonstrates superior potential in computer vision [30]. Various DL architectures exist (convolutional neural networks (CNNs) [31], recurrent neural networks (RNNs) and long short-term memory (LSTM) [32], encoder-decoders [33], and generative adversarial networks (GANs) [34]), among which CNNs are the most successful and widely used for computer visions [35]. CNNs use convolutional filters to scan the whole feature matrix and reduce dimensionality, suitable to deal with spatial problems such as images. For 3D TRUS-based prostate brachytherapy, CNNs can serve as a promising tool for automatic organ segmentation and needle detection to liberate physicians from tedious contouring labors, as well as imaging registration and tissue characterization to provide physiology knowledge of lesions. In this chapter, we will discuss state-of-the-art applications of CNNs in 3D TRUS-based prostate brachytherapy as of July 1, 2022. We have searched the Pubmed and Google Scholar databases with a variety of keywords including, but not limited to, prostate brachytherapy, 3D US, TRUS, TeUS, DL, and CNNs and so on. We collected more than 60 publications from the search and found 27 of them closely relevant to the scope of this chapter, mostly published after 2018. Based on these papers, we group the addressed tasks into four main categories: (1) prostate and OAR segmentation, (2) multi-needle detection, (3) MR to TRUS image registration, and (4) prostate cancer DILs detection with TeUS. With this survey, we aim to:

- Summarize the most recent development in prostate brachytherapy tasks addressed by CNNs
- Comprehensively review variant CNN architectures used for each specific task
- Provide an insightful view of the challenges and outlooks of using CNNs in prostate brachytherapy.

The remainder of the chapter is organized as follows: the second section provides an overview of CNN algorithms and several commonly used CNN architectures detailed in the papers. The third section provides a comprehensive review of the state-of-the-art progress of CNNs in 3D TRUS-based prostate therapy, grouped to four main tasks. In the fourth section, we discuss the challenges and future directions for CNN applications in brachytherapy.

OVERVIEW OF CNN ARCHITECTURES

A convolutional layer is the core element in the CNN concept. It extracts feature maps from the input vectors (such as images or lower-level feature maps) through the convolving of kernels (filters) [36]. Kernels, or filters, are smaller vectors that depict a particular feature pattern with fixed sets of weights and biases for each neuron. Multiple filters can be generated to capture various aspects of image features. For convolution operations of stride length greater than 1, the output higher level feature maps are of reduced dimensionality compared to the input. Rectified linear unit (ReLU) and its variants are now used as most common activation functions for CNNs [37]. ReLU sets all negative elements in the feature map to zeros and preserves the positive elements. In this way, ReLU activation introduces nonlinearity and sparsity to the model, and makes it relatively easy to compute derivatives. The pooling layer is a down-sampling operation that a defined neighborhood zone in the feature map shrinks to a single value, the maximum or averaged value within that neighborhood.

CONVENTIONAL CNNs

Conventional CNNs are a type of variants of the neural networks combining two basic blocks: the convolutional block and fully connected (FC) block (Figure 12.1). The convolutional block consists of convolutional layer and the pooling layer, which essentially functions as a feature-extraction component.

Next to the deepest convolutional block, there is usually a flatten layer to flat out the pooled feature maps/matrices into a one-dimensional (1D) vector of hidden neurons, as the input fed into the subsequent FC block. The FC block is a fully connected simple neural network where each neuron in the previous layer is connected to all neurons in the next layer, and each neuron in the next layer is fully connected to all activations in the previous layer. Dropout layers can be used to nullify some neurons to prevent overfitting. A Softmax layer is added in the last step to generate classification labels, which are compared with ground truth to compute the loss function. Cross entropy loss is among the most widely used loss functions. When training the model, stochastic gradient descent (SGD) and Adam gradient descent optimizations are commonly used to optimize parameters by back-propagation.

A CNN architecture was first developed for document recognition by Lecun et al. namely LeNet in late 1990s [31], consisting of the basic convolutional layers, pooling layers, and FC layers. By extracting multi-resolution pyramid feature layers, higher level features represent larger receptive fields, so that the number of parameters fed into the FC block is much smaller than that of feeding the original input images to a FC NN, dramatically reducing the computational cost. Since the LeNet, several well-known CNNs such as AlexNet [38], VGGNet [39], ResNet [40], GoogLeNet [41], MobileNet [42], and DenseNet [43] have been developed and have achieved great success in various tasks.

FULLY CONVOLUTIONAL NETWORKS

As discussed above, the conventional CNNs take images as input and output classification labels for the whole input patches. But for the main DL tasks in 3D TRUS-based prostate brachytherapy (also in many other medicine fields), the organ/needle segmentation and image registration can be better fulfilled via a dense voxel-wise output instead of a single patch-wise label. The fully convolutional

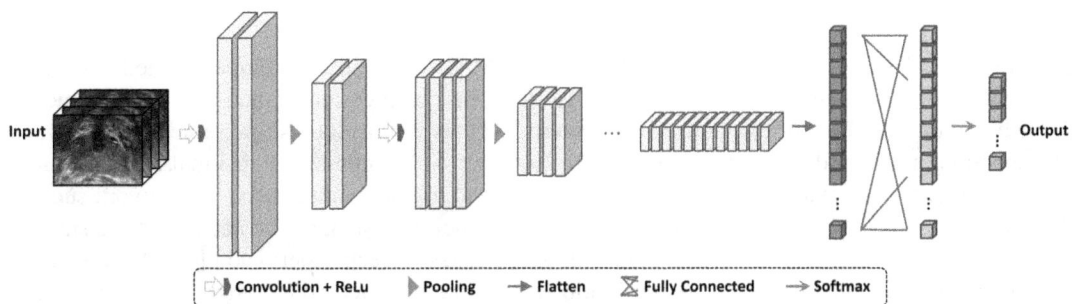

FIGURE 12.1 Schematic diagram of CNNs.

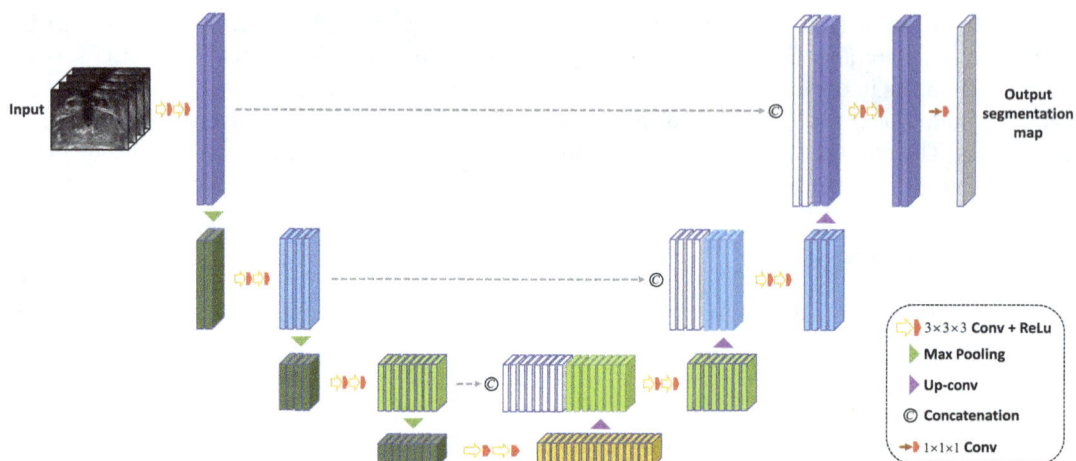

FIGURE 12.2 Schematic diagram of U-Net architecture, depicting a contracting and expanding path.

networks (FCNs) are developed to output a dense voxel-wise prediction map of the same size by replacing the fully connected layers with convolutional networks [44]. In this way, the segmentation of the whole images can be obtained in one forward pass, namely end-to-end segmentation.

U-Net

U-Net is one of the most well-known modified FCN architectures, especially successful in medical image segmentation and registration. U-Net is based on the important concept of deconvolution proposed by Ronneberger et al. in 2015 [45]. The model got the name of "U-Net" due to its elegant symmetry of U-shaped architecture (Figure 12.2). The left branch is a contracting encoding path similar to a typical CNN, and the number of feature channels doubled at each down-sampling step. The right branch is the expanding decoding path that recover feature maps to the same resolution level of input image. At each expansive step, the feature maps experienced up-sampling by deconvolution (up-convolution) that halves the number of channels, concatenation with feature maps at the corresponding level in encoding path, and two following convolutions. Through the carefully designed encoding-decoding path, the final output feature maps are of the same resolution level as the input images, and then a layer of 1×1 convolution is used to assign each voxel in final feature maps to ground truth classes. Moreover, the skip-connection (concatenation) from encoding path brings the high-resolution location information to each up-sampling step, so the U-Net model can rebuild a more precise output. U-Net is the most commonly used backbone architecture in the DL papers for 3D TRUS-based prostate brachytherapy, as we will see in the third section.

V-Net

Inspired by U-Net, V-Net is specifically designed by Milletari et al. for volumetric medical image segmentation [46]. Similar to U-Net, V-Net also has one compression branch and one expanding branch connected with a bridge path at the bottom level, as well as long skip-connections between the encoding and decoding paths (Figure 12.3). However, V-Net contains a residual learning function [40] at each level that ensures convergence compared to non-residual learning network such as U-Net. Besides, V-Net employs convolution to replace pooling layers, so the network has a smaller memory footprint during training. For relatively small dataset, deep supervision [47, 48] was incorporated to compute predictions at each resolution level, and then merged for a hybrid loss regression with respect to the ground truth. In this way, the deeply supervised V-Net directly supervises the hidden layers of multiple resolution levels and increases its discriminative ability for difficult tasks.

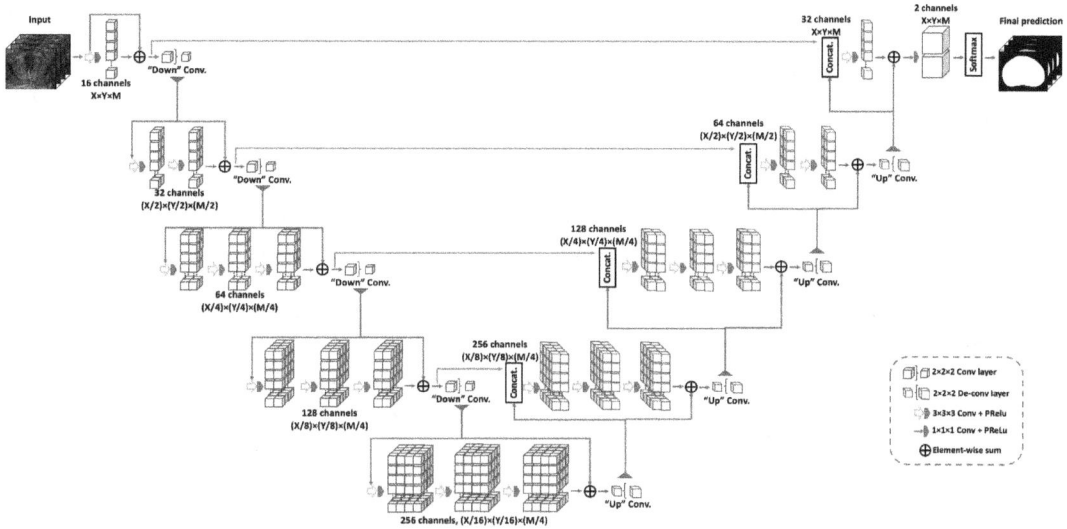

FIGURE 12.3 Schematic diagram of V-Net architecture. Note the element-wise sum operation of the residual block in each level.

Feature Pyramid Network

In both U-Net and V-Net, the classification task is performed at the outermost layer, reconstructed to the highest resolution level. However, multiple layer feature maps are generated in pyramid representations during the encoding path. To leverage the inherent multi-scale, pyramidal hierarchy of CNNs, feature pyramid network (FPN, Figure 12.4) was proposed by Lin et al. to construct feature pyramids with marginal cost [49]. The first branch is a bottom-up CNN architecture that generates feature maps at multiple resolution levels. The second branch is a top-down pathway that reconstructs higher resolution features by up-sampling, merged with lateral connection via element-wise addition. Unlike U-Net/V-Net that only make predictions on the finest level, FPN makes independent predictions for all resolution levels. In this way, FPN reuses the multi-scale features computed in the forward pass without much extra computation, for different levels.

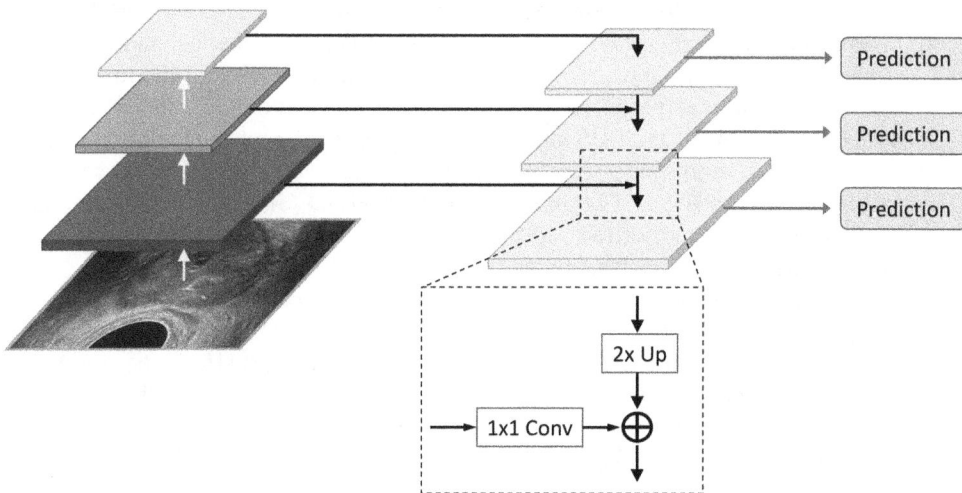

FIGURE 12.4 Schematic diagram of FPN architecture. A building block illustrates the merge of lateral connection and top-down pathway.

THREE-DIMENSIONAL (3D) TRUS-BASED PROSTATE BRACHYTHERAPY

Enormous efforts of using computer or AI tools to aid 3D TRUS-based prostate brachytherapy have been made and various automation algorithms have been developed without the application of DL (CNNs).

Needle detection seems an extensively studied task since 2005 [50]. At early stage, needles were detected by generating difference map between a pre-scan before the needle insertion and a post-scan after the needle insertion, followed by post-processing such as thresholding or a tri-bar model to obtain needle voxel clusters [50, 51]. Yan et al. used the shape information and level set method to segment needles and detect their endpoints [52]. Qiu et al. first cropped out a 3D region of interest (ROI) that contained one needle and converted it to a binary image based on intensity histogram. A 3D Hough transform was then applied to locate the needle axis, and finally determined the endpoint with optimal threshold based on its intensity probability distribution, achieving a needle endpoint localization accuracy <1.43 mm and detection accuracy >84% [53, 54]. In 2017, Hrinivich et al. employed regularized feature point classification and the randomized 3D Hough transform to extract needle trajectory, and the peaks of derivatives of trajectory intensity to define the needle tips. 190 needles from 12 prostate cancer patients were investigated. The algorithm identified 82% and 85% of needle tips with 3D errors ≤3 mm and ≤5 mm [15, 55]. In 2018, Dehghan et al. integrated the electromagnetic (EM) tracking into LDR prostate treatment by implanting seeds with an EM-tracked needle that records the seed drop locations and then fused the EM-tracking and TRUS images to detect seed locations [56]. More recently, Younes et al. detected radioactive seeds in real-time US using a Bayesian classifier and a support vector machine (SVM), achieving a mean error of 1.44 ± 0.45 mm on 10 patients [57].

Other tasks such as prostate segmentation, tissue characterization, and 3D image reconstruction were also studied. In 2016, Li et al. introduced an active band term in the level-set energy functional to distinguish the weak boundary of prostate in US images, by restricting only pixels around prostate boundaries to influence the evolution of level set, with an edge descriptor to avoid unwanted edges [58]. On the other hand, Yang et al. proposed a CT prostate segmentation method based on TRUS contours [59]. The contoured prostate volume on a 3D TRUS image was acquired right after catheter insertion. Then using the reconstructed needles as landmarks, a landmark similarity-based nonrigid registration of TRUS-CT was performed. The prostate volume is thus deformed to the CT images and readily available for treatment planning. Nahlawi et al. employed a probabilistic-temporal framework, namely, hidden Markov models (HMMs), to exploit the temporal features of TeUS to characterize prostate tissue. On an experiment of 284 malignant and 286 benign TeUS signals from 9 patients, they achieved 0.85 accuracy [60], while on another expanded dataset of 12 patients, the malignancy detection demonstrated > 0.85 accuracy and 0.95 AUC [61]. For 3D volume reconstruction, Waine et al. proposed to reconstruct the 3D needle shapes in TRUS by defining needle shape and location in 2D axial US images [62]. Rodgers et al. proposed to obtain a 3D TRUS images by rotation of a 2D transducer, reconstructing 2D image slices into a real-time 3D volume [63]. Hrinivich et al. demonstrated a sagittally reconstructed 3D (SR3D) TRUS needle guidance that shows statistically smaller insertion depth errors, compared to the conventional 2D sagittally assisted axially reconstructed (SAAR) approach to insert needles [16].

In the above summary, most non-DL models used intensity-based methods, such as difference map and binary thresholding [50, 51, 53, 54], shape-based methods such as level-set [52, 58], or were based on simple feature extraction models such as Hough transform [15, 53–55] and SVM [57]. In some studies, manual contours are required as input [56, 59, 62]. Now with the more powerful DL algorithms, we are able to learn deep features of multiple levels to extract a more comprehensive understanding or abstraction of the images. The enormous number of learnable parameters in the sophisticated networks can be optimized to achieve more accurate results. Besides, in DL architectures, the manual labels are only needed during the training stage, but not needed for inference or application stage. Moreover, with the introduction of weakly supervised learning and other techniques, the number of manual labels can be largely reduced for training datasets, or even completely omitted. We will review the emerging use of CNNs to facilitate these tasks in subsequent subsections.

CNNs in Prostate and OAR Segmentation

To carry out a successful radiotherapy, the lesion needs to be irradiated with prescribed doses to achieve desired control rate, but other healthy organs, or OARs, should be spared from doses to limit potential radiation toxicity effects; so, an accurate delineation of organs are of critical interest for treatment planning. For LDR brachytherapy placing radioactive seeds during the surgery, physicians may need to manually delineate prostate and OARs in OR for intraoperative planning [14]. The contouring process can be subjective, tedious and stressful to physicians especially when patient is under anesthesia. High-quality TRUS automatic organ segmentation is thus a great tool to relief the situation. Besides, for HDR brachytherapy, real-time segmentation can help guide the needle placement for treatment plan of better outcome. For example, an intraoperative simulated treatment plan can be generated based on TRUS segmentation in OR and physician can decide the optimal location to insert needles.

Among the DL auto-segmentation papers, some of them focused only on prostate delineation. In 2019, Lei et al. [64] proposed a deep-supervised V-Net based architecture to automatically segment prostate from 3D TRUS images, with manual prostate segmentation contours serving as ground truth. The input TRUS images were first filtered with 3D Gaussian, mean, and median filters to reduce the speckle noises [23]. Together with the original TRUS, these filtered images were combined to a four-channel multi-derivative-based data. Due to computational cost limit, serial 3D image patches were used for training, and then rebuilt the whole image via patch fusion. A 3D patch-based V-Net [24] was then trained with deep supervision. Since their 3D TRUS dataset was considered small, deep supervision [25, 26] was incorporated to compute predictions at each resolution levels, and then merged for a hybrid loss regression with respect to the ground truth. Moreover, they separately trained the transverse, coronal, and sagittal planes for a multi-directional image modeling. Due to the possible segmentation discrepancies between different planes, they merged the segmentation from three planes via multi-direction contour refinement for a final 3D segmentation mask. On a 44-patient dataset, they used leave-one-out cross-validation to evaluate the model performance and achieved a Dice similarity coefficient (DSC) of 0.92 ± 0.03 and mean surface distance (MSD) of 0.60 ± 0.23 mm. The processing time of one prostate segmentation takes 1–2 s on a GPU with 12 GB of memory, which is almost real time. They have also tested the 3D deeply supervised V-Net on a smaller dataset of 30 patients and obtained similar results [65].

Wang et al. [66] performed prostate segmentation by integrating attention modules to FPN architecture. Since shallow layer features carry detailed information but lack semantic learning, while deep layer features capture prostate location cues but lose boundary information, a deep attention module is integrated to each individual layer to generate the refined attentive features by refining each single-layer feature with multi-layer features. So that the shallow single-layer features will gradually capture more semantic saliency regions while deep single-layer features will enhance boundary details. On a dataset of 30 3D TRUS, the algorithm performance was 0.90 ± 0.03 Dice, 0.90 ± 0.06 precision, and 0.91 ± 0.04 recall.

In 2020, Orlando et al. [67] proposed to first predicted prostate segmentations on 2D slices and then reconstructed them into 3D surface. They modified the U-Net by adding 50% dropouts and replaced the standard up-sampling by transpose convolutions. A total of 246 3D TRUS volumes were collected using either end-fire (in prostate biopsy) or side-fire SR (in HDR) scanning protocols [68]. They obtained an overall DSC, recall, precision, and Hausdorff distance (HD) of 0.94, 0.96, 0.93, and 2.89 mm, respectively. By comparison experiments, their proposed 2D method even outperformed some 3D networks such as 3D V-Net, Dense V-Net, and High-resolution 3D-Net. The model computation time takes less than 0.7 s [69]. Later, they tested a modified U-Net and U-Net++

architectures on a dataset of 40 3D TRUS volumes to see the impact of image quality on prostate segmentation performance [70]. They found that image quality had no impact on end-fire (biopsy) images but significantly impacted the side-fire (HDR) images, especially for the boundary visibility.

Girum et al. [71] presented a multi-task deep-learning method to detect clinical target volume (CTV) boundary in TRUS. The method utilized a channel-wise feature calibration for low-level layers to reconstruct CTV shape, which detects the low-contrast and noisy regions around the prostate. The method was tested on a database of 145 patients and achieved an accuracy of 0.96 ± 0.01, and an MSD of 0.10 ± 0.06 mm. They also tried a generative neural network-based architecture [72] on the same dataset. The generative network predicted prior knowledge such as contour proposal and then refined the prostate contours with raw input images. The DSC was 0.97 ± 0.01, 3D HD was 4.25 ± 4.58 mm, and volumetric overlap ratio was 0.94 ± 0.02.

More recently, Xu et al. [73] used a vanilla U-Net equipped with Shadow-AUG and Shadow-DROP in 2022. Shadow-AUG augmented the dataset by adding simulated shadow artifacts to original images, and Shadow-DROP ensured the segmentation was learned from shadow-free pixels. The experiment was performed on a public dataset of 1761 TRUS images and an in-house set of 662 images, achieving DSC, average symmetric surface distance (ASD), and HD of 0.92 ± 0.02, 0.93 ± 0.29 mm, and 5.89 ± 1.93 mm for the public dataset, and 0.89 ± 0.03, 1.26 ± 0.58 mm, and 6.88 ± 3.00 mm for the in-house data.

For OARs segmentation task (Figure 12.5), the algorithms need to simultaneously identify multiple organs which vary in shapes and sizes. In 2021 Lei et al. [14] proposed a novel anchor-free CNN architecture to perform automated pelvic multi-organ segmentation for bladder, prostate, rectum, and urethra in TRUS. The workflow involved three steps: a 3D residual network backbone [40] to extract pyramid feature maps, a fully convolutional one-state object detector (FCOS) [74] to identify the volume of interest (VOI) of organs, and a deep attention U-Net-like [75] mask head of binary segmentation within the selected VOI for each organ. The manual contours from physicians were the ground truth for final binary voxel-level segmentations at the mask head. The center-of-mass of each organ was the ground truth of VOI centers in the FCOS stage, and a 3D bounding box to cover the contoured organ served as the ground truth for VOI bounding index. Compared to the Mask R-CNN which first proposes numerous potential VOIs (anchor boxes), then classifies boxes with or without objects, and later adjusts box boundaries with regression, FCOS is anchor-free and learns to extract the VOIs directly from feature maps obtained from the backbone network. After using FCOS to get the VOIs, all VOIs will be resized to uniform so that the data imbalance problem between large and small organs can be addressed. Since each organ was segmented separately, there was an extra step of consolidation fusion to put all individual masks back to the original whole TRUS. In total 83 prostate cancer patients were included in experiments, by both fivefold cross-validation and a hold-out test. The method achieved DSC >0.9 for rectum and prostate, >0.8 for urethra, and >0.7 for bladder, outperforming U-Net and Mask R-CNN. Since the task is a multi-organ segmentation, the computation time was under 5 s, while for the single prostate segmentation it is about 1 s.

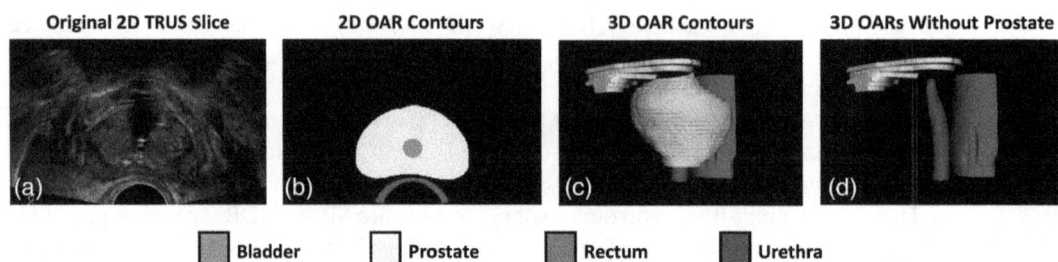

| Original 2D TRUS Slice | 2D OAR Contours | 3D OAR Contours | 3D OARs Without Prostate |

(a) (b) (c) (d)

■ Bladder   □ Prostate   ■ Rectum   ■ Urethra

FIGURE 12.5   2D and 3D pelvic multi-organ segmentation tasks. Note the differences in shapes and sizes of organs.

CNNs in Needle Detection

The needle, or catheter, insertion is a key step in both LDR and HDR brachytherapy, since the radioactive seeds will be implanted via needles in LDR treatment, and the radioactive sources will dwell inside needles for allocated time during HDR treatment. Besides, for HDR brachytherapy the patient needs to be transported for a CT scan after needle insertion to obtain accurate location of needles for treatment planning. But performing extra CT scan is time-consuming which increases patients' holding time and medical cost. Patients have to move on and off the CT couch, potentially altering the placement of inserted needles and impairing treatment quality [76, 77]. If accurate needle localization on TRUS is achieved, physicians can do needle insertion and treatment planning with a single imaging modality. Thus, high accuracy needle localization in 3D TRUS (Figure 12.6) is an important task in the field.

In 2020, Zhang et al. [78] proposed a variant of U-Net [45, 79] architecture to automate multi-needle localization in 3D transrectal US (TRUS). The ground truth was obtained via manual annotations of needles at pivotal slices of TRUS which linearly interpolated to connective needle trajectories. According to the outer diameter of clinical needles, the final needle trajectories were expanded to a diameter of 1.67 mm. The network consists of contracting encoding layers and expanding decoding layers, with skip connections to combine high-level features and low-level localization information. To tackle the difficulty of detecting small objects with CNNs [80] such as needles, they improved the primitive U-Net by integrating attention gates [81] to decrease the response of irrelevant spatial zones. To reduce the image noise of needle shafts, they introduced the total variation (TV) regularization that put penalty on the total gradient of a pixel [82, 83]. The TV regularization is usually defined as the sum of $L1$ norm of pixel gradients and increase for noise pixels. Since the needle shapes elongated in z-direction, so they modified the isotropic TV with deformed weights, of a much larger contribution from z-direction than the x,y-directions, to account for the spatial continuity of needles. The total loss function is to optimize the cross-entropy loss [45] of an attention U-Net under constraint of TV regularization. To alleviate the vanishing gradients problem and better train discriminative features, the deep supervision [84] strategy was employed to compute companion objective function for each resolution level. A fivefold cross-validation of the proposed network on 339 needles from 23 patients yielded an accuracy of 0.96 to detect needles, with 0.290 ± 0.236 mm shaft error and 0.442 ± 0.831 mm tip error. The method could provide needle digitization within 1 s on GPU.

Followed by the above study, they proposed an automated multi-needle detection workflow with two main components: a large-margin mask R-CNN model (LMMask R-CNN) [85] and a needle-based density-based spatial clustering of application with noise algorithm (Needle-DBSCAN) [86]. The LMMask R-CNN uses FPNs to extract feature maps at multiple scales, which were fed into the region proposal networks (RPN) and ROI aligning layer. The many generated ROI proposals (fixed size of 1.67 mm) were binarily classified as with or without needles. A bounding-box center regression was trained to bring the centers of ROI boxes close to the centers of ground truth. Meanwhile, a segmentation task of pixel-level classification of needle was performed for the proposal. The next step was to refine the predicted needle pixels with a modified DBSCAN algorithm.

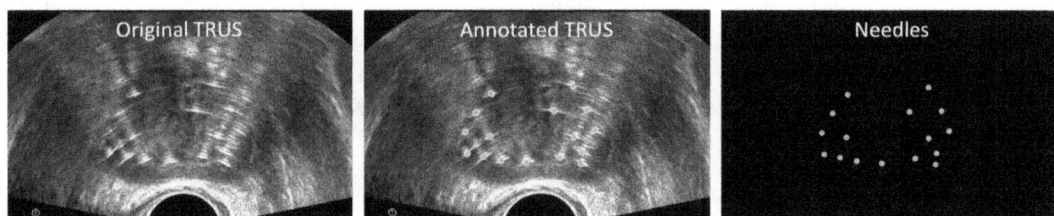

FIGURE 12.6   Multi-needle localization detection.

The needle-based DBSCAN is a density-based clustering algorithm that iteratively refine needle pixel clusters to improve the prediction accuracy of shaft and tip positions. In an experiment on the same dataset of [78], fivefold cross validation was used to evaluate the workflow and it successfully detected 98% needles with 0.09 ± 0.04 mm shaft error and 0.33 ± 0.36 mm tip error, outperforming the methods using Mask R-CNN or LMMask R-CNN alone. The results are also better than the previously proposed U-Net methods [78], with a faster computation time of 0.6 s per patient.

Wang et al. [87] cropped the TRUS into smaller ROIs that mainly contain one needle, and the needle trajectory are defined on each 2D TRUS slices with x and y coordinates, and then stacking 2D trajectories into 3D needle positions, defining the needle tip with x, y, and z coordinates. They employed a deep U-Net backbone but omitted the skip connection in low-resolution levels to train the needle pixel-wise segmentations in x and y directions. Then they used a VGG-16-like network to predict needle tips in z direction. TRUS were collected from 832 HDR treatments, each containing about 15–20 needles. The networks were trained on over 640 treatments and tested on about 183 treatments. For needle trajectory segmentation, 95.4% of the x coordinate predictions and 99.2% of the y directions are within 2 mm data error. The needle tip accuracy is 0.721, 0.369, and 1.877 mm in x, y, and z directions.

With a large dataset of 1,102 brachytherapy treatments for prostate cancer with 24,422 individual needles, Andersen et al. [88] used a 3D CNN U-Net [45] to find needles in TRUS, yielding root-mean-square distance (RMSD) of 0.55 mm. After testing the U-Net to segment 3,500 needles for 242 patients, Liu et al. [89] achieved an overall 0.8 accuracy of catheter reconstruction. They found the individual implantation style would interfere the US artifacts and thus slightly change the accuracy of auto-segmentation. Besides, the distal catheters (relative to the US probe) are usually more difficult to reconstruct due to the distal image degradation such as decreased contrast and increased noise. The network accuracy was 0.89 for easiest reconstructions but dropped to 0.13 for hard cases. Even within the easy catheters, distal reconstruction accuracy was only 0.5.

As a comparison to the above CNNs, in 2020 Zhang et al. propose to use a sparse dictionary learning (SDL)-based method to detect multiple needles in TRUS without the cost of laborious manual labeling [90]. Unlike CNNs that learn discriminative features in layer-by-layer fashion, dictionary learning extracts basis features through matrix factorization. SDL provided an over complete basis that contains numerous features and tries to rebuild images based on a subset of intrinsic features, with a sparse coefficient solution. They utilized the TRUS without needles as the "auxiliary" set and the TRUS with needles as the target set. They trained an order-Graph regularized SDL (ORSDL) on the "auxiliary" set to learn the latent features (basis), which also ensures the spatial continuity of tissues and needles, and then the basis was used to reconstruct the target TRUS. Since the "auxiliary" TRUS contained no needles, so the basis wouldn't recognize the needle patterns in target images, and resulting the rebuilt target TRUS of excellent reappearance of prostate tissues, but without needles. The difference between original target TRUS and the rebuilt images are assumed to be needles. Further clustering and refinement of the obtained needle pixels would produce the final needle patterns. Such method transforms the image recognition problem to a matrix factorization objective and requires no back-propagation updates aiming at ground truth, and thus demands no contouring labors. The model could detect 95% of needles, which is comparable to CNN methods [78, 86], but yield a much larger tip error of 1.01 mm (for CNN methods tip error < 0.5 mm). Moreover, the computation time is about 37.6 ± 4.8 s per patient on a GPU, much slower than CNNs (within 1 s [78, 86]). But in another study in ref. [91], they proposed a novel sparse learning model, dubbed bidirectional convolutional sparse coding (BiCSC), to learn latent features from both TRUS and CT images, and then relate the US features to CT features, thus the needle detection on TRUS was supervised by CT needle location information, rather than manual contours. The experiment was performed on 10 patients and achieved 93.1% detection rate with 0.15 ± 0.11 mm shaft error and 0.44 ± 0.37 tip error, which is much better than [90] and comparable to CNN methods [78, 86].

CNNs in Seed Detection

Radioactive seeds are permanently implanted into patient for prescribed doses. To validate the implantation quality, X-ray/fluoroscopy and CT are commonly used for seed visualization, but these imaging modalities have poor soft-tissue contrast so prostate and other OARs are not easy to delineate. TRUS show good soft-tissue contrast but is hard for seed detection. To address this problem, Golshan et al. [92] used a fully connected CNN to detect brachytherapy seed in 3D TRUS images for 13 patients. The ground truth of seed locations was obtained from the reconstruction of fluoroscopy views. Small patches of sub-regions of the TRUS were used to learn the needle track (seeds were inserted with needles), and then predict the individual seed locations within the needle track. The seed detection method yielded a precision of 0.78 ± 0.08, a recall of 0.64 ± 0.10, and a F1-score of 0.70 ± 0.08.

CNNs in MR-TRUS Image Registration

3D TRUS is widely used in brachytherapy due to its advantage of real-time imaging and good soft-tissue contrast. However, as an intraoperative imaging, TRUS suffer from low image quality such as speckle noise, shadow artifacts, and low signal-to-noise ratio [93]. Multi-parametric MR imaging shows excellent soft-tissue contrast and is suitable for physiology diagnosis. It will be beneficial to fuse the dominant intraprostatic lesion information from high-quality MR imaging with real-time TRUS to better guide needle placement in OR for HDR prostate brachytherapy with a focal boost [24]. The major challenge of such multimodal image registration is that different statistical correlations of anatomy may be generated across different imaging modalities due to their physical acquisition processes [94] such lack robust similarity measure.

Classical pairwise intensity-based image registration methods are generally based on optimizing image similarity. However, multimodal similarity metric is challenging due to different physical acquisition processes, user-dependent variabilities in the intraprocedural imaging, and intraoperative time constraints. To address the problem, Hu et al. [95] proposed that higher level anatomical labels (solid organs, vessels, ducts, structure boundaries and other subject-specific *ad hoc* landmarks, Figure 12.7) highlighted in pairs of images can serve as weak labels for training the prediction of voxel correspondence (dense displacement field (DDF)). Thus, the new image registration (Figure 12.8) is based on label similarity instead of voxel similarity (Intensity-based similarity). The study describes a method to infer voxel-level transformation from higher level correspondence information contained in 3D Gaussian filtered anatomical labels, and they found that such labels are more reliable and practical to obtain for reference sets of image pairs than voxel-level correspondence. In the example of registering T2-weighted MR images to 3D TRUS from prostate patients, a median target registration error (TRE) of 3.6 mm on landmark centroids and a median Dice of 0.87 on prostate glands were achieved.

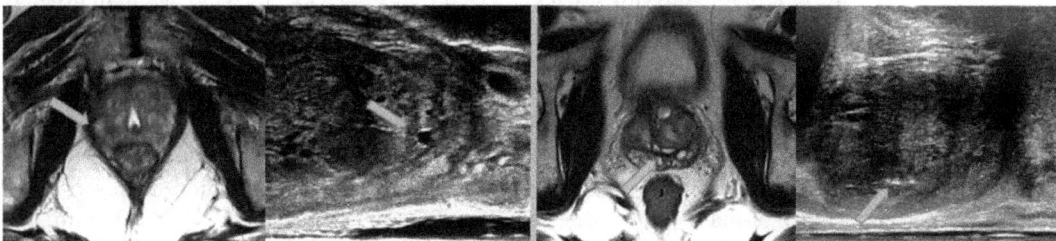

FIGURE 12.7 Examples of corresponding training landmark pairs in MR-TRUS registration: A water-filled cyst (on the left MR-TRUS image pair) and a cluster of calcification deposit (on the right image pair) [95]. ("Weakly-supervised convolutional neural networks for multimodal image registration" by Hu et al. under the terms of the Creative Commons CC-BY license.)

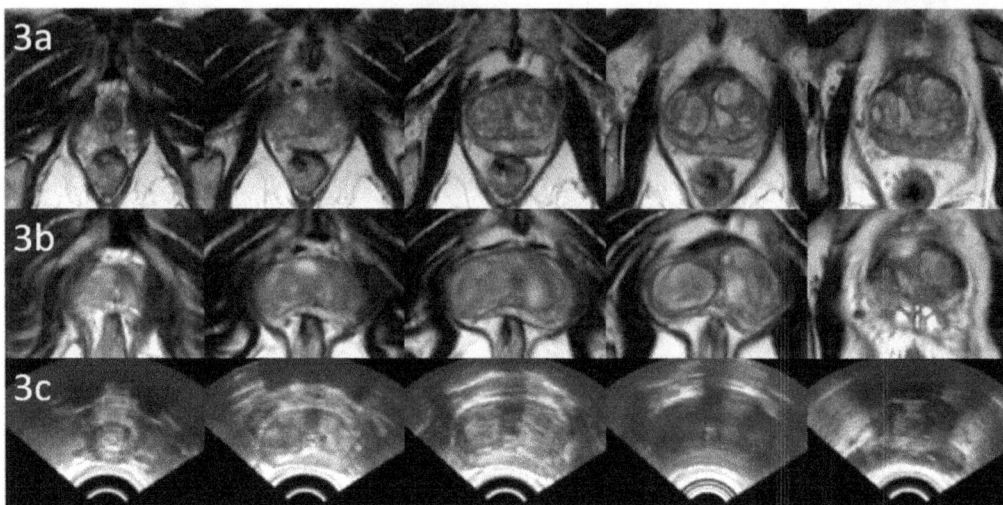

FIGURE 12.8 Examples of MR-TRUS nonrigid registration. The top row are MR images, the bottom row are TRUS images, and the middle row are warped MR images ready to be fused to TRUS [95]. ("Weakly-supervised convolutional neural networks for multimodal image registration" by Hu et al. under the terms of the Creative Commons CC-BY license.)

Zeng et al. [28] attempted to fulfill the registration in three steps: segmenting the landmark labels from both MRI and TRUS images, then predicting an affine transformation, and finally training a nonrigid fine local registration. They trained two separate FCNs for MRI and TRUS auto-segmentations, with a 3D supervision mechanism integrated in the hidden layers to help extracting features. Manual contours of the prostate labels are used as ground truth. The segmented labels served as inputs for registration and played an important role in improving the accuracy of registration. The affine transformation parameters were learned with 2D CNN targeting at high DSC between moving and fixed labels. This step served as rigid registration and provided a reasonable initialization for the subsequent nonrigid registration. A 3D U-Net-like densely connected network containing four down-sampling and four up-sampling blocks was used to train DDFs for fine nonrigid registration. The MRI-TRUS labels were used as input for both affine and nonrigid registration steps. They also tried to use original MRI-TRUS images as input, and in comparison, the segmented labels outperformed images. By combining the three networks, they achieved a successful MRI-TRUS registration with TRE of 2.53 ± 1.39 mm, MSD of 0.88 ± 0.20 mm, HD of 4.41 ± 1.05 mm, and Dice loss of 0.91 ± 0.02, on an experiment of 36 patients. The relatively large TRE may arise from the lack of learned knowledge of the deformation in inner prostate.

Also inspired by the landmark concept, Chen et al. [96] proposed a 3D V-Net to segment prostate labels in MR and TRUS images separately, using the manual prostate contour as ground truth. In this step, the prostate segmentation network was trained on 121 patients for MR image, and 104 patients for US image. Once the two segmentation networks were ready, they were used to generate weak supervision prostate labels for another 63 patients to train a subsequent registration network. Since in the registration phase, the labels in MR and US modalities are automatically generated, this method is free from manual contouring for the pairing images. The method achieved mean DSC of 0.87 ± 0.05, center-of-mass (COM) distance of 1.70 ± 0.89 mm, HD of 7.21 ± 2.07 mm, and averaged symmetric surface distance (ASSD) of 1.61 ± 0.64 mm. Similarly, Fu et al. [97] introduced a deep-learning-based 3D point cloud matching, where the MR and TRUS images were first separately segmented for prostate with CNNs, and then a point cloud matching network was trained for deformation field based on finite element analysis. On a dataset of 50 patients, the mean DSC, MSD, HD, and TRE were 0.94 ± 0.02, 0.90 ± 0.23 mm, 2.96 ± 1.00 mm and 1.57 ± 0.77 mm, respectively.

CNNs in DIL Detection

TeUS is a paradigm that records a temporal sequence of US RF signals or B-mode images at a specific scan location. The information embedded in TeUS reflects the changes in tissue temperatures as well as the micro-vibrational features due to physiology, and thus can be used to effectively classify various in vivo tissue types, e.g., recognizing prostate malignancy from benign tissue. These microstructural differences can be captured by statistical analysis of the scatter intensity variations in TeUS and used for prostate cancer DIL detection [18]. Currently most treatment on prostate gland is of one prescribed dose level to the whole PTV, though DILs account for the majority of tumor burden with less than 10% volume occupation. Besides, DILs are of high recurrence risk post radiotherapy [98–100]. Thus, if DILs can be identified, a focal dose can be delivered to DILs for an optimized boost plan that achieves prescribed coverage of whole prostate when sparing the OARs, as well as improves tumor control probability [24]. Till the time of preparing this manuscript, we found very few papers discussed applying CNNs with TeUS in detection DILs. Sedghi et al. proposed a U-Net-based architecture with attention gates to detect cancer foci in 2020 [101]. They fused the information from both TeUS and MRI to improve the accuracy of lesion detection. The AUC for predicting prostate cancer is 0.76 for all involved cancer foci, and 0.89 for larger foci.

Besides CNNs, other deep-learning techniques have also been applied for TeUS. In 2017, Azizi et al. [102] tried to grade prostate cancer with TRUS RF signals. Delimiting the original image into multiple much smaller ROIs (1×1 mm^2), they first extracted the spectral components of each ROI in TRUS RF signals based on discrete Fourier transforms (DFT). Then they used a deep belief network (DBN) [103, 104] to learn the latent features of RF TeUS, followed by a Gaussian mixture model [105] to assign a cancer grade to each ROI. The tissue biopsy and histopathology analysis were used as the ground truth to grade cancer (Gleason score [106]). Their method achieved 0.7 accuracy, 0.7 sensitivity, 0.71 specificity, and 0.7 AUC on a dataset of 132 patients. Since B-mode images are more readily available in clinics, later they proposed to translate a tissue classification model trained primarily on TeUS RF data to relying on TeUS B-mode images [107]. Applying domain adoption and transfer learning via DBN, they found the common latent feature space between RF and B-mode images, and then a SVM classifier was used to differentiate prostate malignancy with B-mode image alone. 172 benign biopsy cores and 83 cancerous cores from 157 patients were tested and TRUS B-mode images achieved comparable detecting strength compared to RF data. Later, they demonstrated the capability of LSTM networks to detect prostate cancer based on TeUS [108]. In an in vivo study of 157 patients, they got AUC, sensitivity, specificity, and accuracy of 0.96, 0.76, 0.98, and 0.93, respectively.

A summary of the reviewed literature in the above subsections of the third section is listed in Table 12.1.

DISCUSSION

3D TRUS has become a standard imaging modality used for prostate brachytherapy in OR. Since 2019, fusion of DL and 3D brachytherapy TRUS has drawn dramatically increasing interest. We have carefully reviewed 27 publications of close relevance and summarized their architectures and performance as listed in Table 12.1. The prostate segmentation and multi-needle detection are the most popular research topics, followed by MR-TRUS registration and DIL detection. Only one paper addressed the multi-organ (OAR) segmentation, probably due to its difficulty in that different organs vary dramatically in contrast, shape, and sizes. As shown in Figure 12.5, the 3D urethra is a thin tube while the prostate is a much large 3D oval-like volume. When these two organs were to be delineated in the same image, the problem of imbalanced dataset is almost inevitable. One solution is proposed in ref. [14] that VOIs are first selected for each individual organ and then the segmentation of each organ is computed within their corresponding VOI. In this way, small organs will be segmented in small VOIs, and the data imbalance can be greatly reduced. Considering the binary

TABLE 12.1
Summary of the Reviewed Work for 3D TRUS-Based Brachytherapy in the Section Above

Author	Year	Tasks	Architecture	Materials	Results
Azizi et al. [107]	2017	DIL detection	DBN	157 patients	AUC: 0.71 ± 0.02, specificity: 0.73 ± 0.03, and sensitivity: 0.76 ± 0.05
Azizi et al. [102]	2017	DIL detection	DBN	132 patients	Accuracy: 0.7, sensitivity: 0.7, specificity: 0.71, and AUC: 0.7
Azizi et al. [108]	2018	DIL detection	LSTM	157 patients	AUC: 0.96, sensitivity: 0.76, specificity: 0.98, and accuracy: 0.93
Hu et al. [95]	2018	MR-TRUS registration	U-Net	76 patients	Landmark TRE: 3.6 mm and prostate DSC: 0.87
Lei et al. [65]	2019	Prostate segmentation	3D V-Net	30 patients	DSC: 0.92 and MSD: 0.60 mm
Wang et al. [66]	2019	prostate segmentation	FPN	30 3D TRUS	Dice: 0.90 ± 0.03, precision: 0.90 ± 0.06, and recall: 0.91 ± 0.04
Lei et al. [64]	2019	Prostate segmentation	V-Net	44 patients	DSC: 0.92 ± 0.03 and MSD: 0.60 ± 0.23 mm
Fu et al. [79]	2020	MR-TRUS registration	CNNs	50 patients	DSC: 0.94 ± 0.02, MSD: 0.90 ± 0.23 mm, HD: 2.96 ± 1.00 mm and TRE: 1.57 ± 0.77 mm
Girum et al. [71]	2020	Prostate segmentation	U-Net	145 patients	Accuracy: 0.96 ± 0.01, and MSD: 0.10 ± 0.06 mm
Zeng et al. [28]	2020	MR-TRUS registration	3D U-Net	36 patients	TRE: 2.53 ± 1.39 mm, MSD: 0.88 ± 0.20 mm, HD: 4.41 ± 1.05 mm, and Dice: 0.91 ± 0.02.
Orlando et al. [69]	2020	Prostate segmentation	U-Net	206 3D TRUS	DSC: 93.5% and DSC and VPD: 5.89%
Zhang et al. [90]	2020	Needle detection	Sparse Dictionary Learning	70/21 patients without/with needles	Accuracy: 95%, tip error: 1.01 mm
Golshan et al. [92]	2020	Seed detection	FCN	13 patients	Precision: 0.78 ± 0.08, Recall: 0.64 ± 0.10, and F1-score: 0.70 ± 0.08.
Zhang et al. [91]	2020	Needle detection	Convolutional dictionary learning	10 patients	Accuracy: 93.1%, shaft error: 0.15 ± 0.11 mm, and tip error: 0.44 ± 0.37.
Orlando et al. [67]	2020	Prostate segmentation	U-Net	246 3D TRUS	DSC: 0.94, Recall: 0.96, Precision: 0.93, MSD: 0.89 mm, and HD: 2.89 mm.
Zhang et al. [78]	2020	Needle detection	Attention U-Net	23 patients, 339 needles	Accuracy: 0.96, shaft error: 0.290 ± 0.236 mm, tip error: 0.442 ± 0.831 mm.

(Continued)

TABLE 12.1 (Continued)
Summary of the Reviewed Work for 3D TRUS-Based Brachytherapy in the Section Above

Reference	Year	Task	Network	Dataset	Results
Sedghi et al. [101]	2020	DIL detection	U-Net	107 patients with 145 biopsy cores	AUC: 0.76 (all foci) and 0.89 (larger foci)
Girum et al. [72]	2020	Prostate segmentation	generative NN	145 3D TRUS	DSC: 0.97 ± 0.01, HD: 4.25 ± 4.58 mm, and volumetric overlap ratio: 0.94 ± 0.02
Wang et al. [87]	2020	Needle detection	U-Net	832 HDR treatments	Tip accuracy: 0.721, 0.369, and 1.877 mm in x, y, and z directions
Zhang et al. [86]	2020	Needle detection	LMMask R-CNN	23 patients, 339 needles	Accuracy: 98%, shaft error: 0.09 ± 0.04 mm, and tip error: 0.33 ± 0.36 mm
Anderse'n et al. [88]	2020	Needle detection	3D U-Net	1102 HDR treatments, 24,422 needles	RMSD: 0.55 mm
Zhang et al. [85]	2021	Needle detection	LMMask R-CNN	23 patients, 339 needles	Accuracy: 98%, shaft error: 0.0911 ± 0.0427 mm, and tip error: 0.3303 ± 0.3625 mm
Chen et al. [96]	2021	MR-TRUS registration	3D V-Net	63 patients	DSC: 0.87 ± 0.05, COM distance: 1.70 ± 0.89 mm, HD: 7.21 ± 2.07 mm, and ASSD: 1.61 ± 0.64 mm
Lei et al. [14]	2021	OAR segmentation	U-Net	83 patients	DSC >0.9 for rectum and prostate, >0.8 for urethra, and >0.7 for bladder
Orlando et al. [70]	2022	Prostate segmentation	U-Net and U-Net++	40 3D TRUS	Image quality had no impact on end-fire (biopsy) images but significantly impacted the side-fire (HDR) images
Liu et al. [89]	2022	Needle detection	U-Net	242 patients, 3,500 needles	Accuracy: 0.89 for easiest reconstructions, and 0.13 for hard cases
Xu et al. [73]	2022	Prostate segmentation	Vanilla U-Net	1761 + 662 3D TRUS	DSC: 0.92 ± 0.02/0.89 ± 0.03, ASD: 0.93 ± 0.29/1.26 ± 0.58 mm, and HD: 5.89 ± 1.93/6.88 ± 3.00 mm

TRE, target registration error; MSD, mean surface distance; HD, Hausdorff distance; DSC, Dice similarity coefficient; RMSD, root-mean-square distance; ASSD, averaged symmetric surface distance; ASD, average symmetric surface distance.

segmentation CNN models are relatively well-established, the efficient and accurate VOI selection for each organ becomes the major concern in OAR segmentation tasks, e.g., Mask R-CNN [109, 110] addresses this problem by region proposals. In ref. [14], the VOI selection method was further optimized to achieve better performance. Besides the four major tasks discussed in the first to fourth sections, other research topics are also open to investigations. For example, in clinical TRUS-guided brachytherapy, due to the limited mechanically scanning precision and limited intraoperative time allowing for image acquisition, 2D TRUS images with large spatial interval are generally acquired, resulting a much worse through-plane resolution than the in-plane directions. To deal with this issue, Dai et al. [111, 112] propose a self-supervised learning framework based on cycle-consistent cycleGAN to reconstruct high-resolution 3D TRUS images that only relies on the acquired sparsely distributed 2D slices. Their work provides an efficient solution to obtain high-resolution 3D TRUS volumes in clinics, which helps catheter placement and treatment planning.

The most frequently used deep-learning architectures are U-Net based, while V-Net and FPN are also popular choices. These CNNs-based architectures all compute feature maps at different resolution levels to extract a comprehensive abstraction of image. The ability of learning both high-level semantic features and low-level edge/location information makes CNNs a powerful image recognition model compared to traditional ML methods. Taking needle detection as an example, traditional ML (AI) methods usually detect a single needle in a small VOI, while CNNs can simultaneously detect multiple needles present in the whole TRUS by learning richer information. On the other hand, though CNN architectures like U-Net can reconstruct the outmost layer feature maps with the same size as input image to fulfill end-to-end classification in one pass, the input images are often decomposed into overlapped patches before being fed into the network. In those cases, patch-fusion is needed to stack patches back into the final whole image in the inference stage. One reason to use the patch-based method is due to the limitation of GPU memory if one pushes the whole image in one pass. Another reason is that multiple smaller patches can be obtained from one whole image, and the number of trainable data is thus greatly increased. So, patch-based methods are now widely adopted, though the patches allow smaller inception field compared to the whole image.

CHALLENGES AND OPPORTUNITIES

A long-standing obstacle for deep-learning projects has been the limited size of datasets, especially the lack of manually annotated data. In the reviewed 27 papers, for the needle detection task the largest dataset involves 1,102 HDR treatments with 24,422 needles [88]; for prostate segmentation the largest dataset contains 1,761 public and 662 in-house 3D TRUS images [73]; and for MR-TRUS registration, the largest dataset has 76 patients [95]. For most other papers focusing on these tasks, the numbers of patients involved are 10~50. Though each patient can generate multiple TRUS slices from the 3D volume imaging, the variation between each individual patient may not be fully investigated with small number of individuals. To address the lack of manually labeled data, weakly supervised learning and transfer learning have been utilized. The underlying mechanism of weakly supervised learning is enabling the models to automatically generate labels to replace the manual labels as ground truth. The transfer learning uses the pre-trained network for a new dataset that shares some similarities in nature with the pre-trained on dataset.

Technically, the evaluation of DL model performance is quantified as geometric metrics, such as commonly used DSC, MSD, and TRE, that reflect the accuracy of the specific task. However, the clinical application of brachytherapy consists of a series of steps involving pre-planning, needle insertion, intraoperative/postoperative treatment planning, and delivery of the treatment. Automation of any step in the chain may introduce some error and induce variations in following steps, which could result in considerable accumulated errors and degrade the final delivered treatment. To ensure the reliability and quality of treatment, any replacement of human works with AI substitutes needs to undergo extensive investigation to validate its application in clinics.

The inherent complexity of CNNs poses considerate computational costs for applications. Though the reported computation times are usually within 1 s per process [69, 78, 86] or 5 s per patient for the more difficult multi-OAR segmentation task [14], these methods were tested on high-performance GPUs. When performed on CPUs, it could take much longer time. Not to mention that many papers didn't report the computation time at all. For feasible application in OR, the AI applications need to be high fidelity, real time, and not too resource demanding. Thus, the computation efficiency of CNN-based models may need to be validated on not only high-performance GPUs but also on more readily available CPUs in OR.

OUTLOOKS

The application of CNNs in 3D TRUS-based prostate brachytherapy has achieved encouraging results since 2019. Major tasks of automatic organ segmentation, needle detection, and multimodal (MR-TRUS for now) image registration have all been addressed with properly designed solutions. Thanks to the modern technology of computer science, the computation time is within few seconds, close to the requirement of being real time in OR. For both HDR and LDR, an intraoperative planning can be generated to guided needle/seed placement if high-fidelity organ segmentations are available. A real-time MR-TRUS image fusion is also beneficial to guide needle/seed insertion by providing dominant lesion information that is not readily discernable in TRUS. In 2021, considering the difficulty in needle placement, de Vries et al. [113] introduced a robotic steerable instrument for needle insertion in HDR brachytherapy, using US images for visualization during insertion. In the same year, Smith et al. [114] reported the commissioning and workflow development of a real-time US image-guided treatment planning system, assembled along with a stepper and a US unit. With the possibility of integrating AI (CNNs) architectures into the robotic instruments and real-time treatment planning system, a more automated and intelligent 3D TRUS-based brachytherapy may be around the corner.

REFERENCES

1. Ferlay J, et al., *Cancer statistics for the year 2020: An overview.* International Journal of Cancer, 2021. **149**(4): p. 778–789.
2. Yoshioka Y, et al., *High-dose-rate brachytherapy as monotherapy for prostate cancer: Technique, rationale and perspective.* Journal of Contemporary Brachytherapy, 2014. **6**(1): p. 91.
3. Martinez AA, et al., *Phase II prospective study of the use of conformal high-dose-rate brachytherapy as monotherapy for the treatment of favorable stage prostate cancer: A feasibility report.* International Journal of Radiation Oncology, Biology, Physics, 2001. **49**(1): p. 61–69.
4. Demanes DJ, et al., *High-dose-rate monotherapy: Safe and effective brachytherapy for patients with localized prostate cancer.* International Journal of Radiation Oncology, Biology, Physics, 2011. **81**(5): p. 1286–1292.
5. Barkati M, et al., *High-dose-rate brachytherapy as a monotherapy for favorable-risk prostate cancer: A phase II trial.* International Journal of Radiation Oncology, Biology, Physics, 2012. **82**(5): p. 1889–1896.
6. Ghilezan M, et al., *High-dose-rate brachytherapy as monotherapy delivered in two fractions within one day for favorable/intermediate-risk prostate cancer: Preliminary toxicity data.* International Journal of Radiation Oncology, Biology, Physics, 2012. **83**(3): p. 927–932.
7. Yoshioka Y, et al., *Monotherapeutic high-dose-rate brachytherapy for prostate cancer: Five-year results of an extreme hypofractionation regimen with 54 Gy in nine fractions.* International Journal of Radiation Oncology, Biology, Physics, 2011. **80**(2): p. 469–475.
8. Yoshioka Y, et al., *Monotherapeutic high-dose-rate brachytherapy for prostate cancer: A dose reduction trial.* Radiotherapy and Oncology, 2014. **110**(1): p. 114–119.
9. Rogers CL, et al., *High dose brachytherapy as monotherapy for intermediate risk prostate cancer.* The Journal of Urology, 2012. **187**(1): p. 109–116.
10. Hoskin P, et al., *High-dose-rate brachytherapy alone for localized prostate cancer in patients at moderate or high risk of biochemical recurrence.* International Journal of Radiation Oncology, Biology, Physics, 2012. **82**(4): p. 1376–1384.

11. Zamboglou N, et al., *High-dose-rate interstitial brachytherapy as monotherapy for clinically localized prostate cancer: Treatment evolution and mature results.* International Journal of Radiation Oncology, Biology, Physics, 2013. **85**(3): p. 672–678.

12. Shen F, et al., *Three-dimensional sonography with needle tracking: Role in diagnosis and treatment of prostate cancer.* Journal of Ultrasound in Medicine, 2008. **27**(6): p. 895–905.

13. Pfeiffer D, et al., *AAPM task group 128: Quality assurance tests for prostate brachytherapy ultrasound systems.* Medical Physics, 2008. **35**(12): p. 5471–5489.

14. Lei Y, et al., *Male pelvic multi-organ segmentation on transrectal ultrasound using anchor-free mask CNN.* Medical Physics, 2021. **48**(6): p. 3055–3064.

15. Hrinivich WT, et al., *Accuracy and variability of high-dose-rate prostate brachytherapy needle tip localization using live two-dimensional and sagittally reconstructed three-dimensional ultrasound.* Brachytherapy, 2017. **16**(5): p. 1035–1043.

16. Hrinivich WT, et al., *Three-dimensional transrectal ultrasound guided high-dose-rate prostate brachytherapy: A comparison of needle segmentation accuracy with two-dimensional image guidance.* Brachytherapy, 2016. **15**(2): p. 231–239.

17. Crivianu-Gaita D, et al., *3D reconstruction of prostate from ultrasound images.* International Journal of Medical Informatics, 1997. **45**(1-2): p. 43–51.

18. Bayat S, et al., *Investigation of physical phenomena underlying temporal-enhanced ultrasound as a new diagnostic imaging technique: Theory and simulations.* IEEE Transactions on Ultrasonics, Ferroelectrics, and Frequency Control, 2017. **65**(3): p. 400–410.

19. Feng Y, et al., *A deep learning approach for targeted contrast-enhanced ultrasound based prostate cancer detection.* IEEE/ACM Transactions on Computational Biology and Bioinformatics, 2018. **16**(6): p. 1794–1801.

20. Ukimura O, *Evolution of precise and multimodal MRI and TRUS in detection and management of early prostate cancer.* Expert Review of Medical Devices, 2010. **7**(4): p. 541–554.

21. Schalk SG, et al., *3-D quantitative dynamic contrast ultrasound for prostate cancer localization.* Ultrasound in Medicine and Biology, 2018. **44**(4): p. 807–814.

22. Nath R, et al., *AAPM recommendations on dose prescription and Reporting methods for permanent interstitial brachytherapy for prostate cancer: Report of task group 137.* Medical Physics, 2009. **36**(11): p. 5310–5322.

23. Crook J, Marbán M, Batchelar D, *HDR Prostate Brachytherapy.* In Seminars in Radiation Oncology. 2020. Elsevier.

24. Wang T, et al., *Multiparametric MRI-guided dose boost to dominant intraprostatic lesions in CT-based high-dose-rate prostate brachytherapy.* The British Journal of Radiology, 2019. **92**(1097): p. 20190089.

25. Pinto PA, et al., *Magnetic resonance imaging/ultrasound fusion guided prostate biopsy improves cancer detection following transrectal ultrasound biopsy and correlates with multiparametric magnetic resonance imaging.* The Journal of Urology, 2011. **186**(4): p. 1281–1285.

26. Rastinehad AR, et al., *Improving detection of clinically significant prostate cancer: Magnetic resonance imaging/transrectal ultrasound fusion guided prostate biopsy.* The Journal of Urology, 2014. **191**(6): p. 1749–1754.

27. Siddiqui MM, et al., *Comparison of MR/ultrasound fusion–guided biopsy with ultrasound-guided biopsy for the diagnosis of prostate cancer.* JAMA, 2015. **313**(4): p. 390–397.

28. Zeng Q, et al., *Label-driven magnetic resonance imaging (MRI)-transrectal ultrasound (TRUS) registration using weakly supervised learning for MRI-guided prostate radiotherapy.* Physics in Medicine and Biology, 2020. **65**(13): p. 135002.

29. Santos MK, et al., *Artificial intelligence, machine learning, computer-aided diagnosis, and radiomics: Advances in imaging towards to precision medicine.* Radiologia Brasileira, 2019. **52**: p. 387–396.

30. LeCun Y, Bengio Y, Hinton G, *Deep learning.* Nature. 2015. **521**(7553): p. 436–444.

31. LeCun Y, et al., *Gradient-based learning applied to document recognition.* Proceedings of the IEEE, 1998. **86**(11): p. 2278–2324.

32. Hochreiter S, Schmidhuber J, *Long short-term memory.* Neural Computation. 1997. **9**(8): p. 1735–1780.

33. Badrinarayanan V, Kendall A, Cipolla R, *Segnet: A deep convolutional encoder-decoder architecture for image segmentation.* IEEE Transactions on Pattern Analysis and Machine Intelligence, 2017. **39**(12): p. 2481–2495.

34. Goodfellow I, et al., *Generative adversarial nets.* Advances in Neural Information Processing Systems, 2014. **27**: p. 2672–2680.

35. Minaee S, et al., *Image segmentation using deep learning: A survey.* IEEE Transactions on Pattern Analysis and Machine Intelligence, 2021. **44**: p. 3523–3542.

36. Zhou T, Ruan S, Canu S. *A review: Deep learning for medical image segmentation using multi-modality fusion*. Array, 2019. **3**: p. 100004.
37. He K, et al., *Delving deep into rectifiers: Surpassing human-level performance on imagenet classification*. In *Proceedings of the IEEE International Conference on Computer Vision*. 2015.
38. Krizhevsky A, Sutskever I, Hinton GE, *Imagenet classification with deep convolutional neural networks.*, Communications of the Acm 2017. **60**(6): p. 84–90.
39. Simonyan K, Zisserman A, *Very deep convolutional networks for large-scale image recognition*. arXiv preprint arXiv:1409.1556, 2014.
40. He K, et al, *Deep residual learning for image recognition*. In *Proceedings of the IEEE Conference on Computer Vision and Pattern Recognition*. 2016.
41. Szegedy C, et al, *Going deeper with convolutions*. In *Proceedings of the IEEE Conference on Computer Vision and Pattern Recognition*. 2015.
42. Howard AG, et al., *Mobilenets: Efficient convolutional neural networks for mobile vision applications*. arXiv preprint arXiv:1704.04861, 2017.
43. Huang G, et al, *Densely connected convolutional networks*. In *Proceedings of the IEEE Conference on Computer Vision and Pattern Recognition*. 2017.
44. Long J, Shelhamer E, Darrell T, *Fully convolutional networks for semantic segmentation*. in *Proceedings of the IEEE Conference on Computer Vision and Pattern Recognition*. 2015.
45. Ronneberger O, Fischer P, Brox T, *U-Net: Convolutional networks for biomedical image segmentation*. In *International Conference on Medical Image Computing and Computer-Assisted Intervention*. 2015. Springer.
46. Milletari F, Navab N, Ahmadi S-A, *V-Net: Fully convolutional neural networks for volumetric medical image segmentation*. In *2016 Fourth International Conference on 3D Vision (3DV)*. 2016. IEEE.
47. Zhu Q, et al., *Deeply-supervised CNN for prostate segmentation*. In *2017 International Joint Conference on Neural Networks (IJCNN)*. 2017. IEEE.
48. Dou Q, et al., *3D deeply supervised network for automated segmentation of volumetric medical images*. Medical Image Analysis, 2017. **41**: p. 40–54.
49. Lin T-Y, et al. *Feature pyramid networks for object detection*. In *Proceedings of the IEEE Conference on Computer Vision and Pattern Recognition*. 2017.
50. Wei Z, et al., *Oblique needle segmentation and tracking for 3D TRUS guided prostate brachytherapy*. Medical Physics, 2005. **32**(9): p. 2928–2941.
51. Ding M, et al., *Needle and seed segmentation in intra-operative 3D ultrasound-guided prostate brachytherapy*. Ultrasonics, 2006. **44**: p. e331–e336.
52. Yan P, Cheeseborough JC III, Chao KC, *Automatic shape-based level set segmentation for needle tracking in 3-D TRUS-guided prostate brachytherapy*. Ultrasound in Medicine & Biology, 2012. **38**(9): p. 1626–1636.
53. Qiu W, et al., *Needle segmentation using 3D Hough transform in 3D TRUS guided prostate transperineal therapy*. Medical Physics, 2013. **40**(4): p. 042902.
54. Qiu W, Yuchi M, Ding M, *Phase grouping-based needle segmentation in 3-D trans-rectal ultrasound-guided prostate trans-perineal therapy*. Ultrasound in Medicine & Biology, 2014. **40**(4): p. 804–816.
55. Hrinivich WT, et al., *Simultaneous automatic segmentation of multiple needles using 3D ultrasound for high-dose-rate prostate brachytherapy*. Medical Physics, 2017. **44**(4): p. 1234–1245.
56. Dehghan E, et al., *EM-enhanced US-based seed detection for prostate brachytherapy*. Medical Physics, 2018. **45**(6): p. 2357–2368.
57. Younes H, Troccaz J, Voros S, *Machine learning and registration for automatic seed localization in 3D US images for prostate brachytherapy*. Medical Physics, 2021. **48**(3): p. 1144–1156.
58. Li X, et al., *Segmentation of prostate from ultrasound images using level sets on active band and intensity variation across edges*. Medical Physics, 2016. **43**(6Part1): p. 3090–3103.
59. Yang X, et al., *Prostate CT segmentation method based on nonrigid registration in ultrasound-guided CT-based HDR prostate brachytherapy*. Medical Physics, 2014. **41**(11): p. 111915.
60. Nahlawi L, et al., *Stochastic modeling of temporal enhanced ultrasound: Impact of temporal properties on prostate cancer characterization*. IEEE Transactions on Biomedical Engineering, 2017. **65**(8): p. 1798–1809.
61. Nahlawi L, et al., *Stochastic sequential modeling: Toward improved prostate cancer diagnosis through temporal-ultrasound*. Annals of Biomedical Engineering, 2021. **49**(2): p. 573–584.
62. Waine M, et al., *Three-dimensional needle shape estimation in TRUS-guided prostate brachytherapy using 2-D ultrasound images*. IEEE Journal of Biomedical and Health Informatics, 2015. **20**(6): p. 1621–1631.

63. Rodgers JR, et al., *Toward a 3D transrectal ultrasound system for verification of needle placement during high-dose-rate interstitial gynecologic brachytherapy.* Medical Physics, 2017. **44**(5): p. 1899–1911.

64. Lei Y, et al., *Ultrasound prostate segmentation based on multidirectional deeply supervised V-Net.* Medical Physics, 2019. **46**(7): p. 3194–3206.

65. Lei Y, et al., *Ultrasound Prostate Segmentation Based on 3D V-Net With Deep Supervision.* In Medical Imaging 2019: Ultrasonic Imaging and Tomography. 2019. SPIE.

66. Wang Y, et al., *Deep attentive features for prostate segmentation in 3D transrectal ultrasound.* IEEE Transactions on Medical Imaging, 2019. **38**(12): p. 2768–2778.

67. Orlando N, et al., *Automatic prostate segmentation using deep learning on clinically diverse 3D transrectal ultrasound images.* Medical Physics, 2020. **47**(6): p. 2413–2426.

68. Fenster A, Parraga G, Bax J, *Three-dimensional ultrasound scanning.* Interface Focus, 2011. **1**(4): p. 503–519.

69. Orlando N, et al., *Deep Learning-Based Automatic Prostate Segmentation in 3D Transrectal Ultrasound Images from Multiple Acquisition Geometries and Systems.* In *Medical Imaging 2020: Image-Guided Procedures, Robotic Interventions, and Modeling.* 2020. SPIE.

70. Orlando N, et al., *Effect of dataset size, image quality, and image type on deep learning-based automatic prostate segmentation in 3D ultrasound.* Physics in Medicine and Biology, 2022. **67**(7): p. 074002.

71. Girum KB, et al., *A deep learning method for real-time intraoperative US image segmentation in prostate brachytherapy.* International Journal of Computer Assisted Radiology and Surgery, 2020. **15**(9): p. 1467–1476.

72. Girum KB, et al., *Fast interactive medical image segmentation with weakly supervised deep learning method.* International Journal of Computer Assisted Radiology and Surgery, 2020. **15**(9): p. 1437–1444.

73. Xu X., et al., *Shadow-consistent semi-supervised learning for prostate ultrasound segmentation.* IEEE Transactions on Medical Imaging, 2021. **41**(6): p. 1331–1345.

74. Tian Z, et al., *Fully Convolutional One-Stage Object Detection2019.* In *Conference: 2019 IEEE/CVF International Conference on Computer Vision (ICCV).* 2019.

75. Dong X, et al., *Synthetic MRI-aided multi-organ segmentation on male pelvic CT using cycle consistent deep attention network.* Radiotherapy and Oncology, 2019. **141**: p. 192–199.

76. Holly R, et al., *Use of cone-beam imaging to correct for catheter displacement in high dose-rate prostate brachytherapy.* Brachytherapy, 2011. **10**(4): p. 299–305.

77. Milickovic N, et al., *4D analysis of influence of patient movement and anatomy alteration on the quality of 3D U/S-based prostate HDR brachytherapy treatment delivery.* Medical Physics, 2011. **38**(9): p. 4982–4993.

78. Zhang Y, et al., *Multi-needle localization with attention u-Net in US-guided HDR prostate brachytherapy.* Medical Physics, 2020. **47**(7): p. 2735–2745.

79. Falk T, et al., *U-Net: Deep learning for cell counting, detection, and morphometry.* Nature Methods, 2019. **16**(1): p. 67–70.

80. Roth HR, et al., *Spatial aggregation of holistically-nested convolutional neural networks for automated pancreas localization and segmentation.* Medical Image Analysis, 2018. **45**: p. 94–107.

81. Oktay O, et al., *Attention U-Net: Learning where to look for the pancreas.* arXiv preprint arXiv:1804.03999, 2018.

82. Vishnevskiy V, et al., *Isotropic total variation regularization of displacements in parametric image registration.* IEEE Transactions on Medical Imaging, 2016. **36**(2): p. 385–395.

83. Ehrhardt MJ, Betcke MM, *Multicontrast MRI reconstruction with structure-guided total variation.* SIAM Journal on Imaging Sciences. 2016. **9**(3):1084–1106.

84. Lee C-Y, et al., *Deeply-Supervised Nets.* In *Artificial Intelligence and Statistics.* 2015. PMLR.

85. Zhang Y, et al., *Multi-Needle Detection in Ultrasound Image Using Max-Margin Mask R-CNN.* In *Medical Imaging 2021: Ultrasonic Imaging and Tomography.* 2021. SPIE.

86. Zhang Y, et al., *Automatic multi-needle localization in ultrasound images using large margin mask RCNN for ultrasound-guided prostate brachytherapy.* Physics in Medicine and Biology, 2020. **65**(20): p. 205003.

87. Wang F, et al., *Deep learning applications in automatic needle segmentation in ultrasound-guided prostate brachytherapy.* Medical Physics, 2020. **47**(9): p. 3797–3805.

88. Andersén C, et al., *Deep learning-based digitization of prostate brachytherapy needles in ultrasound images.* Medical Physics, 2020. **47**(12): p. 6414–6420.

89. Liu D, et al., *The challenges facing deep learning-based catheter localization for ultrasound guided high-dose-rate prostate brachytherapy.* Medical Physics, 2022. **49**(4): p. 2442–2451.

90. Zhang Y, et al., *Multi-needle detection in 3D ultrasound images using unsupervised order-graph regularized sparse dictionary learning.* IEEE Transactions on Medical Imaging, 2020. **39**(7): p. 2302–2315.

91. Zhang Y, et al., *Weakly Supervised Multi-Needle Detection in 3D Ultrasound Images With Bidirectional Convolutional Sparse Coding*. In *Medical Imaging 2020: Ultrasonic Imaging and Tomography*. 2020. SPIE.

92. Golshan M, et al., *Automatic detection of brachytherapy seeds in 3D ultrasound images using a convolutional neural network*. Physics in Medicine and Biology, 2020. **65**(3): p. 035016.

93. Ding M, et al., *3D TRUS image segmentation in prostate brachytherapy*. In *2005 IEEE Engineering in Medicine and Biology 27th Annual Conference*. 2006. IEEE.

94. Zöllei L, Fisher JW, Wells WM, *A unified statistical and information theoretic framework for multimodal image registration*. In *Biennial International Conference on Information Processing in Medical Imaging*. 2003. Springer.

95. Hu Y, et al., *Weakly-supervised convolutional neural networks for multimodal image registration*. Medical Image Analysis, 2018. **49**: p. 1–13.

96. Chen Y, et al., *MR to ultrasound image registration with segmentation-based learning for HDR prostate brachytherapy*. Medical Physics, 2021. **48**(6): p. 3074–3083.

97. Fu Y, et al., *Biomechanically constrained non-rigid MR-TRUS prostate registration using deep learning based 3D point cloud matching*. Medical Image Analysis, 2021. **67**: p. 101845.

98. Chapman CH, et al., *Phase I study of dose escalation to dominant intraprostatic lesions using high-dose-rate brachytherapy*. Journal of Contemporary Brachytherapy, 2018. **10**(3): p. 193–201.

99. Wise AM, et al., *Morphologic and clinical significance of multifocal prostate cancers in radical prostatectomy specimens*. Urology, 2002. **60**(2): p. 264–269.

100. Arrayeh E, et al., *Does local recurrence of prostate cancer after radiation therapy occur at the site of primary tumor? Results of a longitudinal MRI and MRSI study*. International Journal of Radiation Oncology, Biology, Physics, 2012. **82**(5): p. e787–e793.

101. Sedghi A, et al., *Improving detection of prostate cancer foci via information fusion of MRI and temporal enhanced ultrasound*. International Journal of Computer Assisted Radiology and Surgery, 2020. **15**(7): p. 1215–1223.

102. Azizi S, et al., *Detection and grading of prostate cancer using temporal enhanced ultrasound: Combining deep neural networks and tissue mimicking simulations*. International Journal of Computer Assisted Radiology and Surgery, 2017. **12**(8): p. 1293–1305.

103. Azizi S, et al., *Ultrasound-based detection of prostate cancer using automatic feature selection with deep belief networks*. In *International Conference on Medical Image Computing and Computer-Assisted Intervention*. 2015. Springer.

104. Bengio Y, et al., *Introduction to the special issue on neural networks for data mining and knowledge discovery*. IEEE Transactions on Neural Networks, 2000. **11**(3): p. 545–549.

105. Xu L, Jordan MI, *On convergence properties of the EM algorithm for Gaussian mixtures*. Neural Computation, 1996. **8**(1): p. 129–151.

106. Epstein JI., et al., *Upgrading and downgrading of prostate cancer from biopsy to radical prostatectomy: Incidence and predictive factors using the modified Gleason grading system and factoring in tertiary grades*. European Urology, 2012. **61**(5): p. 1019–1024.

107. Azizi S, et al., *Transfer learning from RF to B-mode temporal enhanced ultrasound features for prostate cancer detection*. International Journal of Computer Assisted Radiology and Surgery, 2017. **12**(7): p. 1111–1121.

108. Azizi S, et al., *Deep recurrent neural networks for prostate cancer detection: Analysis of temporal enhanced ultrasound*. IEEE Transactions on Medical Imaging, 2018. **37**(12): p. 2695–2703.

109. Jeong J, et al., *Brain tumor segmentation using 3D Mask R-CNN for dynamic susceptibility contrast enhanced perfusion imaging*. Physics in Medicine and Biology, 2020. **65**(18): p. 185009.

110. He K, et al., *Mask R-CNN*. In *Proceedings of the IEEE International Conference on Computer Vision*. 2017.

111. Dai X, et al., *Self-supervised learning for accelerated 3D high-resolution ultrasound imaging*. Medical Physics, 2021. **48**(7): p. 3916–3926.

112. Yang X, et al., *Self-supervised learning-based high-Resolution ultrasound imaging for prostate brachytherapy*. International Journal of Radiation Oncology, Biology, Physics, 2021. **111**(3): p. e119–e120.

113. de Vries M, et al., *Axially rigid steerable needle with compliant active tip control*. PloS One, 2021. **16**(12): p. e0261089.

114. Smith BR, et al., *Implementation of a real-time, ultrasound-guided prostate HDR brachytherapy program*. Journal of Applied Clinical Medical Physics, 2021. **22**(9): p. 189–214.

13 Freehand 3D Reconstruction in Fetal Ultrasound

Mingyuan Luo, Xin Yang, and Dong Ni

INTRODUCTION

Ultrasound (US) imaging is one of the main diagnostic tools in clinical practice due to its safety, portability, and low cost. Three-dimensional (3D) US is increasingly used in clinical diagnosis [1–3] because of its rich context information that is not offered in two-dimensional (2D) US. However, routine 3D US probe, such as electronic phased array and mechanical system, are constrained by the limited field of view and high cost. Thus, freehand 3D US, which explores 3D freehand US reconstruction from a series of 2D frames, has great application benefits, such as large field of view, relatively high resolution, low cost, and ease of use [4].

TRADITIONAL 3D ULTRASOUND RECONSTRUCTION

Developing automatic freehand US reconstruction methods is a challenging task due to the complex in-plane and out-of-plane shifts among adjacent frames, which are caused by the diverse scan strategies and paths. Freehand 3D US reconstruction has been studied for over half a century [5]. Early solutions mainly relied on complex and expensive external positioning systems to accurately calculate image locations [4–6]. The non-sensor scheme is mainly speckle decorrelation [7], which uses speckle correlation between adjacent images to estimate relative motion and decomposes it into in-plane and out-of-plane parts. However, the reconstruction quality is susceptible to scan direction and speed [4].

DNN-BASED 3D ULTRASOUND RECONSTRUCTION

Recently, the resurgence of deep neural networks (DNNs) profoundly promotes the 3D reconstruction performance over traditional machine learning-based methods. Interleaving convolution layer, pooling layer, and nonlinearity layer in a bionic connection fashion, DNNs discard traditional hand-crafted features and seamlessly learn the hierarchical features of an image with the training of the classifier. Prevost et al. [8] first used a convolutional neural networks (CNN) to estimate the relative motion of US images. Guo et al. [9] proposed a deep contextual learning network (DCL-Net) to mine the correlation information of US video clips for 3D reconstruction. Although effective, the reconstruction performance based on deep learning relies on the training data heavily and ignores the reconstruction robustness on the complex sequences during the test phase and still faces challenges such as difficult elevational displacement estimation and large cumulative drift. In this chapter, we will introduce our recent efforts in modifying the design of deep-learning-based method for 3D fetal US reconstruction. We concentrate on three problems that prove to be closely related to the network performance.

> **Temporal network**: How to digest the temporal information from 2D image sequence input can fundamentally affect the 3D reconstruction performance. With our work, we prove that DNNs can benefit more from the spatial information that is inherently contained in the 2D image sequence by using temporal-based deep-learning model, i.e., long short-term memory (LSTM) [10].

DOI: 10.1201/9781003299462-16

Training data generation: The learning process of DNN relies on training on a large amount of data. For freehand 3D US reconstruction, the training data consists of 2D US image sequences with position information. However, it is very difficult to obtain the position information during the US scanning. Our work designs a method that generates 2D US sequences and corresponding position information from 3D US volumes, which proves to be effective in greatly facilitating the training process.

Online learning: Most previous deep-learning-based 3D reconstruction methods [8, 9, 11] relied solely on offline network training and direct online inference. This strategy often fails to handle scans with a different distribution than the training data. Online self-supervised learning can address this challenge by leveraging adaptive optimization and valuable prior knowledge.

END-TO-END 3D RECONSTRUCTION

In this section, we will introduce the online learning framework (Figure 13.1) to combat the challenge of freehand 3D US reconstruction.

CONVOLUTIONAL LSTM NETWORK

Considering that extracting representations across temporal context is vitally important for 3D US volume reconstruction, we first implement a temporal CNN, i.e., LSTM. Compared with traditional CNN structure, the LSTM is capable of encoding representations from temporal sequence, and therefore extracting more discriminative features via richer temporal information. The main components of the LSTM are the memory gates and forget gates which are successively stacked as a temporal architecture.

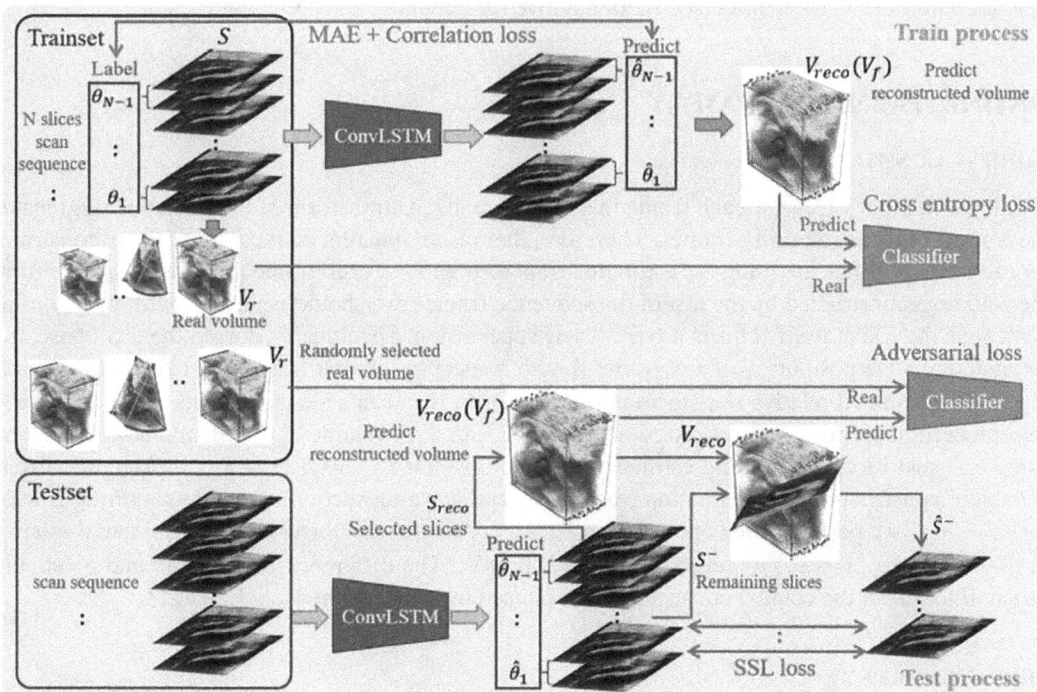

FIGURE 13.1 Overview of our proposed online learning framework.

To make LSTM better handle temporal information, we input 2D US image sequences into ResNet to extract features, and then input the output features into LSTM. ResNet is a powerful feature extractor mainly composed of convolutional layer, normalization layer, nonlinear activation function, and fully connected layer. As shown in Figure 13.1, ResNet and LSTM are connected to form a convolutional LSTM which learns by stochastic gradient descent in a data-driven manner. It is a key advancement in DNNs compared to the traditional predefined transformation of hand-crafted features.

END-TO-END LEARNING FOR 3D RECONSTRUCTION

We define the N-length US scan as $S = \{S_i \mid i = 1, 2, \ldots, N\}$. The goal of 3D reconstruction is to calculate the position of each US image frame in 3D space. Instead of directly computing the 3D position, we compute the position transformation between two consecutive US image frames, which eliminates the effect of the 3D position coordinate system. We use the rigid transformation θ_i to represent the position transformation between frame S_i and frame S_{i+1}, containing three translation and three rotational degrees. The 3D relative transformation parameters of the entire sequence S are denoted as $\Theta = \{\theta_i \mid i = 1, 2, \ldots, N\}$.

In the training phase, the ConvLSTM predicts relative transform parameters $\hat{\Theta}$ between all adjacent frames in the scan sequence. Its loss function includes two items, the first item is the mean absolute error (MAE) loss (i.e., L1 normalization), and the second item is the case-wise correlation loss from [9], which is beneficial to improve the generalization performance.

$$L = \| \hat{\Theta} - \Theta \|_1 + \left(1 - \frac{\mathbf{Cov}(\hat{\Theta}, \Theta)}{\sigma(\hat{\Theta})\sigma(\Theta)} \right), \tag{13.1}$$

where $\| \cdot \|_1$ indicates L1 normalization, $\mathbf{Cov}$ gives the covariance, and σ calculates the standard deviation. In the test phase, we construct an online learning process to optimize the model's estimation of complex scan sequences (see section above for details).

ONLINE LEARNING STRATEGY

CONTEXT CONSISTENCY

As shown in Figure 13.1, for each frame in a sequence, its 3D transformation estimation aggregates the context of its neighboring frames. Therefore, there is an inherent context consistency constraint for each frame. When applying the estimated transformation of each frame to extract the slice from the volume reconstructed by the rest of the sequence frames, we should get a slice with very similar content as the frame itself. This is a typical self-supervision. Specifically, during the test phase, the estimated relative position $\hat{p}_i$ of any frame s_i with respect to the first frame s_1 is calculated according to the estimated relative transform parameter $\hat{\Theta}$. In the scan sequence S, we firstly uniformly select a certain proportion (we set 0.5) of frames $S_{reco} \subset S$. A volume V_{reco} is then reconstructed by using S_{reco} and its corresponding estimated relative position $\hat{P}_{reco} = \{\hat{p}_i \mid s_i \in S_{reco}\}$ through a differentiable reconstruction approximation (see end-to-end learning section). For the remaining frames $S^- = S - S_{reco}$, we perform slice operation in V_{reco} according to the corresponding estimated relative position $P^- = \{\hat{p}_i \mid s_i \in S^-\}$ to get the generated slice $\hat{S}^-$. The difference between S^- and $\hat{S}^-$ should be small to reflect the context consistency and is used to optimize the ConvLSTM.

SHAPE CONSTRAINT

Shape prior is a strong constraint to regularize the volume reconstruction. Exposing the models to shape prior under the online learning scheme, rather than the offline training, provides us more chances to better explore the constraint for more generalizability, especially in handling the complex scan sequences where unplausible reconstructions often occur. Specifically, as shown in Figure 13.1, we propose to encode the shape prior with an adversarial module [12]. The volume reconstruction V_r from the training set is randomly selected as a sample with real anatomical structures. While the volume V_f reconstructed from the testing sequence S with relative estimated position parameters $\hat{P}$ is taken as a sample of fake structure. A classifier C pre-trained to distinguish between the volume V_r and V_f reconstructed from all training sequences serves as the adversarial discriminator. It adversarially tunes the structure contained in the reconstructed volume via the proposed differentiable reconstruction approximation.

DIFFERENTIABLE RECONSTRUCTION APPROXIMATION

The end-to-end optimization of our online learning framework is inseparable from the differentiable reconstruction. In general, most of the unstructured interpolation operations in the reconstruction algorithm (such as the Delaunay algorithm) are not differentiable and they block our online learning. As a foundation of our online design, we firstly propose a differentiable reconstruction approximation to mimic the interpolation. As shown in Figure 13.2, the volume V is reconstructed using the N slices $\{s_j \mid j = 1, \cdots, N\}$. For any pixel $v_i \in V$, the distance d_{ij} from any slice s_j and the gray value G_{ij} of the projection points on any slice s_j are calculated. Then the reconstructed gray value G_{v_i} at pixel v_i is calculated as:

$$G_{v_i} = \sum_j W(d_{ij})G_{ij}. \tag{13.2}$$

Among them, $W(\cdot)$ is a weight function. Its purpose is to encourage small d_{ij}. The core formulation is softmax operation on the reciprocal of d_{ij}:

$$W(d_{ij}) = \frac{\exp(1/(d_{ij} + \epsilon))}{\sum_j \exp(1/(d_{ij} + \epsilon))}, \tag{13.3}$$

where ϵ prevents division by 0. If only the nearest slice is weighted, the approximation can be expressed as follows:

$$W(d_{ij}) = \frac{1}{Z} \left(\frac{1}{d_{ij} + \epsilon} \frac{\exp(1/(d_{ij} + \epsilon))}{\sum_j \exp(1/(d_{ij} + \epsilon))} \right), \tag{13.4}$$

where $1/Z$ is the normalization coefficient such that $\sum_j W(d_{ij}) = 1$.

Figure 13.2 shows two reconstruction examples. The left volume of each example directly puts the slice into the volume according to the position without interpolation, and the right is the result of differentiable reconstruction approximation.

LOSS FUNCTION

In the test phase, SSL and ADL are jointly iteratively optimized. The first two items of L_d are the adversarial loss used to optimize the classifier C, and the third item is the quadratic potential

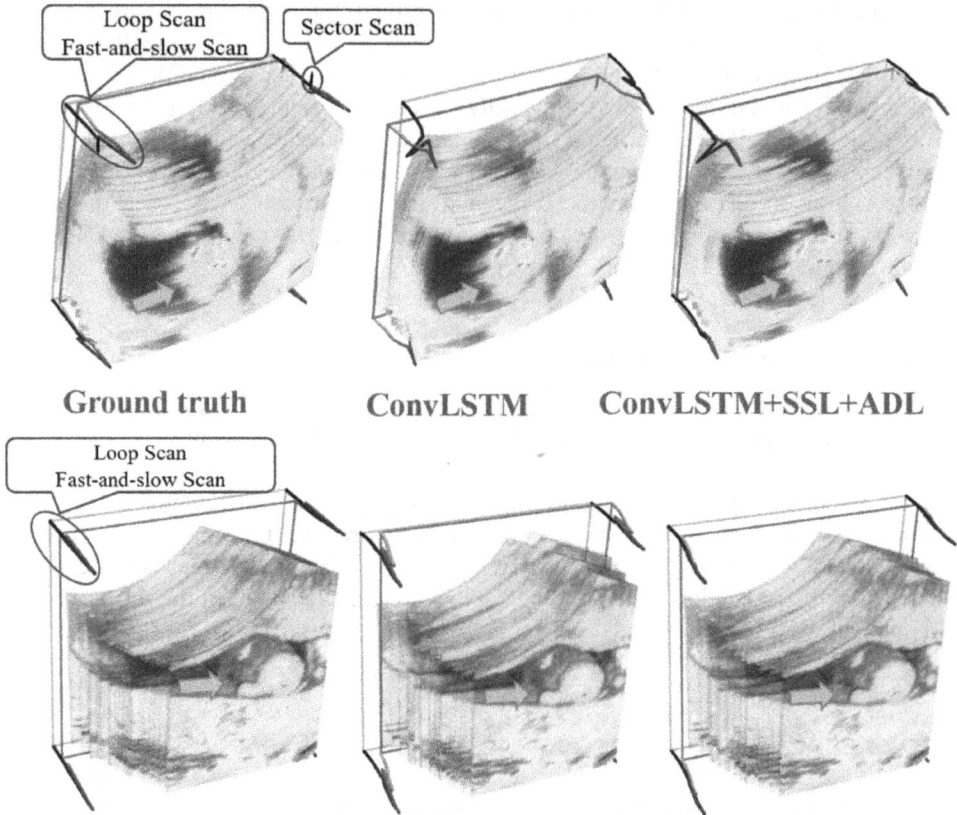

FIGURE 13.2 Two reconstruction examples of differentiable reconstruction approximation.

divergence from ref. [13], which helps to prevent the gradient vanishing without additional Lipschitz constraint. The first term of L_g is the adversarial loss used to optimize the ConvLSTM, and the second term is the self-supervised loss.

$$L_d = E_{V_f \sim P_{V_f}, V_r \sim P_{V_r}} \left[C(V_f) - C(V_r) + \frac{\| C(V_f) - C(V_r) \|_2^2}{2 \| V_f - V_r \|_1} \right], \tag{13.5}$$

$$L_g = -E_{V_f \sim P_{V_f}}[C(V_f)] + \| \hat{S}^- - S^- \|_1, \tag{13.6}$$

where $\| \cdot \|_1$ and $\| \cdot \|_2$ indicate L1 and L2 normalizations, respectively.

EXPERIMENTS

DATASETS

Our experiments involve 3D fetus US dataset. The fetal dataset contains 78 fetal US volumes from 78 volunteers. The gestational age ranges from 10 to 14 weeks. Its average volume size is $402 \times 535 \times 276$, the voxel resolution is $0.3 \times 0.3 \times 0.3 \text{mm}^3$. All data collection is anonymous. The collection and use of data are approved by the local IRB.

It is time-consuming and expensive to collect a large number of real freehand sequences with complex scan strategies. In order to verify the proposed framework on the complex scan strategies, the real 3D volumes are used to generate massive complex scan strategies for our training and evaluation. Multiple complex scan sequences are dynamically simulated with diverse scan strategies from each US volume to form our final corpus. Specifically, the scan sequences are a complex combination of loop scan, fast-and-slow scan, and sector scan with the aim to simulate the loop movement, uneven speed, and anisotropic movement of probe.

EXPERIMENTAL SETTINGS

We randomly divide the 3D fetal US dataset into 65/13 volumes for training/testing according to the patients, and the sequence length is 90. The number of generated sequences is 100 and 10 for training and testing, respectively. All generated slices are of size 300×300 pixel. In the training phase, the Adam optimizer is used to iteratively optimize our ConvLSTM. The epoch and batchsize are 200 and 4, respectively. The learning rate is 10^{-3}, and the learning rate is reduced by half every 30 epochs. The ConvLSTM has one layer and convolutional kernel size is 3. For classifier C, the epoch, batchsize, and learning rate is 50, 1, and 10^{-4}, respectively. During the test phase, for each testing sequence, we iterate the online learning with adversarial and self-supervised losses for 30 iterations, and the learning rate is set as 10^{-6}.

EVALUATION METRICS

The current commonly used evaluation indicator is the final drift [8, 9], which is the drift of final frame of a sequence, and the drift is the distance between the center points of the frame according to the real relative position and the estimated relative position. On this basis, a series of indicators is used to evaluate the accuracy of our proposed framework in estimating the relative transform parameters among adjacent frames. Final drift rate (FDR) is the final drift divided by the sequence length. Average drift rate (ADR) is the cumulative drift of all frames divided by the length from the frame to the starting point of the sequence, and finally, the average value is calculated. Maximum drift (MD) is the maximum accumulated drift of all frames. Sum of drift (SD) is the sum of accumulated drifts of all frames. The bidirectional Hausdorff distance (HD) emphasizes the worst distances between the predicted positions (accumulatively calculated by the relative transform parameters) and the real positions of all the frames in the sequence.

QUALITATIVE AND QUALITATIVE RESULTS

Table 13.1 summarizes the overall comparison of the proposed online learning framework with other existing methods and ablation frameworks on simulated fetus complex scan strategies, respectively. CNN-OF [8] and DCL-Net [9] are considered in this study for comparison. SSL and ADL are our proposed online self-supervised learning and adversarial learning methods, respectively. As can be seen from Table 13.1, our proposed methods get consistent and better improvements on complex scan strategies. The comparison with the ablation frameworks fully proves the necessity of introducing online learning in the test phase. In particular, our proposed online learning methods help the ConvLSTM improve the accuracy by 6.77%/1.67mm for fetal sequences in terms of the FDR/HD indicators. Under the complementary effects of mining context consistency and shape constraint, the proposed framework has significantly robust the estimation accuracy of complex scan strategies.

Figure 13.3 shows the indicator declining curves for online iterative optimization of the ConvLSTM on fetus US datasets. In each part of Figure 13.3, the abscissa and ordinate represent the

TABLE 13.1

The Mean (Std) Results of Different Methods on the Fetus Sequences. The Best Results Are Shown in Bold

Methods	FDR(%)↓	ADR(%)↓	MD(mm)↓	SD(mm)↓	HD(mm)↓
CNN-OF [8]	16.59(9.67)	31.82(14.02)	24.50(10.43)	1432.89(698.88)	19.16(8.21)
DCL-Net [9]	12.47(8.49)	30.58(16.05)	17.74(8.56)	882.00(452.00)	15.88(7.71)
ConvLSTM	16.71(5.22)	31.33(12.68)	20.57(8.33)	938.42(391.39)	15.79(6.78)
ConvLSTM+SSL	15.86(5.10)	30.78(12.20)	20.01(8.27)	909.95(382.43)	15.59(6.53)
ConvLSTM+ADL	10.49(4.86)	28.15(9.86)	17.49(7.68)	762.79(292.92)	14.18(5.99)
ConvLSTM+SSL+ADL	**9.94(4.41)**	**27.08(9.31)**	**16.84(7.56)**	**730.11(303.20)**	**14.12(5.89)**

number of iterations and the extent of indicators decline, respectively. The blue dot curve is the average indicator declining curve of all data. It is obvious that the indicators curve of almost all single sequences is declining significantly, and the downward trend is from rapid decline to gentle convergence. Among the various indicators, FDR has the largest decline, which is reduced to 59.48% of the ConvLSTM on the fetus US dataset. Figure 13.4 shows four specific cases of online iterative optimization on fetus complex scan strategies. In each part of Figure 13.4, the orange and red boxes represent the first and final frame, respectively and the pipeline that change color represent the scan path. It can be observed that through online iterative optimization, the final reconstruction result is significantly improved compared with the ConvLSTM, with smaller drift and more stable trajectory estimation.

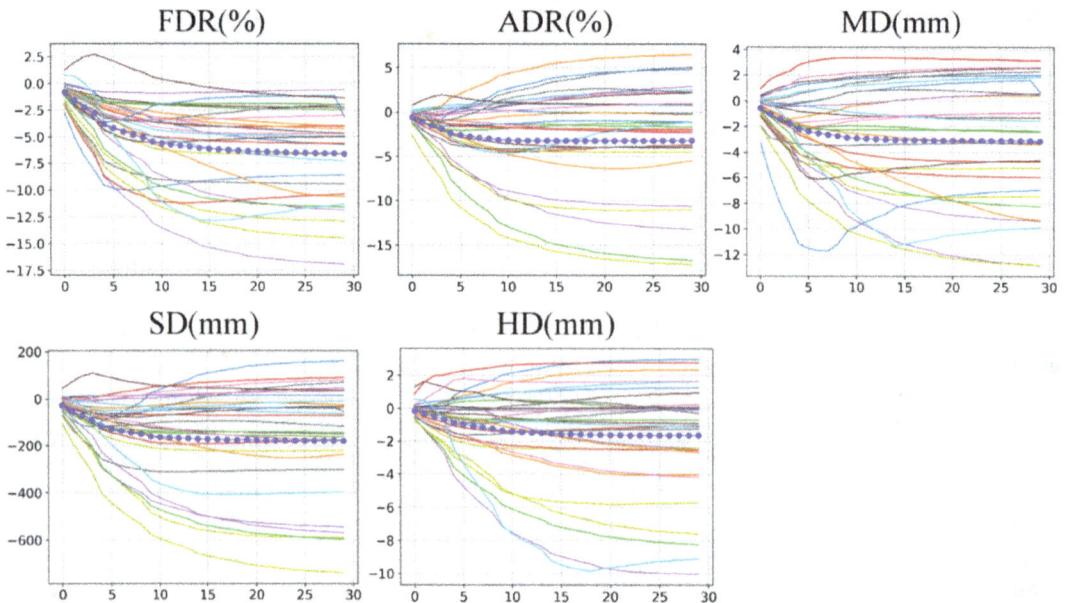

FIGURE 13.3   Indicator declining curves for online iterative optimization on fetus US datasets.

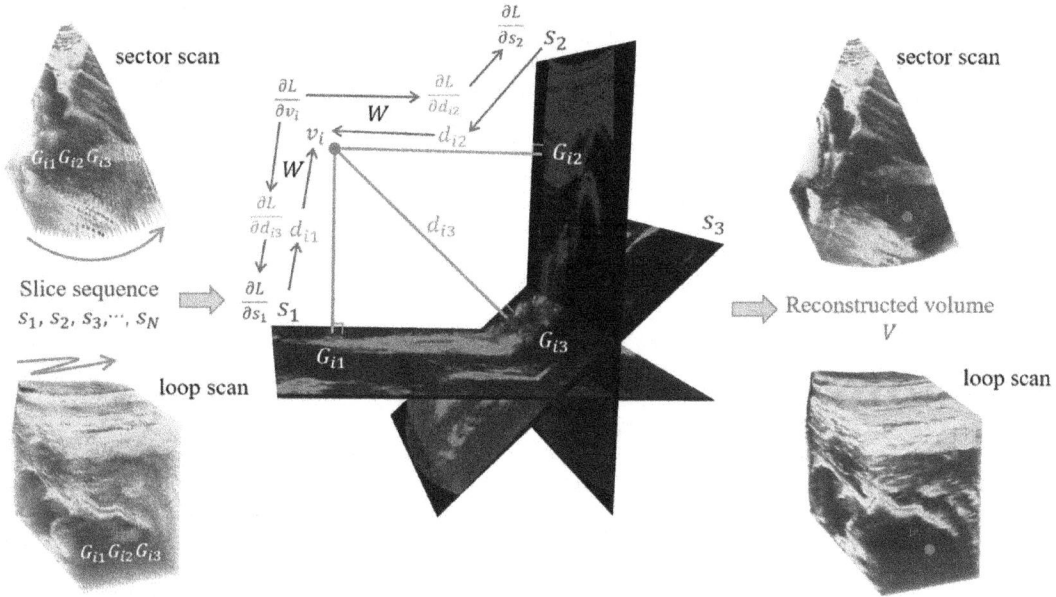

FIGURE 13.4 Specific case of online iterative optimization on fetus cases. The shaded arrows indicate the anatomic structure.

CONCLUSION

With this chapter, we present our investigation about freehand complex scan 3D US volume reconstruction. Benefiting from self-context, shape prior, and differentiable reconstruction approximation, the proposed online learning framework can effectively solve the challenge of 3D US reconstruction under complex scan strategies. Experiments on fetal US dataset prove the effectiveness and versatility of the proposed framework.

ACKNOWLEDGMENTS

This work was supported by the National Key R D Program of China (No. 2019YFC0118300), Shenzhen Peacock Plan (No. KQTD2016053112051497, KQJSCX20180328095606003), Royal Academy of Engineering under the RAEng Chair in Emerging Technologies (CiET1919/19) scheme, EPSRC TUSCA (EP/V04799X/1), and the Royal Society CROSSLINK Exchange Programme (IES/NSFC/201380).

REFERENCES

1. Yuhao Huang, Xin Yang, Rui Li, Jikuan Qian, Xiaoqiong Huang, Wenlong Shi, Haoran Dou, Chaoyu Chen, Yuanji Zhang, and Huanjia Luo, et al. Searching collaborative agents for multi-plane localization in 3d ultrasound. In *International Conference on Medical Image Computing and Computer-Assisted Intervention*, pages 553–562. Springer, 2020.
2. Shengfeng Liu, Yi Wang, Xin Yang, Baiying Lei, Li Liu, Shawn Xiang Li, Dong Ni, and Tianfu Wang. Deep learning in medical ultrasound analysis: A review. *Engineering*, 5(2):261–275, 2019.
3. Pádraig Looney, Gordon N Stevenson, Kypros H Nicolaides, Walter Plasencia, Malid Molloholli, Stavros Natsis, and Sally L Collins. Fully automated, real-time 3D ultrasound segmentation to estimate first trimester placental volume using deep learning. *JCI Insight*, 3(11):e120178, 2018.
4. Farhan Mohamed, and C Vei Siang. A survey on 3D ultrasound reconstruction techniques. Artificial Intelligence: Applications in Medicine and Biology. IntechOpen, 2019.

5. Mohammad Hamed Mozaffari, and Won-Sook Lee. Freehand 3-D ultrasound imaging: A systematic review. *Ultrasound in Medicine and Biology*, 43(10):2099–2124, 2017.

6. Laurence Mercier, Thomas Langø, Frank Lindseth, and Louis D Collins. A review of calibration techniques for freehand 3-D ultrasound systems. *Ultrasound in Medicine and Biology*, 31(2):143–165, 2005.

7. Theresa A Tuthill, JF Krücker, J Brian Fowlkes, and Paul L Carson. Automated three-dimensional us frame positioning computed from elevational speckle decorrelation. *Radiology*, 209(2):575–582, 1998.

8. Raphael Prevost, Mehrdad Salehi, Simon Jagoda, Navneet Kumar, Julian Sprung, Alexander Ladikos, Robert Bauer, Oliver Zettinig, and Wolfgang Wein. 3D freehand ultrasound without external tracking using deep learning. *Medical Image Analysis*, 48:187–202, 2018.

9. Hengtao Guo, Sheng Xu, Bradford Wood, and Pingkun Yan. Sensorless freehand 3D ultrasound reconstruction via deep contextual learning. In *International Conference on Medical Image Computing and Computer-Assisted Intervention*, pages 463–472. Springer, 2020.

10. Xingjian Shi, Zhourong Chen, Hao Wang, Dit-Yan Yeung, Wai-kin Wong, and Wang-chun Woo. Convolutional LSTM network: A machine learning approach for precipitation nowcasting. In C. Cortes, N. Lawrence, D. Lee, M. Sugiyama, and R. Garnett, editors, *Advances in Neural Information Processing Systems*, volume 28. Curran Associates, Inc., 2015.

11. Raphael Prevost, Mehrdad Salehi, Julian Sprung, Alexander Ladikos, Robert Bauer, and Wolfgang Wein. Deep learning for sensorless 3D freehand ultrasound imaging. In *International Conference on Medical Image Computing and Computer-Assisted Intervention*, pages 628–636. Springer, 2017.

12. Xin Yi, Ekta Walia, and Paul Babyn. Generative adversarial network in medical imaging: A review. *Medical Image Analysis*, 58:101552, 2019.

13. Jianlin Su. Gan-qp: A novel gan framework without gradient vanishing and lipschitz constraint, 2018.

14 Localizing Standard Plane in 3D Fetal Ultrasound

Yuhao Huang, Yuxin Zou, Houran Dou,
Xiaoqiong Huang, Xin Yang, and Dong Ni

INTRODUCTION

Ultrasound (US) is the primary screening method for assessing fetal health and development due to its advantages of being low-cost, real-time, and radiation-free [1]. Generally, during the diagnosis, sonographers first scan pregnant women to obtain US videos or volumes, then manually localize standard planes (SPs) from them. Next, they measure biometrics on the planes and make the diagnosis [2]. Of these, SP acquisition is vital for subsequent biometric measurement and diagnosis. However, it is very time-consuming to acquire dozens of SPs during the diagnosis and the process often requires extensive experiences due to the large difference in fetal posture and the complexity of SP definitions. Thus, automatic SP localization is highly expected to improve the diagnostic efficiency and decrease operator dependency. However, manually localizing SPs in 3D US is challenging because of the huge search space and large anatomical diversity. Therefore, an automatic approach to localize SPs in 3D US is desired to simplify the clinical process and reduce observer dependency.

As shown in Figure 14.1, there remain several challenges for localizing SPs in the 3D US automatically. First, the same SPs often have high intra-class variation. In the fetal brain, the high intra-class variation of SPs comes from the scale, shape, and contrast difference caused by fetal brain development, and also from the varying fetal postures. Second, the inter-class variation among different types of planes is often low. The differences between SPs and non-SPs, and even different SPs are often negligible, since they can contain the same or different anatomical structures with a similar appearance. For example, both cavum septum pellucidum and lateral ventricle appear on TT, TV, and other non-SPs, which may cause a high confusion for the algorithms. Third, SPs have different appearance patterns and fine definitions, which makes designing a general machine learning algorithm difficult. The fourth challenge lies in the varied spatial relationship among SPs. In our review of the related work on SP localization, we first introduce the approaches based on 2D US, and then we summarize the 3D US methods.

STANDARD PLANES DETECTION IN 2D US IMAGES

The early works of [3–6] selected SPs based on conventional machine learning methods (i.e., adaboost, random forest, support vector machine) through detecting key anatomical structures or landmarks of each frame in the video. Recent approaches made use of the convolutional neural network (CNN) due to its powerful ability in automatically learning hierarchical representations. The first two studies [7, 8] built the CNN model with transfer learning technology to detect fetal SPs. Chen et al. [9] then equipped the CNN with recurrent neural network (RNN) to capture the spatial-temporal information to detect three fetal SPs. Similar design can also be found in refs. [10, 11]. Baumgartner et al. [12] further proposed a weakly supervised approach to detect 13 fetal SPs and locate region of interest in each plane. Inspired by ref. [12], Schlemper et al. [13] incorporated the gated attention mechanism into the CNN to contextualize local information for detecting SPs. More recently, some works [14–16] proposed to assess US image quality automatically. Wu et al. [14] first

DOI: 10.1201/9781003299462-17

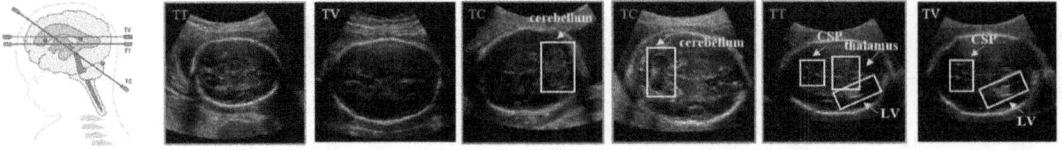

FIGURE 14.1 Visualization of the targeted fetal brain SPs in 3D US (left to right). Schematic diagram of anatomy (showing the genu of corpus callosum, splenium of corpus callosum, and cerebellar vermis from left to right). One typical example of the transthalamic (TT), transventricular (TV), and the transcerebellar (TC) plane. We show the TC plane of another subject. It proves that the fetal brain can have 180° orientation difference (note the cerebellum can appear on both sides of the image.) Also, note that the TT and TV planes are spatially close, their appearance can be extremely similar (e.g., the two right-most subfigures) and add more difficulty to the task.

introduced the quality assessment system of fetal abdominal plane by a cascade CNN. Luo et al. [15] and Lin et al. [16] then proposed to assess the quality of fetal brain, abdomen, and heart SPs by multi-task learning. These above methods showed the efficacy of detecting SPs and assessing image quality by transfer learning, spatial-temporal information, attention mechanisms, and multi-task learning. However, automatic SP detection in 2D US still suffers from the high dependence on clinicians' scanning skills. Although the aforementioned methods are effective in detecting SPs in 2D US, their tasks are essentially different from SPs localization in 3D US. In 2D US, plane localization is considered as a frame search task in video sequences with countable frames and thus the classification-based methods can handle this well. However, compared with 2D US, the 3D US contains countless slices, which makes the classification-based approaches intractable and brings exponentially growing false positives. Besides, localizing planes in 3D US require pinpointing the exact locations and navigating to the correct orientation in the huge space simultaneously, which is more complex than just locating the frame index in 2D US.

STANDARD PLANES DETECTION IN 3D US IMAGES

Automatic solutions for plane localization in 3D volumes can be roughly classified into two types:

1. Supervised learning (SL)-based methods such as registration, regression, and classification, and
2. Reinforcement learning (RL)-based approaches.

SL-Based Methods

Lu et al. proposed a boosting-based classification approach for automatic view planning in 3D cardiac MRI volumes [17]. Chykeyuk et al. developed a random forest-based regression method to extract cardiac planes from 3D echocardiography [18]. Ryou et al. proposed a cascaded algorithm to localize SPs of fetal head and abdomen in 3D US [19]. An SL-based landmark-aware registration solution was also employed to detect fetal abdominal planes [20]. Two regression methods were applied to localize the TT and TV planes in the fetal brain [21], and the abdominal circumference planes [22].

RL-Based Methods

In an RL framework, an agent interacts with the environment to learn the optimal policy that can maximize the accumulated reward. It is similar to the behavior of sonographers as they manipulate (*action*) the probe (*agent*) to scan the organ (*environment*) while visualizing the intermediate planes on the screen (*state*) till the SP is acquired (*reward* and *terminal state*). Alansary et al. [23] were the first to utilize RL in SP localization for 3D MRI volumes. These approaches achieved high accuracy and showed great potential in addressing the issues. Authors of [24] proposed to equip the RL framework with a landmark-aware alignment module for warm start and an RNN-based termination module for stopping the agent adaptively during inference. Ref. [25] explored the neural architecture search (NAS) in designing multi-agent collaborate framework, achieving light-weight model while with good plane localization performance. Ref. [26] restructured the action space and designed auxiliary task learning strategy and spatial-anatomical reward to further improve the model efficacy. These approaches achieved high accuracy and showed great potential in addressing the plane localization issues.

DEEP REINFORCEMENT LEARNING FRAMEWORK

In the classical single-agent RL (SARL) framework, one agent interacts with the environment $\mathcal{E}$ by making successive actions $a \in \mathcal{A}$ at states $\mathcal{S}$ to maximize the expectation of accumulated reward $\mathcal{R}$, where $\mathcal{A}$ is the action space. While in a multi-agent RL (MARL) system, multiple agents are created and the system endeavors to maximize the total reward obtained by all the agents [27, 28]. The pipelines for SARL-based plane localization methods are shown in Figures 14.2 and 14.3, and Figure 14.4 visualizes the MARL-based plane localization approach. In the following subsections, we will first introduce the key elements of simple-version SARL in detail. Then, we will extend the introduction to vallina MARL. Last, the classic RL algorithms will be presented.

SARL-BASED FRAMEWORK

We defined the *Environment* as the 3D US volume, and the plane in 3D space is typically modeled as $cos(\alpha)x + cos(\beta)y + cos(\gamma)z = d$, where $n = (cos(\alpha), cos(\beta), cos(\gamma))$ denotes the unit normal vector of the plane, and d is its Euclidean distance from the origin. The origin is set as the center of an US volume. We therefore define the main elements of this plane-localization RL framework as follows:

State: The state is defined as the reconstructed 2D US image from the volume given the current plane parameters. Since the reconstructed image size may change, we pad the image to a square and resize it to fixed size (e.g., 224×224). In addition, we concatenate the two images obtained from the previous two iterations with the current plane to enrich the state information, which is similar to study in ref. [31].

Action: The action is defined as incremental adjustment to the plane parameters. The complete action space is defined as $\mathcal{A} = \left\{ \pm a_\alpha, \pm a_\beta, \pm a_\gamma, \pm a_d \right\}$. Given an action, the plane parameters

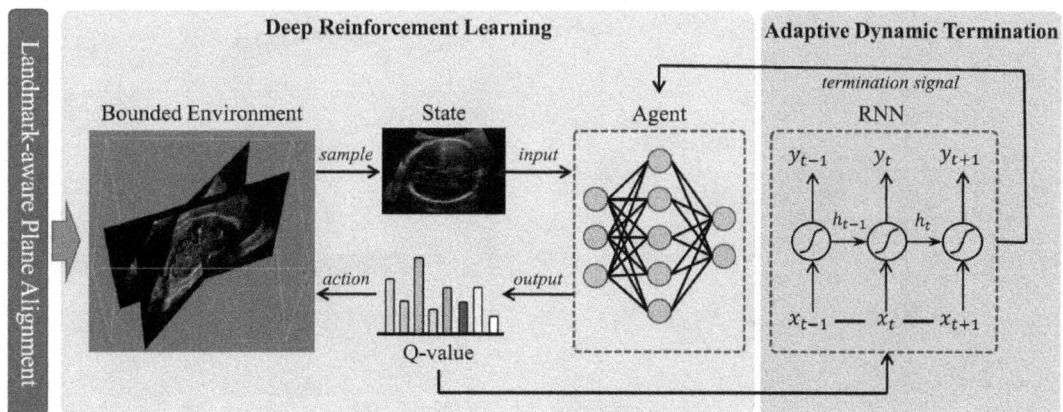

FIGURE 14.2 Schematic view of the proposed SARL framework in ref. [29].

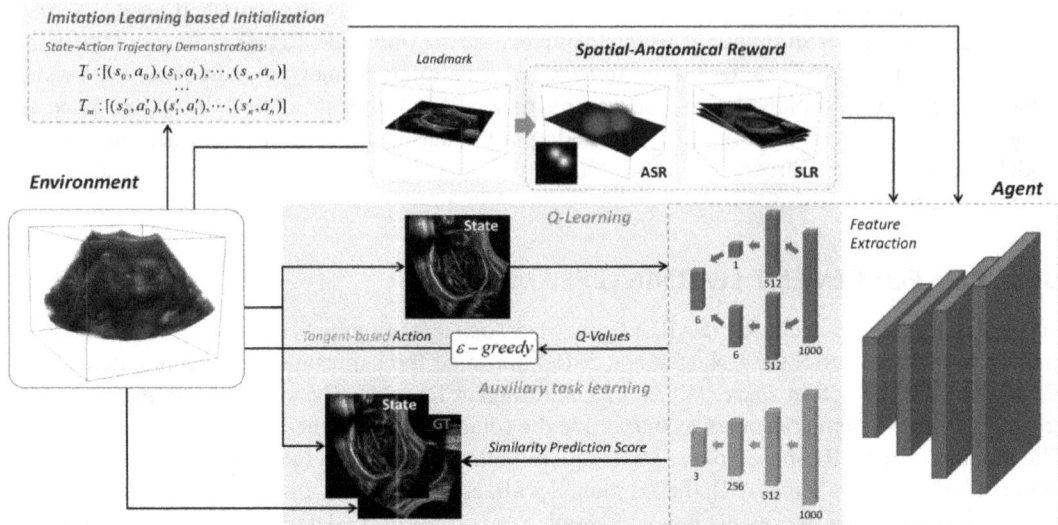

FIGURE 14.3 Schematic view of the proposed SARL framework in ref. [26].

are modified accordingly (e.g., $\alpha_i = \alpha_{i-1} + a_\alpha$). We perform one action to adjust only one plane parameter with the others unchanged for each iteration.

Reward: The reward signal defines the goal in an RL problem. It instructs the agent what policy should be taken to select the proper action. In this study, the reward is defined as whether the agent approaches or moves away from the target, which can be obtained by:

$$r = sgn(\| P_{i-1} - P_g \|_2 - \| P_i - P_g \|_2),\tag{14.1}$$

where P_i, P_g indicate the plane parameters of the predicted plane and the ground truth in iteration i, $sgn(\cdot)$ is the sign function. The universal set of the calculated reward signal

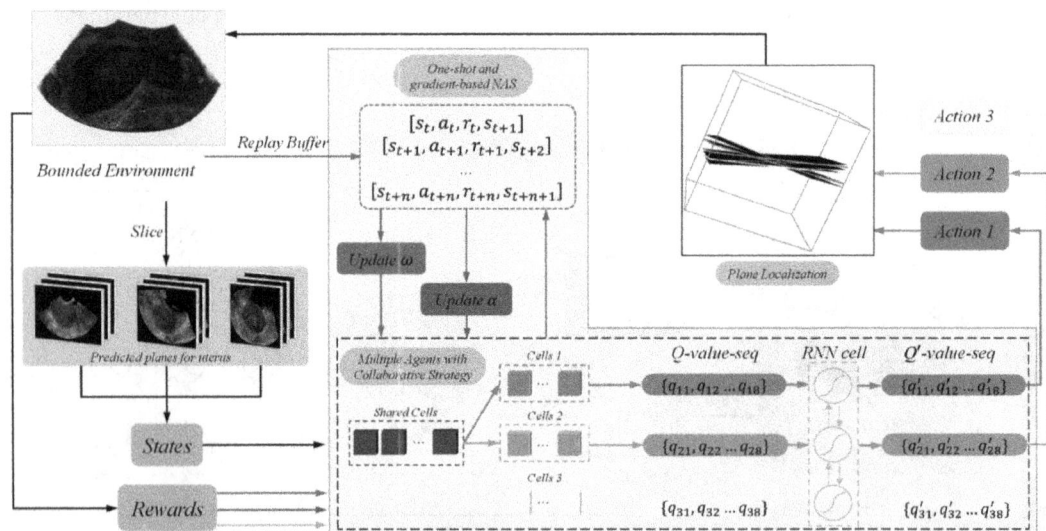

FIGURE 14.4 Schematic view of the proposed MARL framework in ref. [30].

is: $\{+1,0,-1\}$, where $+1$ and -1 indicate the positive and negative movement, respectively, and 0 refers to no adjustment.

Terminal step: In RL, a decision needs to be made by the agent as to whether to terminate the interaction with environment. Several works terminated the agent searching by setting the fixed steps [24], detecting oscillation [32], or the lowest Q-value [23].

MARL-Based Framework

Assume that there are three agents $agent_{k,k=1,2,3}$ searching three planes (P_1, P_2, P_3) in the MARL system. The *States* are defined as the last nine planes predicted by three agents, with each agent obtaining three planes. Like SARL, this setting can provide rich state information for the agents while keep the learning speed. The plane parameters of the P_n can be updated accordingly to obtain new planes (e.g., $\zeta_n \pm \alpha_{\zeta_n}$). The *Reward*, $R_k \in \{-1,0,+1\}$, for each $agent_k$ can be calculated by Eq. 14.1. For the *Terminal States*, the strategies to stop the agent–environment interaction are the same with SARL. The last step predictions will be treated as the final outputs.

RL Algorithms

Deep Q Network (DQN) and Double DQN

One could imagine that there exist numerous combinations of states and actions before the agents reach the final states. Therefore, instead of storing all the state-action values (Q-values) in a table in classical Q-learning [33], we opt for a strategy like DQN [31] where a CNN is used to model the relationship among states, actions, and Q-values. In specific, we adopt the double DQN (DDQN) [34] method for mitigating the upward bias caused by DQN and better stabilizing the training process. The training loss in DDQN can be defined as:

$$\mathcal{L} = E[(r + \gamma Q^*(s', \underset{a'}{argmax}\, Q(s', a'; \omega); \tilde{\omega}) - Q(s, a; \omega))^2], \tag{14.2}$$

where states s, actions a, rewards r and the states of the next step s' sampled from the naive replay buffer. It can remove data correlations and guarantee the training data matches the requirement of independent identically distributed (i.i.d). $\gamma \in [0,1]$ is a discount factor to weight future rewards. The Q network $Q(\omega)$ is used to obtain the Q-values of the current step and the actions a' in the next step. $Q^*(\tilde{\omega})$ represents the target Q network leveraged to estimate the Q-values of the next step. ω and $\tilde{\omega}$ are the network parameters of Q network and target Q network. $\varepsilon - greedy$ algorithms are utilized for the action selection to balance the exploration and exploitation during training [31]. Normally, the Q-values will be output by the fully connected layers and guide the actions of agents to update the plane parameters.

Dueling DQN

To further improve the RL performance, the dueling learning [35] is utilized to encourage the agent to learn which states should be weighted more and which are redundant in choosing proper actions. Specifically, the Q-value function is decomposed into a state value function and a state-dependent action advantage function, respectively.

As shown in Figure 14.5, the deep duel neural network takes the state as input and outputs the action. The extracted high-level features are fed into two independent streams of fully connected layers to estimate the state value and the state-dependent action advantage value. The hidden units of the last fully connected layers are 1 and 8 in the state value estimation and state-dependent action value estimation streams, respectively. The outputs of the two streams are fused to predict the final Q-value.

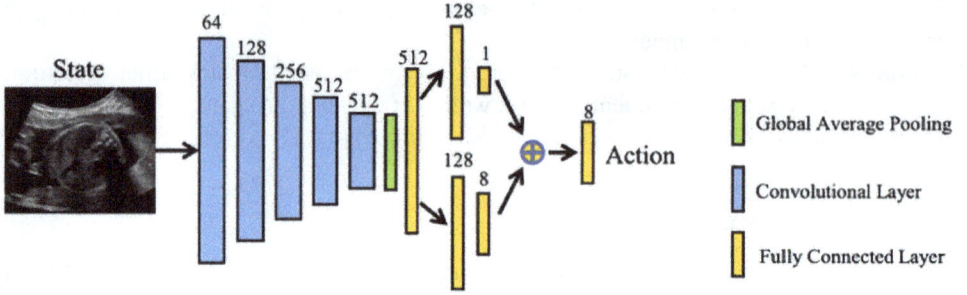

FIGURE 14.5 Architecture of our agent neural network. Blue, green, and yellow rectangles represent convolutional layers global average pooling and fully connected layers, respectively.

As explained above, the Q-value function, $Q(s,a;w)$, is decomposed into two separate estimators including the state value function $V(s;w_c,w_v)$ and the state-dependent action advantage function $A(s,a;w_c,w_s)$, where s is the input state of the agent, a is the action, w_c, w_v, w_s represent the parameters of the convolution layers and the two streams of fully connected layers, respectively, and $w = \{w_c, w_v, w_s\}$. The Q-value function of the agent is calculated as:

$$Q(s,a;w) = V(s;w_c,w_v) + A(s,a;w_c,w_s)$$
$$-\frac{1}{|A|}\sum_a A(s,a;w_c,w_s) \tag{14.3}$$

where $|A| = 8$ denotes the size of the action space. The loss function is then defined as:

$$L(w) = E_{s,a,r,\hat{s}\sim U(M)}[(r + \gamma \max_{\hat{a}} Q_{target}(\hat{s},Q(\hat{s},\hat{a};w);\tilde{w})$$
$$- Q(s,a;w))^2] \tag{14.4}$$

where γ is a discount factor to weight future rewards; $\hat{s}$ and $\hat{a}$ are the state and the action in next step; $U(M)$ represents uniform data sampling from the experience replay memory M; w and $\tilde{w}$ are the parameters of Q network ($Q(w)$) and target Q network ($Q_{target}(\tilde{w})$).

Prioritized Replay Buffer

For further improving the sampling efficiency of uniform-based replay buffer, prioritized replay buffer [36] was proposed to store the data sequences containing states s, actions a, rewards r and the states of the next step s' in the agent–environment interaction. Specifically, the i-th sequence element with high error will be sampled from the buffer preferentially, and its sampling probability can be calculated as $P_i = \frac{e_i^p + \delta}{\Sigma_k(e_k^p + \delta)}$, where $p = 0.6$ controls how much prioritization is used and $\delta = 0.05$ is set to adjust the error e_i. Besides, we also adopted importance-sampling weights for correcting the bias caused by the change of data distribution in prioritized replay similar to [36].

IMPROVEMENTS ON RL PLANE LOCALIZATION FRAMEWORK

RL SYSTEM INITIALIZATION

Background

Due to the low image quality, large data size and diverse fetal postures, it is very challenging to localize SPs in 3D US. Random state and model initialization used in ref. [23] often fail in localizing

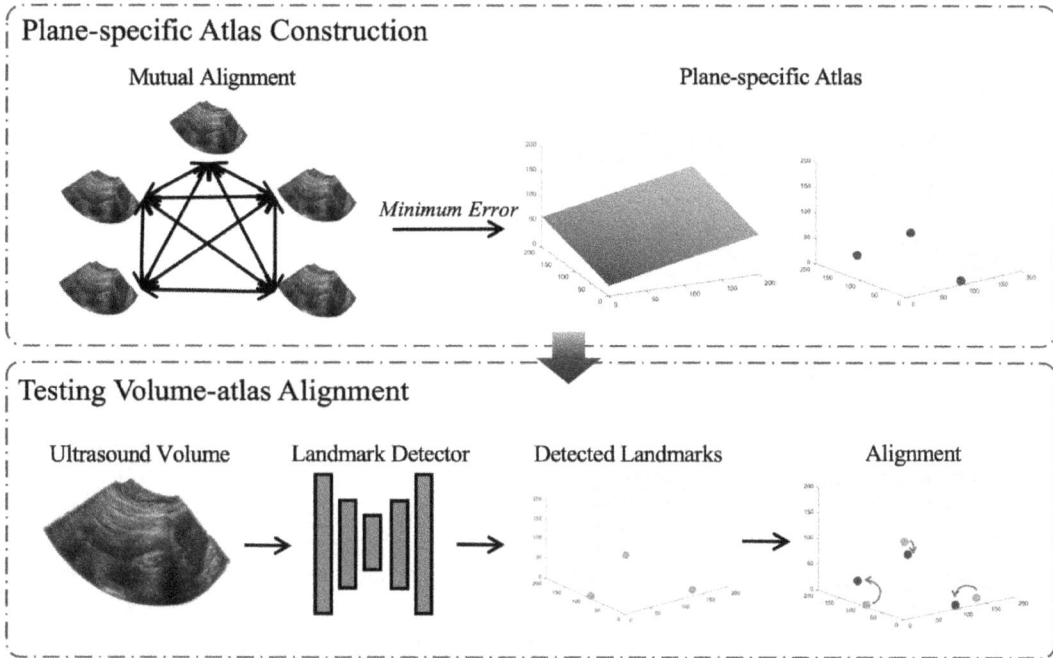

FIGURE 14.6 Pipeline of our landmark-aware alignment module. We first select the plane-specific atlas from the training dataset. Then, a pre-trained detection network is employed to predict the landmarks. Finally, we obtain the bounded environment by aligning the testing volume with the atlas based on the landmarks. The bounded environment includes the aligned volume and the initial position of the standard plane.

SPs because of the noisy 3D US environment. Thus, it is highly desired to provide good initialization for the RL-based localization framework. This section presents two options for solving the initialization problem. One is providing warm starts for the initial planes based on the landmark-aware registration module [24, 25, 29, 30]. The alternative one is to initialize the agent in RL framework using imitation learning module [26].

Landmark-aware Plane Alignment for Warm Start

The landmark-aware module aligns US volumes to the atlas space, thus reducing the diversity of fetal posture and US acquisition. As shown in Figure 14.6, the alignment module consists of two steps, namely plane-specific atlas construction and testing volume-atlas alignment. The details are described as follow.

ALGORITHM 14-1: ATLAS SELECTION

1: **for all** $i \in \{1,\ldots,N\}$ **do**
2:    Refer to the US volume V_i as the proxy atlas.
3:    **for all** $j \in \{1,\ldots,N\}$ **do**
4:      **if** $i == j$ **then**
5:       *continue*
6:      **else**
7:       $T_j^i \leftarrow$ transformation matrix from Vj to Vi.
8:       Perform rigid registration from Vj to Vi.

9:      $n_P^{\vec{j}} \leftarrow$ compute registered normal vector of plane P.

10:     $d_P^j \leftarrow$ compute the distance from the origin to the plane P in registered Vj.

11:     $error_{i,j} = \Theta^1(T_j^i \times \vec{n}_P^j, \vec{n}_P^i) + d_P^j - d_{P1}^i$

12:   **end if**

13:   **end for**

14:   $error_i = \dfrac{1}{N-1}\Sigma_j^{N-1} error_{i,j}$

15: **end for**

16: **return** The US volume with the minimum error.

- *Plane-specific atlas construction:* In this study, the atlas is constructed to initialize the SP localization in the testing volume through landmark-based registration. Hence, the atlas selected from the training dataset must contain both reference landmarks for registration and SP parameters for plane initialization. As shown in Figure 14.6, instead of selecting a common anatomical model for all SPs [37, 38], we propose to select specific atlas for each SP to improve the localization accuracy. In order to ensure the initialization effectiveness, ideally, the specific SP of the selected atlas should be as close to the SPs of other training volumes as possible.

  Algorithm 14.1 shows the determination of the plane-specific atlas volume from the training dataset based on minimum plane error (i.e., sum of the angle and distance between two planes). During the training stage, each volume is first taken as an initialized proxy atlas, then performing landmark-based rigid registration with the remaining volumes. According to the mean plane error measured between the linear-registered planes and ground truth for each proxy atlas, volume with the minimum error is chosen as the final atlas.

- *Testing volume-atlas alignment:* Our alignment module is based on landmark detection and matching. Unlike the direct regression, we convert the landmark detection as a heatmap regression task [39] to avoid learning a highly abstract mapping function (i.e. feature representations to landmark coordinates). We trained a customized 3D U-net [40] with the L2-norm regression loss, denoted as:

$$L = \frac{1}{N}\sum_{n=1}^{N}\left(\mathcal{H}_i - \hat{\mathcal{H}}_i\right)^2 \tag{14.5}$$

where $N = 3$ denotes the number of landmarks, and $\mathcal{H}_i$, $\hat{\mathcal{H}}_i$ represent the *i-th* predicted landmark heatmap and ground truth landmark heatmap, respectively. These ground truth heatmaps are created by placing a Gaussian kernel at the corresponding landmark location. During inference, we pass the test volumes to the landmark detector to get predicted landmark heatmaps. The coordinates with the highest value in the landmark heatmap are selected as the final prediction. We map the volume to the atlas space through the transform matrix calculated by the landmarks to create a bounded environment for the agent. Furthermore, we utilize the annotated target plane function of the atlas as the initial starting plane function for the agent.

Imitation Learning for Model Weight Initialization

It is difficult for the agent to obtain effective samples during interacting with the unaligned US environment because the replay buffer will store a number of futile data during the agent exploring the 3D space. It might reduce the agent learning efficiency even harm the performance.

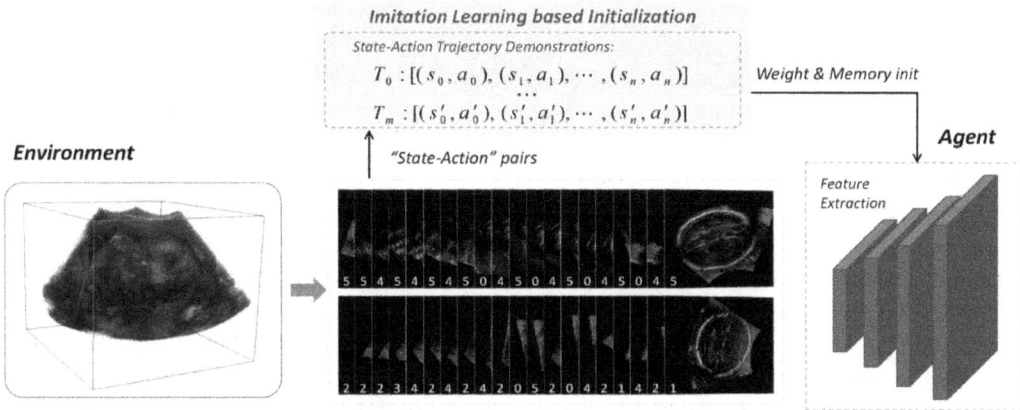

FIGURE 14.7 Pipeline for the imitation learning module.

[41] pointed out that imitation learning could effectively address this issue by pre-training the agent to ensure it could gain enough knowledge before exploring the environment, thus boosting the learning efficiency. Therefore, we adopt imitation learning as an initialization of the agent, as shown in Figure 14.7. Specifically, we first randomly select 20 initial tangent points in each training data and then approach the target plane by taking the optimal action in terms of the distance to the target tangent point. This can imitate the expert's operation and obtain a number of effective demonstrated state-action trajectories (e.g., $(s_0,a_0),(s_1,a_1),\ldots,(s_n,a_n)$). After that, we perform the supervised learning with the cross-entropy loss on the agent based on the randomly sampled state-action pair. With a well-trained agent based on imitation learning initialization, the learning of the RL framework can be eased and accelerated.

AGENT NETWORK CONSTRUCTION

Background

Constructing the agent(s) should include the communication modules in MARL, and the network architecture design (e.g., CNN, RNN) in both SARL and MARL, etc. **First**, in a MARL system, communication among multiple agents will influence the model learning a lot [42]. Recently, authors of [43] proposed a collaborative MARL framework for detecting multiple landmarks in MRI volumes. In their framework, the agents can only communicate with each other by sharing the low-level representations in a transfer learning manner, which can not explicitly represent the spatial relationships among SPs, and also, the decision policies of agents. Thus, an RNN-based module was proposed to collaborate the agents to catch the high-level information for strengthening the communication among agents. **Second**, in deep RL, designing a suitable neural network architecture of the agent is crucial in achieving good learning performance [44]. Previous methods used the same CNNs (like VGG) to localize different planes [23, 24]. One optimal strategy is to design plane-specific, task-adapting, and inter-tasks balanced (for MARL) neural network architectures (i.e., CNNs) for different planes via NAS, enabling more flexibility that might benefit the learning. Similarly, the architectures of the RNN in the agent collaborative module may also influence its ability to extract the shared spatial information and action policies. Hence, the similar technique can be utilized in designing a suitable RNN model. **Third**, localizing SPs in 3D US is challenging due to the low inter-class variability between SPs and non-SPs in the searching procedure and high intra-class variability of SPs. Most approaches[21, 29, 30] lack the proper strategy to use the image-level content information (e.g., anatomical priors), resulting in inefficient data utilization and agent learning. Auxiliary tasks for RL [45, 46] could improve learning efficiency and boost the performance by learning the fine-grained representations. In this section, we will first introduce the RNN-based agent collaborative module proposed in ref. [30]. Then, we will present the topic about NAS in automatically designing both CNN and RNN [25, 30]. Last, the auxiliary task and the designed network branch will be described [26].

RNN-Based Agent Collaborative Module

Figure 14.1 shows the most typical spatial relationships among planes in the fetal brain, and these relationships vary within certain ranges. Thus, learning such spatial information could boost the robustness and the accuracy of an automatic multi-plane localization model. However, the previous methods have not considered such latent spatial information [23, 43].

We aim at modeling such spatial relations explicitly using a novel RNN based agent collaborative module to improve the accuracy of plane localization. In specific, instead of directly outputting the Q-values through fully connected layers without communication like the established MARL frameworks [42] (dashed arrows at bottom of Figure 14.8), an RNN cell with bi-directional structure is proposed to be inserted among the agents (see Figures 14.4 and 14.8). The main reason for applying a bi-directional structure is that it can combine forward and backward information, and thus realizes information flowing and policy sharing among multiple agents. Therefore, it can enhance communication among agents.

Agents can collaborate to benefit each other in the learning of *plane-wise* spatial features. The Q-values in recurrent form are the collaborate media, which encode the *states* and *actions*, and inherently reflect the plane spatial positions and trajectory policy information. Specifically, at step t, the Q-value sequence set is denoted as $Q_t = (Q_{P_1}, Q_{P_2}, Q_{P_3})^T$, where the Q-value sequence

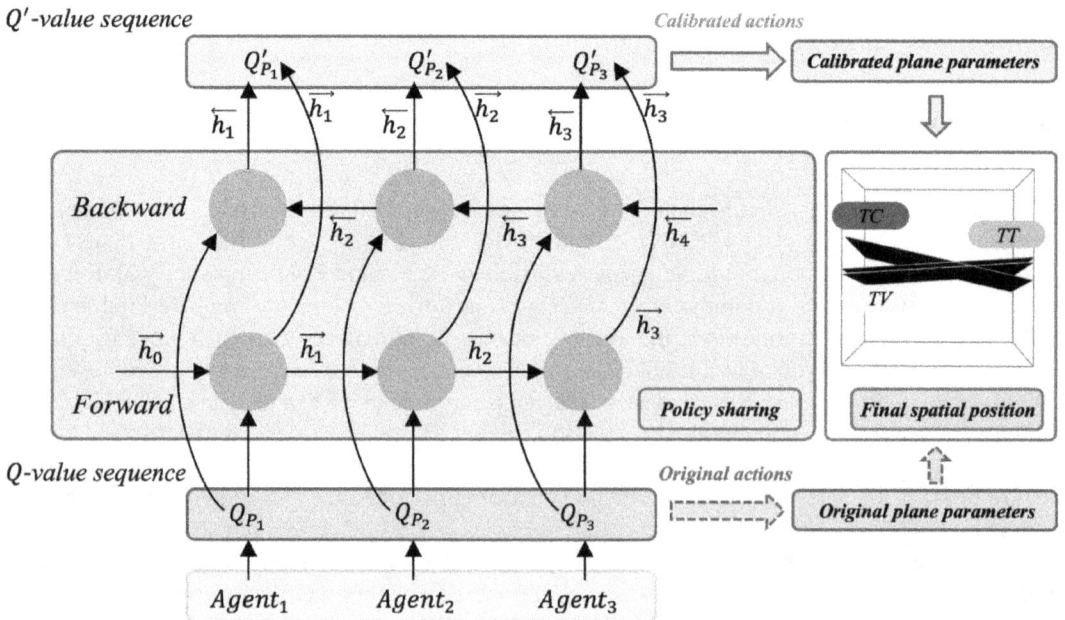

FIGURE 14.8 The proposed RNN collaborative module. The bottom block and dotted arrows indicate the plane-updating process which is based on the independently predicted Q-value sequence. The TT block is the RNN cell that makes the agents share policies. It takes the original Q-value sequence as inputs and outputs the calibrated Q'-values (top block). The proposed method uses the updated Q'-values to choose actions and update plane parameters (solid arrow) to decide the final spatial position of planes (center block).

Q_{P_i} for plane P_i, which contains 8 Q-values (according to the *Actions* space), can be written as: $Q_{P_i} = \{q_{i1}, q_{i2}, q_{i3}, ..., q_{i8}\}, i = 1, 2, 3$. The Q-value sequence Q_{P_i} of each plane is then passed to an RNN with bi-directional structure as its hidden-state of each time-step. Then the RNN module outputs a calibrated Q-value sequence set $Q'_t = (Q'_{P_1}, Q'_{P_2}, Q'_{P_3})^T$ and thus exploits full knowledge of all the detected planes, formally, we defined:

$$Q'_t = \mathcal{H}(Q_t, \overrightarrow{h^{i-1}}, \overleftarrow{h^{i+1}}; \theta), \tag{14.6}$$

where $\overrightarrow{h^i}$ and $\overleftarrow{h^i}$ are the forward and backward hidden sequence, respectively. $\mathcal{H}$ is the hidden layer function including linear and different activation operations, and θ represents the model parameters of the RNN. Then the agents take calibrated actions based on Q'_t to update the plane parameters (solid green arrow in Figure 14.8). In other words, the actions taken by each agent are based on not only its own action policy and spatial information, but also the policies and information shared by other agents.

NAS-Based Automatic Network Design

We explore to take advantage of NAS methods to design plane-specific networks automatically. To the best of our knowledge, it is the first time that NAS method is adopted to search the network architecture of agent in RL. It is noted that both RL and NAS require long training time and are hard to train even training them separately. Jointly training them thus faces more serious challenges. To address this issue, we explore to adopt one-shot and gradient-based NAS, i.e. GDAS, for searching both the CNN and the RNN to save training time and achieve a satisfactory performance. Here, one-shot and gradient-based methods represent that instead of training countless candidate sub-networks from scratch, we only need to train one supernet which is updated by gradient descent. Then, we can sample lots of sub-graph from the supernet for subsequent performance estimation.

Figure 14.4 illustrates the whole designed architecture. The details of its components are shown in Figure 14.9. The search process of both the agent (CNN) and the collaborative module (RNN) can be defined by the following elements: search space, search strategy and performance estimation strategy. Simply, search space represents the whole pre-defined network structure to be searched (in this paper, i.e. supernet) and the candidate operations. The search strategy defines how to search and

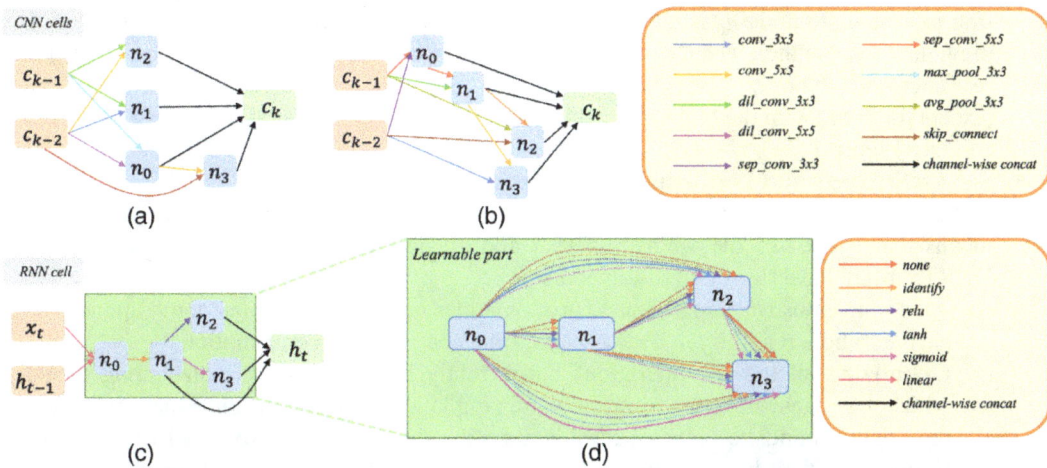

FIGURE 14.9 Details of CNN and RNN cells. (a) and (b) are two typical examples of CNN cells. (c) and (d) show one typical example of RNN cell and its learnable part, respectively. Different colored arrows represent different operations defined in the set of candidate operations. In the learnable part, the solid arrows form the sub-graph sampled from the supernet (represented by dotted arrows).

update the network structure. Besides, a performance estimation strategy is defined as the method for evaluating the performance of searched architectures.

- **Search Space:**

  Designing a large search space, including the number of layers, network branches, connection rules, candidate operations, hyper-parameters, etc., is time-consuming and may obtain an over-complex search space. In order to improve design efficiency and simplify the search space, we use the cell-based structure (like VGG, ResNet, etc.). It only needs to define several types of cells (including the number of nodes, filters and channels, the connection rules between nodes, etc.), which can be stacked to form the final search structure according to the connection rules of cells. The cell is a convolutional block and nodes in the cell represent the features maps. The basic design of the cells to construct the agent (i.e., CNN) and the RNN are shown in Figure 14.9, where the arrows represent different operations.

  Similar to DARTS [47], the CNN has two kinds of convolutional cells including normal cell and reduction cell. In definition, the input and the output of normal cells have the same size, while reduction cells output pooled feature maps with doubled channels. Each CNN cell consists of 7 nodes, including two input nodes, four intermediate nodes and one output node. c_{k-1} and c_{k-2} are equal to the outputs of the forward two cells, and the output c_k are defined as the channel-wise concatenation of nodes n_0 to n_3 (see Figure 14.9a,b). In the proposed framework, as shown in Figure 14.4, 3 agents share 8 CNN cells (5 normal cells and 3 reduction cells) while each of them has 4 unique cells (3 normal cells and 1 reduction cell). Thereby, the agents can share the common low-level features while extracting unique high-level representations for their own tasks. The CNN cells consist of 10 candidate operations, including none, 3×3 and 5×5 convolutions, separable convolutions and dilated convolutions, 3×3 max pooling, 3×3 average pooling, and skip-connection. In the validation stage, operations (except for *none*) with the top two weights are chosen from all previous nodes for each intermediate node in CNN cells (see Figure 14.9a,b).

  The RNN module consists of only one cell with a bi-direction structure. The recurrent cell contains 5 candidate operations, including none, identity mapping and three activation functions (i.e. *tanh*, *relu* and *sigmoid*). As shown in Figure. 14.9d, our recurrent cell also consists of 7 nodes: two input nodes, four intermediate nodes and one output node. The first intermediate node n_0 is calculated based on two input nodes x_t (input) and h_{t-1} (the previous hidden state). The output current hidden state h_t is defined as the average of the rest intermediate nodes (nodes n_1 to n_3). Different from the CNN cells, only one operation (except for *none*) with the highest weight is select for intermediate node n_1 to n_3 in the RNN cell (see Figure 14.9c).

- **Search Strategy and Performance Estimation Strategy:**

  For the search strategy, in DARTS, the sub-operations are optimized jointly, which means that it updates the whole supernet and all parameters in every iteration. Thus, DARTS might have two main disadvantages:
  i. All connections (sub-operations) between nodes need to be calculated and updated in each iteration, resulting in heavy search time and memory consumption.
  ii. Optimizing all the sub-operations concurrently leads to competition among them, which may cause unstable training. For example, sometimes, different sub-operations may output opposite values with a summation that tends to zero. It will hinder the information flow between two connected nodes and thus make the learning process unstable.

To accelerate and stabilize the learning process, GDAS uses the differentiable sampler to obtain the sub-graph from the supernet and only updates the sub-graph in each iteration. Specifically,

GDAS uses the *argmax* during the forward pass, and *softmax* with *Gumbel − Max* trick during the back propagation. The *Gumbel − Max* trick used on *softmax* can better approximate the realistic sampling operations. Specifically, when the training epoch increases and the softmax temperature parameter τ drops until $\tau \to 0$, the *Gumbel − softmax* tends to *argmax*. We refer readers to [48, 49] for more details. During search, the network weights ω and architecture parameters α are updated by gradient (Eq. 14.7 and Eq. 14.8) using prioritized weight based batch (M_p) and random-way based batch (M_r) data sampled from the replay buffer, respectively.

$$\omega = \omega - \nabla_\omega \mathcal{L}_{s,a,r,s' \sim M_p}(q,q^*;\omega,\alpha), \tag{14.7}$$

$$\alpha = \alpha - \nabla_\alpha \mathcal{L}_{s,a,r,s' \sim M_r}(q,q^*;\omega,\alpha), \tag{14.8}$$

where $\mathcal{L}$ is the mean square error (MSE) loss function. q is the direct output of the Q network $Q(\cdot)$, and q^* is calculated based on both the Q network and the target Q network $Q^*(\cdot)$, as shown below:

$$q^* = r + \gamma Q^*(s', \underset{a'}{argmax} \, Q(s',a')) \tag{14.9}$$

For the performance estimation strategy, when NAS is applied in non-RL tasks, the selection of the searched model is usually decided on the loss function [50]. Specifically, after the loss becomes converged, one of the searched architectures will be chosen as the final designed model according to their performances on the validation set. However, the loss in RL is oscillated and difficult to converge. Therefore, we select the optimal architecture parameters α^*_{CNN} and α^*_{RNN} for both the CNN and RNN using the maximum of the accumulated rewards on the validation set in all the training epochs. Using these, we can then construct bespoke CNN for each agent and an ideal RNN for the collaborative module.

Auxiliary Task of State-Content Similarity Prediction

To facilitate the agent to learn the content representations, we design an auxiliary task of state-content similarity prediction (SCSP). As shown in Figure 14.3, we utilize an additional regression branch in the agent network to predict the similarity of the current state to the target state. The content similarity is measured by normalized cross-correlation (NCC) [51]. The loss function for the auxiliary task part of our framework is defined as:

$$\mathcal{L}_A(\omega) = E_{s \sim U(\mathcal{M})} \, \| \, Score_{gt}(s, s_{gt}) - Score_{pre}(s; \omega) \, \|_2 \tag{14.10}$$

where $Score_{gt}(s, s_{gt})$ is the NCC between the current state s and the target state s_{gt}; $Score_{pre}(s; \omega)$ is the NCC score predicted by SCSP. Overall, the total loss function of our proposed RL framework is $\mathcal{L} = \mathcal{L}_Q + \delta \mathcal{L}_A$, where $\delta = 0.5$ is the weight to balance the importance of RL and auxiliary task losses.

ACTION SPACE AND TASK FORMULATION

Formulation of the SP localization is essential for optimizing the RL framework. Typically, previous works [23, 24, 29, 30] modeled the plane movement by adjusting the plane function in terms of normal and distance (see Figure 14.10). However, the coupling among the directional cosines ($cos^2(\alpha) + cos^2(\beta) + cos^2(\gamma) = 1$) makes actions dependent and unable to reflect the model's objective accurately, resulting in obstacles to agent learning. Furthermore, the RL training can easily fail without the pre-registration processing [24] to limit orientation variability and search space [29]. This study proposes a novel tangent-point-based formulation for SP localization in RL. Our formulation builds a simplified and mutual-independent action space to improve the optimization of the RL framework, enabling accurate SP localization even within the unaligned US environment.

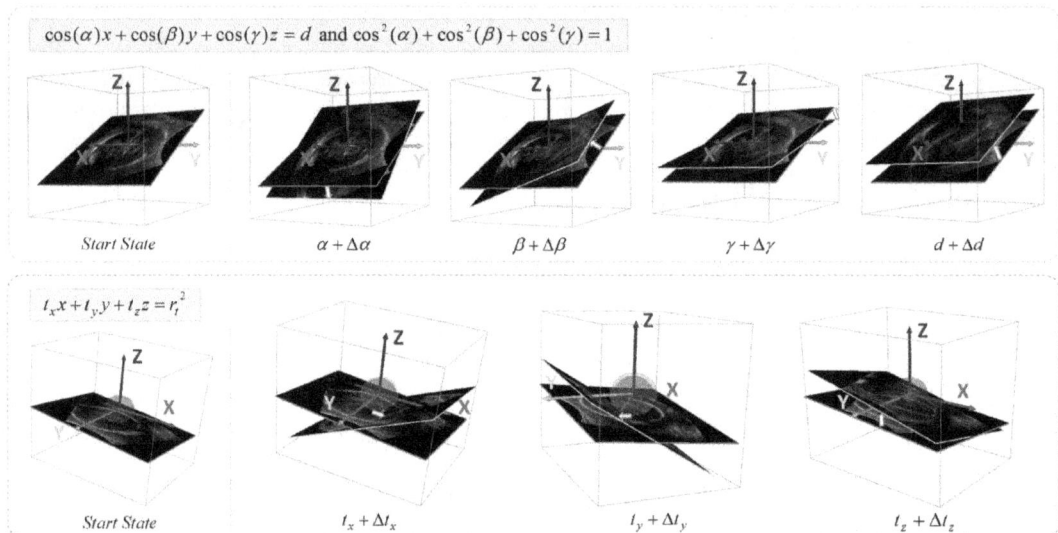

FIGURE 14.10 Comparison of our formulation (bottom) and previous one (top). Previous formulation controls the plane movement by adjusting directional cosines (α, β, γ) and the distance to the origin (d). Instead, our design modifies the plane movement by translating the coordinate (t_x, t_y, t_z) of the tangent point.

Figure 14.10 illustrates the comparison of our formulation and that of previous works. We discovered that any view plane in the 3D space can be defined uniquely as its tangent point (t_x, t_y, t_z) on the sphere centering in the origin with the radius of r_t, where $r_t^2 = t_x^2 + t_y^2 + t_z^2$. The plane function can be written as $t_x x + t_y y + t_z z = r_t^2$. Therefore, SP localization can be re-formulated into the tangent point searching task, where the action space only contains the translation of the coordinate of the tangent point. The proposed formulation is unrestricted by directional cosines coupling with less action space than the previous ones ($6 < 8$), facilitating agent learning.

Based on our formulation, the action space is defined as $\left\{ \pm a_{t_x}, \pm a_{t_y}, \pm a_{t_z} \right\}$. Given an action in step i, the tangent point coordinate can be modified accordingly, e.g., $t_x^{i+1} = t_x^i + a_{t_x}$. We noticed experimentally that the image content is sensitive to the step size when the corresponding sphere radius is small. To address this issue, we model the agent–environment interaction as a multi-stage motion process by progressively scaling down the step size from 1.0 to 0.01 when the agent appears to oscillate for three steps.

SPATIAL-ANATOMICAL REWARD DESIGN

The reward function instructs the agent on the optimal searching policy with the proper action. Recent works by authors of refs. [29, 30] calculated the reward function based on the differences of parameters in the defined plane function between adjacent iterations. Although effective, we argue that such design may cause the agent to lack anatomical perception and guidance, which may affect the agent's performance on the abnormal data. In this study [26], we design a spatial-anatomical reward, involving 1) spatial location reward (SLR) and 2) anatomical structure reward (ASR). Specifically, SLR motivates the agent to approach the target location by minimizing the Euclidean distance of the plane parameters between the current plane and target plane, while ASR encourages the agent to perceive anatomical information. We construct the heatmap with a Gaussian kernel at anatomical landmarks to calculate the ASR. The reward can be defined as:

$$ r = sgn(\| P_{t-1} - P_g \|_2 - \| P_t - P_g \|_2) + sgn\left(|I_t - I_g| - |I_{t-1} - I_g| \right) \tag{14.11} $$

where $sgn(\cdot)$ is the sign function, P_t and P_g indicate the tangent point parameters of the prediction plane and the target plane, respectively. Likewise, I_t and I_g represent the sum of the heatmap value corresponding to the prediction plane and the target plane, respectively.

ADAPTIVE DYNAMIC TERMINATION STRATEGY

In RL, a decision needs to be made by the agent as to whether to terminate the inference. The termination conditions are usually pre-set, such as reaching to the destination in *MountainCar* [52], pole's falling up in *CartPole* [53], etc. However, the termination conditions are often indistinct and can not be precisely determined in many tasks. Specifically, in the SP localization, the agent might fail to catch the target SP and continue to explore without termination condition. One solution was to extend the action space with a further terminate action [54]. However, enlarging the action space will result in insufficient training. Several works terminated the agent searching by detecting oscillation [32] or the lowest Q-value [23]. Our previous work [24] indicated that a learning-based termination strategy can improve the planning performance in deep RL. Although no additional action was introduced, these approaches still required the agent to complete inference with maximum step and obtain the whole Q-value sequence, which is inefficient. Therefore, an adaptive dynamic termination strategy to ensure the efficacy and efficiency of the SP localization is highly desirable in the SP localization task.

We update the active termination strategy into the adaptive dynamic termination, which is proposed in deep RL framework for the first time [29]. Specifically, considering the sequential characteristics of the iterative interaction, as shown in Figure 14.2, we model the mapping between the Q-value sequence and optimal step with an additional RNN model. The Q-value is defined as $\mathbf{q}_t = \{q^1, q^i, ..., q^8\}$, consisting of 8 action candidates at the iteration t; and the Q-value sequence refers to a time-sequential matrix $\mathbf{Q} = [\mathbf{q}_1, \mathbf{q}_2, ..., \mathbf{q}_n]$, where n denotes the index of iteration step. Taking the Q-value sequence as input, the RNN model can learn the optimal termination step based on the highest Angle and Distance Improvement (ADI). During training, we randomly sampled the sub-sequences from the Q-value sequence as the training data and denoted the highest ADI during the sampling interval as the ground truth. Unlike the previous studies [24], we design a dynamic termination strategy to improve the inference efficiency of the reinforcement framework. Specifically, our RNN model performs one inference every two iterations based on the current zero-padding Q-value sequence, enabling an early stop at the iteration step having the first three repeated predictions.

Previous study [24] used mean absolute error loss function to train the RNN in the termination module. However, it has constant gradient of back-propagation and lack of measuring the fine-grained error. This study replaces it with the MSE loss function to relive this and target a more stable training procedure. Since ground truth, i.e. optimal termination step, is usually larger than 1 (e.g., 10~75), the conventional MSE loss function may struggle to converge in training due to the excessive gradient. We adopt the MSE loss function with a balance hyper-parameter, and defined as:

$$L_{MSE}(w) = \left\| f(x;w) - \delta G_i \right\|_2^2 \tag{14.12}$$

where w is the RNN parameters, x is the input sequence of the RNN, $f(x;w)$ represents the RNN network, and G denotes the optimal termination step. The balance hyper-parameter $\delta = 0.01$ can normalize the value range of learned steps to [0, 0.75] approximately, thus simplifying the training process. The RNN model is trained using inference results obtained from training volumes.

EXPERIMENTS

DATASETS

The fetal brain dataset consists of 432 volumes acquired from different healthy subjects under trans-abdomen scanning protocol. The fetuses have a broad gestational age range: 19 to 31 weeks. As a

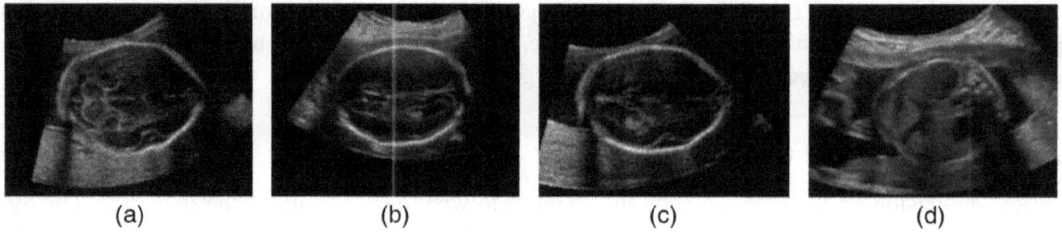

(a)   (b)   (c)   (d)

FIGURE 14.11 Sample cases of the studied four target planes (a–d) correspond to the TT, TV, TC, and AM planes, respectively. Particularly, planes (a–c) are in the fetal brain; plane (d) is in the fetal abdomen.

result, the fetal brain structures have large variations in size, shape and appearance. For the fetal abdomens, we have 519 volumes. The average volume size is $270 \times 207 \times 235$, and with isotropic voxel size of $0.5 \times 0.5 \times 0.5$ mm^3. Four sonographers with 5-year experience provided manual annotations of landmarks and SPs for every volume. Specifically, the annotated SPs are the TV, the TT and the TC planes in the fetal brain, and the fetal abdominal (AM) SP in the fetal abdomen (see Figure 14.11). For fetal brain data, the three landmarks are the genu and the splenium of the corpus callosum, and the center of cerebellar vermis. For the fetal abdomen volumes, the umbilical vein entrance, the centrum, and the neck of the gallbladder.

IMPLEMENTATION DETAILS

In this section, the introduced improvement mainly involve three works [26, 29, 30]. Thus, in this subsection, we will summarize the implementation details for the three works respectively.

In the work of [29], We implemented our framework in PyTorch [55], using a standard PC with an NVIDIA Titan XP GPU. We trained the whole framework through Adam optimizer [56] with a learning rate of 5e-5 and a batch size of 4 for 100 epochs, which cost about 4 days. We set the discount factor γ in the loss function (Eq. 14.4) as 0.9. The size of the Replay Buffer was set as 15000. The target Q network copied the parameters of the Q network every 1500 iterations. The maximum number of iterations in one episode was set as 75 in fetal dataset to reserve enough moving space for agent exploration. The initial ϵ for $\epsilon - greedy$ action selection strategy [10] was set as 0.6 at first and multiplied by 1.01 every 10000 iterations until 0.95 during training. The RNN variants, i.e. vanilla RNN and long short-term memory (LSTM), were trained for 100 epochs, using mini-batch stochastic gradient descent optimizer with a learning rate of 1e-4 and batch size of 100, which costed about 45 mins. The number of hidden units was 64 and that of the RNN layers is 2. The starting plane function for training the framework was randomly initialized around the ground truth plane within an angle range of $25°$ and distance range of $10\,mm$ to ensure the agent can explore enough space within the US volume. For landmark detection, we trained the network using Adam optimizer with a batch size of 1 and a learning rate of 0.001 for 40 epochs.

In the work of [30], the proposed framework is trained with Adam optimizer on an NVIDIA TITAN 2080 GPU. In the first stage of searching network architecture (about 3 days), we set the learning rates to 5e-5 for learning network weights ω and to 0.05 for learning the architecture parameters α, respectively. In the second stage of training the RL system for SP localization (about 3 days), we retrained the agents and the RNN module with fixed architectures for 100 epochs using a learning rate of 5e-5. The batch sizes is set as 24. The size of the RL replay buffer is set as 15000 and the target network copies the parameters of the current network every 1500 steps. During training, the initialized planes were randomly set around the target planes within an angle and distance range of $\pm25°/\pm5mm$. The step sizes in each update iteration for angles (a_ζ, a_β and a_ϕ) are set as $\pm1.0°$, and the step size of distance a_d is set as $\pm0.1mm$.

In the work of [26], we implemented the method by PyTorch using a standard PC with an NVIDIA RTX 2080Ti GPU. We trained the model through Adam optimizer with a learning rate of 5e-5 and a batch size of 32 for 100 epochs, which costs about 2.5 days. We set the discount factor γ equal to 0.85. The size of the prioritized Replay Buffer was set as 15000. The initial ϵ for $\epsilon - greedy$ exploration strategy was set as 0.6 at first and multiplied by 1.01 every episode until 0.95 during training. We calculated the mean (μ) and standard deviation (σ) of target tangent point locations in the training dataset and randomly initialized start points for training within $\mu \pm 2\sigma$ to capture 95% variability approximately. For testing, the origin was set as the initial tangent point. We terminate the agent searching at 60 steps.

EVALUATION CRITERIA

In [29], we used three main criteria to evaluate the localization accuracy of the predicted planes compared with the target plane. First, the spatial metrics, including angle and distance deviation between the two are estimated. Formally, we defined:

$$Ang = \arccos \frac{n_p \cdot n_g}{|n_p||n_g|} \tag{14.13}$$

$$Dis = |d_p - d_g| \tag{14.14}$$

where the n_p, n_g represent the normal of the predict plane and target plane, the d_p, d_g represent the distance from the volume origin to the predicted plane, and that to the ground truth plane. It is noted that the Ang and Dis are evaluated based on the plane sampling function, i.e., $cos(\alpha)x + cos(\beta)y + cos(\gamma)z = d$, with an effective voxel size of $0.5\,mm^3/voxel$. Moreover, it is also important to examine whether these two planes are visually alike. Therefore, Peak Structural Similarity (SSIM) [57] was leveraged to measure the image similarity of the planes. Besides, the ADI in iteration t is defined as the sum of the cumulative changes of distance and angle from the start plane, which is as follows:

$$ADI = (Ang_t - Ang_0) + (Dis_t - Dis_0) \tag{14.15}$$

Moreover, in ref. [30], we used the Sum of Angle and Distance (SAD) to reveal how close is the target to the ground truth. In ref. [26], instead of using SSIM only, we further added NCC to evaluate the content similarity between two images.

QUALITATIVE AND QUALITATIVE RESULTS

To analyze the experimental results in [26, 29, 30] respectively, this subsection mainly includes three parts. We first present a brief introduction to these three works. In [29], we proposed a SARL-based plane localization framework with landmark-aware plane alignment module and adaptive dynamic termination strategy (See Figure 14.2). The framework was validated on both fetal brain and abdomen datasets. In ref. [30], we introduced a collaborative MARL-based framework to localize three SPs in the fetal brain volume simultaneously. Meanwhile, the agent and the RNN-based agent collaborate module were automatically designed via the GDAS-based NAS method (See Figure 14.4). In ref. [26], we first restructured the action space and adopted imitation learning to initialize the model parameters. Then, we leveraged auxiliary task learning to ensure better quality control and spatial-anatomical reward to drive the moving of the agent efficiently (See Figure 14.3). The proposed framework used fetal brain US dataset for validation. The following parts will provide a detailed and thorough introduction to the results of the three studies.

Qualitative and Qualitative Results: PART 1

This section will present the results of the study discussed in ref. [29]. To examine the effectiveness of our proposed method in standard plane localization, we conducted a comparison experiment with the classical learning-based regression method, denoted as *Regression*, the current state-of-the-art automatic view planning (AVP) method [23], and our previous method [24], denoted as *RL-US*. To achieve a fair comparison, we used the default plane initialization strategy of the *Regression* and *AVP*, and re-trained all the two compared models using the public implementations. We also adjusted the training parameters to obtain the best localization results. As shown in Table 14.1, it can be observed that our method achieves the highest accuracy compared with the alternatives in almost all of the metrics. This indicates the superior ability of our method in standard plane localization tasks.

To verify the impact of the landmark-aware alignment module of the proposed approach, we compared the performance of the proposed framework with and without this module. In the *Pre-Regist* method, we set the agent with random starting plane function like [23] and choose the lowest Q-value [23] as the termination step. The *Regist* method represents the framework equipped with the alignment module but without agent searching. The *Post-Regist* method denotes the searching result of the agent with a warm-up initialization with the alignment module. We also chose the lowest Q-value termination strategy to implement the *Post-Regist* for a fair comparison. As shown in the Table 14.2, the accuracy of the *Pre-Regist* method is significantly lower than that of the *Regist* and the *Post-Regist* method. This proves that the landmark-aware alignment module can improve the plane detection accuracy consistently and substantially. Figure 14.12 provides visualization of the 3D spatial distribution of the fetal brain landmarks pre-/post-alignment. It can be observed that all the landmarks are mapped to a similar spatial position, which indicates that all the fetal postures are roughly aligned.

To demonstrate the impact of the proposed adaptive dynamic termination (ADT) strategy, we performed comparison experiments with existing popular strategies such as the termination with max iterations (*Max-Step*), the lowest Q Value (*Low Q-Value* [23]), and the active termination [24] with LSTM (*AT-LSTM*). We also compared with our proposed *ADT* with different backbone network including Multi-layer Perceptron (*ADT-MLP*), vanilla RNN (*ADT-RNN*) and LSTM (*ADT-LSTM*). The superscript * represents the model was trained with the normalized MSE loss function (L_{MSE}, Eq. 14.12). As shown in Table 14.4, equipped with the adaptive dynamic termination strategy, the agent was able to avoid being trapped into an inferior local minimum and achieved better performance. Furthermore, from Table 14.5, we can observe that our proposed dynamic termination can save approximately 67% inference time at most, thus improving the efficiency of the reinforcement framework.

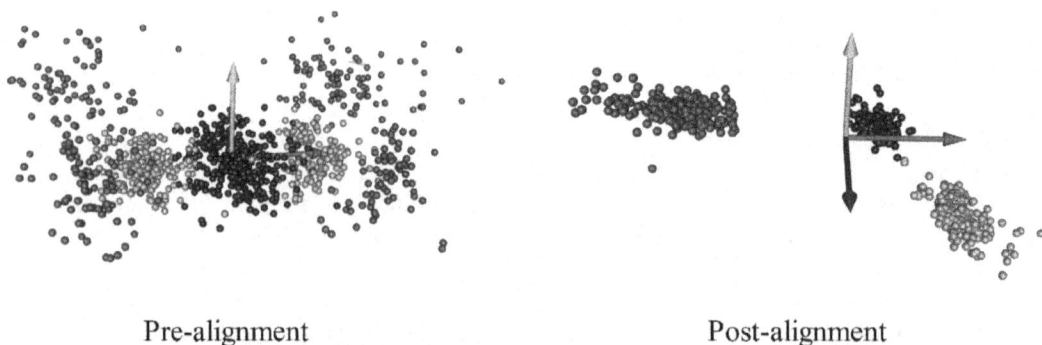

Pre-alignment   Post-alignment

FIGURE 14.12 The 3D visualization of fetal brain landmarks distribution of pre- and post-alignment. Different color points represent different category landmarks.

TABLE 14.1

Comparison Results of Our Proposed Method and Other Existing Methods in Fetal US (Mean±Std, Best Results Are Highlighted in Bold)

Method	TC			TV			TT			AM		
	Ang(°)	Dis(mm)	SSIM	Ang(°)	Dis(mm)	SSIM	Ang(°)	Dis(mm)	SSIM	Ang(°)	Dis(mm)	SSIM
Regression	12.44±7.78	**2.18±2.12**	0.634±0.157	13.62±4.98	5.52±4.02	0.636±0.136	13.87±11.77	2.81±2.16	0.760±0.141	17.48±12.45	6.24±4.77	0.758±0.081
AVP [58]	48.42±12.45	10.55±7.46	0.580±0.047	48.31±18.25	14.64±10.20	0.586±0.054	57.35±12.31	9.92±6.29	0.554±0.088	46.52±13.54	7.71±7.01	0.649±0.071
RL–US [59]	10.54±9.45	2.55±2.45	0.639±0.131	10.40±8.46	2.65±1.62	0.655±0.131	**10.37±8.08**	3.46±2.89	0.769±0.087	14.84±8.22	2.42±1.96	0.784±0.080
Ours	**10.26±7.25**	2.52±2.13	**0.640±0.144**	**10.39±4.03**	**2.48±1.27**	**0.659±0.135**	10.48±5.80	**2.02±1.33**	**0.783±0.060**	**14.57±8.50**	**2.00±1.64**	**0.790±0.074**

TABLE 14.2

Comparison Results of the Ablation Study for Analysis of Warm Start in Fetal US (Mean±Std, Best Results Are Highlighted in Bold)

Method	TC			TV			TT			AM		
	Ang(°)	Dis(mm)	SSIM	Ang(°)	Dis(mm)	SSIM	Ang(°)	Dis(mm)	SSIM	Ang(°)	Dis(mm)	SSIM
Pre-Regist	48.42±12.45	10.55±7.46	0.580±0.047	48.31±18.25	14.64±10.20	0.586±0.054	57.35±12.31	9.92±6.29	0.554±0.088	46.52±13.54	7.71±7.01	0.649±0.071
Regist	14.28±7.62	3.48±2.41	0.609±0.129	13.96±4.33	**2.39±1.34**	0.642±0.120	14.36±13.41	**2.11±1.41**	0.767±0.079	17.31±12.04	2.57±2.34	0.773±0.080
Post-Regist	**10.98±9.86**	**2.88±2.46**	**0.636±0.144**	**11.30±10.80**	2.66±1.69	**0.649±0.128**	**12.28±8.77**	2.62±2.50	**0.769±0.071**	**16.05±8.93**	**2.24±2.10**	**0.776±0.079**

TABLE 14.3

p-values of Pairwise *t*-tests between the Results of Each Method and Our Method for the Three Performance Metrics in the Fetal Dataset. The Bolded Results Represent Significant Difference

Metric	TC			TV			TT			AM		
	Ang(°)	Dis(mm)	SSIM	Ang(°)	Dis(mm)	SSIM	Ang(°)	Dis(mm)	SSIM	Ang(°)	Dis(mm)	SSIM
Ours vs. Regression	10^{-4}	**0.003**	0.138	**0.003**	10^{-15}	0.282	10^{-4}	**0.008**	0.301	**0.006**	10^{-25}	0.161
Ours vs. AVP [58]	10^{-56}	10^{-45}	**0.003**	10^{-57}	10^{-56}	**0.001**	10^{-64}	10^{-44}	10^{-18}	10^{-49}	10^{-32}	10^{-10}
Ours vs. Registration	10^{-4}	**0.003**	0.162	**0.001**	0.454	0.399	10^{-4}	**0.049**	0.329	**0.005**	**0.048**	0.399

To investigate if the difference between methods were statistically significant, we performed paired *t*-tests between the results of our methods and *Regression*, *AVP* [24], *Registration*. These tests were conducted for all of the performance metrics including *Angle*, *Distance* and *SSIM*. We set the significance level as 0.05. The results are shown in the Table 14.3. It indicate that our method performed best among the state-of-art methods (*Regression*, *AVP* [23]) and *Registration*. Although our method outperforms the *AT-LSTM* [24] without significant difference, our method could save at most 67% inference time as shown in Table 14.5.

Figure 14.13 provides visualization results of the proposed method. It shows the prediction plane, the ground truth, the termination curve, and the 3D spatial visualization of four randomly selected cases. It can be observed that the predictions are spatially close and visually similar to the ground truth. Furthermore, the proposed method can reach an ideal stopping point consistently. Both the maximum iteration and lowest Q-values termination strategies fail in spotting the optimal termination step.

Qualitative and Qualitative Results: PART 2

This section will present the results in discussed in ref. [29]. Extensive experiments were conducted to compare the proposed framework with the classical SL methods and the state-of-the-art RL methods. The registration results obtained by transformation using the Atlas-based registration [24] are considered as the baseline localization accuracy. We first compared the SL-based methods, i.e., single-plane regression (S-Regression) and multi-plane regression (M-Regression), which take the whole 3D volume as input and output the regressed plane parameters (ζ, β, ϕ, d in plane function $\cos(\zeta)x + \cos(\beta)y + \cos(\phi)z + d = 0$). RL-based methods are also compared, including SARL and MARL. To further validate each component in the proposed framework, we also conduct ablation studies. In specific, to validate the effectiveness of the landmark-alignment module, we reported the results when the module is not applied, i.e., SARL without alignment module (SARL-WA). To test the contribution of the proposed RNN-based collaborative module, we implemented the MARL method without NAS while equipped it with the proposed RNN (MARL-R). To test whether NAS could further help the MARL and compare the performance of different NAS measures, we added DARTS and GDAS to the MARL framework (D-MARL and G-MARL). Furthermore, to validate the compound effect of the NAS and the RNN collaborative module, we implemented D-MARL-R and G-MARL-R. Note that the RNN modules in MARL-R, D-MARL-R, and G-MARL-R are classical hand-crafted BiLSTM [60]. We used 3D ResNet18 as backbone for S-Regression and M-Regression, and 2D ResNet18 served as the network backbone for SARL-WA, SARL, MARL and MARL-R.

TABLE 14.4

Comparison Results of the Ablation Study for Analysis of Termination Strategy in Fetal US (Mean±Std, Best Results are Highlighted in Bold)

Method	TC			TV			TT			AM		
	Ang(°)	Dis(mm)	SSIM	Ang(°)	Dis(mm)	SSIM	Ang(°)	Dis(mm)	SSIM	Ang(°)	Dis(mm)	SSIM
Max-Step	12.73±11.33	3.74±3.33	0.619±0.135	12.17±11.54	2.80±1.85	0.645±0.127	17.29±14.56	4.76±5.17	0.727±0.091	18.97±11.93	2.90±3.10	0.755±0.095
Low Q-Value [58]	10.98±9.86	2.88±2.46	0.636±0.144	11.30±10.80	2.66±1.69	0.649±0.128	12.28±8.77	2.62±2.50	0.769±0.071	16.05±8.93	2.24±2.10	0.776±0.079
AT-LSTM [59]	10.54±9.45	2.55±2.45	0.639±0.131	10.40±8.46	2.65±1.62	0.655±0.131	10.37±8.08	3.46±2.89	0.769±0.087	14.84±8.22	2.42±1.96	0.784-0.080
ADT-MLP	10.39±7.33	2.55±2.15	0.640±0.145	10.97±4.56	2.57±1.49	0.653±0.133	11.90±7.50	3.20±3.43	0.764±0.074	15.30±8.27	2.34±2.18	0.779±0.081
ADT-RNN	10.63±7.24	2.66±2.19	0.640±0.142	10.90±4.46	2.55±1.47	0.655±0.134	11.45±7.18	2.78±3.12	0.774±0.068	15.05±0.77	2.26±2.06	0.781±0.077
ADT-LSTM	10.49±7.33	2.56±2.14	0.639±0.144	10.60±4.30	2.49±1.44	0.657±0.136	10.84±6.23	2.64±2.88	0.775±0.066	14.92±8.08	2.28±2.15	0.784±0.076
ADT-LSTM*	**10.26±7.25**	**2.52±2.13**	**0.640±0.144**	**10.39±4.03**	**2.48±1.27**	**0.659±0.135**	**10.48±5.80**	**2.02±1.33**	**0.783±0.060**	**14.57±8.50**	**2.00±1.64**	**0.790±0.074**

TABLE 14.5

Average Termination Step of the Adaptive Dynamic Termination and Active Termination

Plane	Termination Step	
	AT	ADT
TC	75	39.0
TV	75	24.8
TT	75	18.8
AM	75	20.3

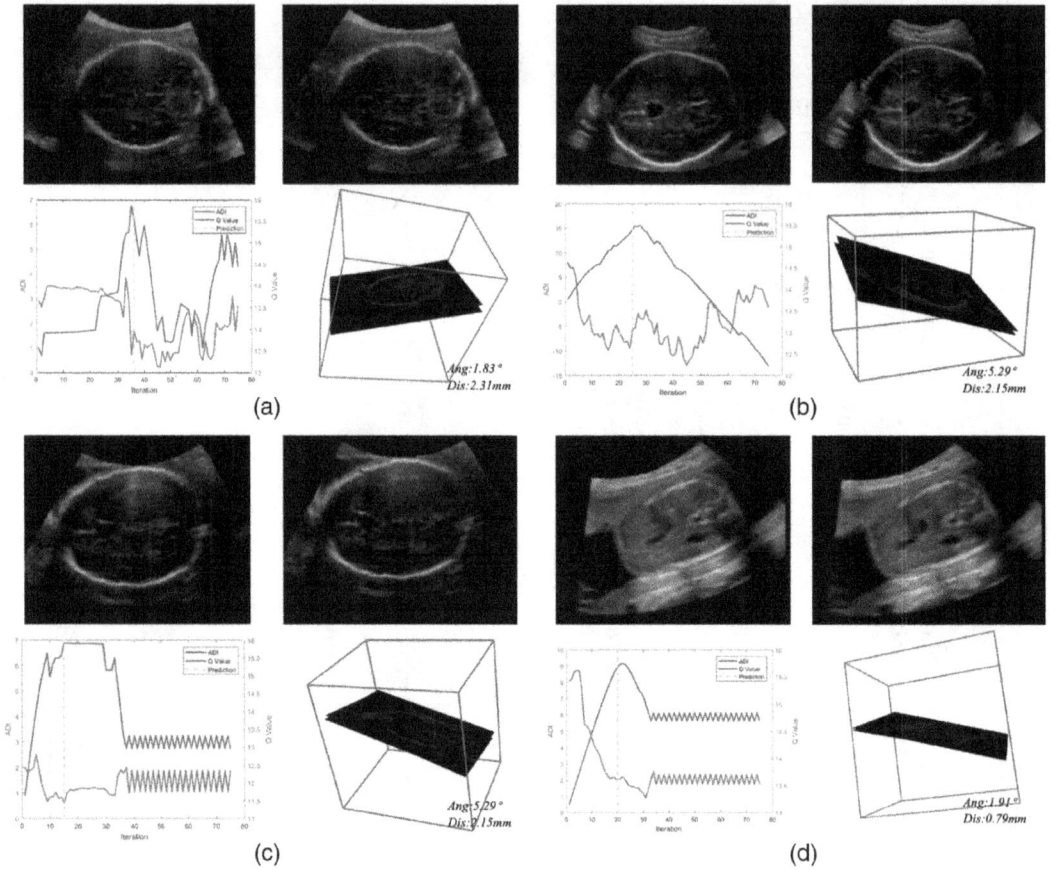

FIGURE 14.13 Visualization of our method in sampled SPs of the US fetal dataset. (a) is the TC SP, (b) is the TV SP, (c) is the TT SP, and (d) is the AM SP. For each case, the upper left is the predicted standard plane, the upper right is the ground truth, the bottom left is the inferring curve of the termination module, and the bottom right is the 3D spatial position of the predicted plane and ground truth.

Table 14.6 shows the results of our baseline method (i.e., Registration), SL- and RL-based methods (i.e., S-Regression vs. SARL and M-Regression vs. MARL). Note that for a fair comparison, we carefully designed the network structure (3D&2D ResNet18) and training strategy (Adam with learning rate = 5e-5 and batch size = 1) for SL- and RL-based methods. It can be seen that the RL-based methods can achieve better performance than the SL-based algorithms in almost all the evaluation metrics of different planes. This proves that directly learning the mapping from high-dimension volumetric data to low-dimension abstract features (i.e., plane parameters) is more difficult than learning that from 2D plane data in a sonographers-like RL system. Thus, RL-based methods are more reasonable for our task of plane localization in 3D US. Table 14.7 reports the RL results fetal brain datasets, respectively. The results of "SARL-WA" and "SARL" show that without the landmark-alignment module, the localization error will be large, and the agents fail to find their target planes. On comparing the "SARL" with the "MARL" column, it can be seen that the MARL-based methods achieved better performance than the SARL. This validates previous conjecture that multiple agents can share their useful information and improve the performance when detecting multiple SPs. Considering the necessity of NAS, both the "D-MARL" and the

TABLE 14.6

Localization Results of SL and RL-based Methods (Batch Size = 1 for Fair Comparion). The best results of S-Regression vs. SARL, and M-Regression vs. MARL are shown in bold.

Plane Metrics		TT	TV	TC
Registration	Ang (°)	19.75±13.83	17.52±13.46	18.25±16.65
	Dis (mm)	1.41±1.31	2.69±2.27	2.22±2.30
	SSIM	0.68±0.11	0.66±0.10	0.68±0.11
S-Regression	Ang (°)	16.63±15.77	15.92±18.21	15.33±13.35
	Dis (mm)	2.33±1.44	3.55±2.19	**2.31±1.72**
	SSIM	0.71±0.12	0.74±0.11	0.78±0.12
SARL	Ang (°)	**16.23±14.31**	**15.32±14.99**	**14.77±15.21**
	Dis (mm)	**1.49±1.28**	**3.47±2.34**	2.42±1.96
	SSIM	**0.76±0.08**	**0.75±0.06**	**0.80±0.07**
M-Regression	Ang (°)	16.77±14.21	15.89±16.02	15.21±14.99
	Dis (mm)	2.24±1.82	3.15±2.44	2.19±1.77
	SSIM	0.72±0.10	0.75±0.09	0.79±0.11
MARL	Ang (°)	**16.08±14.55**	**13.66±12.58**	**13.02±14.79**
	Dis (mm)	**1.41±1.55**	**1.55±1.42**	**1.88±1.99**
	SSIM	**0.75±0.08**	**0.79±0.07**	**0.81±0.08**

"G-MARL" outperformed plain "MARL" in both scenarios. Furthermore, "G-MARL" scored higher accuracy than "D-MARL," while searching the network architecture faster than the latter one. Another interesting phenomenon is that the RNN-based collaborative module is able to boost the model performance consistently (i.e., MARL-R, D-MARL-R and G-MARL-R). It proves that interaction and communication among different tasks are beneficial, as it provides knowledge on the spatial relationship between SPs. The last column of Table 14.7 shows the accuracy of the proposed framework, it can be seen that it outperformed its counterparts in almost all metrics in both datasets.

In addition, the computational cost and the size of a model are also a matter of practical concern. Previous methods [23, 49] adopted the VGG model as the backbone, which has been widely adopted in many computer vision tasks. In this work, we opt for ResNet18 which is also popular but relatively smaller, thus reducing both FLOPs and Params. Table 14.8 shows the FLOPs (computational costs) and the Params (model sizes) of all the tested methods. First, it can be seen that "SARL" and "MARL" are more resource-saving than "S-Regression" and "M-Regression," respectively. Furthermore, compared with "SARL," "MARL" has fewer FLOPs and Params. Intuitively, the more planes are located, the more network parameters and the FLOPs will drop. Besides, the DARTS-based and GDAS-based methods ("D-MARL," "D-MARL-R," "G-MARL," "G-MARL-R", and "Ours") are even more lightweight compared to the hand-crafted networks ("SARL," "MARL. and 'MARL-R'). The "G-MARL-R" model has comparable FLOPs and parameters with the proposed method. However, the proposed method achieved better performance in both datasets. It shows that the automatically discovered RNN is superior to the classical hand-crafted ones.

Furthermore, we performed the expert evaluation for the predicted planes from the clinical perspective. In specific, we invited two sonographers to score the predicted planes based on (1) the existence of anatomical structures and (2) visual plausibility. The score of a plane larger than 0.6

TABLE 14.7

Quantitative Evaluation of Plane Localization in Fetal Brain US Volume. The Best Results Are Shown in Bold

	Metrics	SARL-WA	SARL	MARL	MARL-R	D-MARL	D-MARL-R	G-MARL	G-MARL-R	Ours
TT	Ang (°)	57.35±12.31	16.11±13.42	15.94±11.01	13.60±15.30	13.34±12.11	11.76±13.71	12.51±14.14	11.34±13.11	**11.06±11.77**
	Dis (mm)	9.92±6.29	1.37±1.27	1.34±1.27	1.21±1.03	1.32±1.24	**0.92±1.01**	1.32±0.89	0.93±1.06	0.94±1.01
	SSIM	0.55±0.07	0.77±0.10	0.73±0.10	0.74±0.18	0.77±0.12	0.78±0.12	0.78±0.14	0.82±0.12	**0.83±0.11**
TV	Ang (°)	48.42±12.45	15.27±13.91	9.75±11.86	9.72±14.00	9.83±10.61	9.61±11.48	9.89±12.56	9.54±12.89	**8.59±12.51**
	Dis (mm)	14.64±10.20	3.40±1.97	1.68±2.10	**1.30±1.74**	1.72±2.05	1.46±1.17	1.66±1.93	1.50±2.13	1.43±1.98
	SSIM	0.59±0.05	0.76±0.09	0.80±0.10	0.80±0.14	0.79±0.12	0.80±0.12	0.79±0.11	0.83±0.11	**0.84±0.11**
TC	Ang (°)	48.31±18.25	14.43±17.32	12.23±18.03	12.48±17.90	11.73±15.35	11.02±14.79	11.26±15.66	10.04±15.03	**9.62±14.34**
	Dis (mm)	10.55±7.46	2.20±2.00	1.62±2.19	1.25±1.74	1.49±1.51	1.23±1.09	1.41±1.89	1.27±1.98	**1.20±1.76**
	SSIM	0.58±0.05	0.80±0.08	0.78±0.10	0.81±0.12	0.79±0.09	0.80±0.11	0.80±0.10	0.83±0.11	**0.84±0.10**
Avg	Ang (°)	51.36±15.06	15.27±15.11	12.64±14.14	11.93±15.90	11.63±14.67	10.80±13.91	11.22±14.22	10.31±13.74	**9.75±12.92**
	Dis (mm)	11.70±6.57	2.32±1.97	1.55±1.91	1.25±1.55	1.51±1.44	1.20±1.43	1.32±1.67	1.23±1.81	**1.19±1.66**
	SSIM	0.57±0.06	0.78±0.09	0.77±0.12	0.78±0.15	0.78±0.13	0.79±0.13	0.79±0.12	0.83±0.10	**0.84±0.11**

TABLE 14.8
Model Information of Compared Methods

	S-Regression	M-Regression	SARL	MARL	MARL-R	D-MARL	D-MARL-R	G-MARL	G-MARL-R	Ours
FLOPs (G)	337.23*3	337.26	1.82*3	1.82	1.82	0.72	0.72	0.70	0.70	0.70
Params (M)	33.16*3	33.17	12.21*3	12.22	12.26	4.69	4.73	4.52	4.56	4.56

will be considered as *small error*. The *score* of one plane is calculated by averaging two experts' ratings, and the *qualified rate* is denoted as the percentage of planes with *score* >0.6 to the total. As shown in Table 14.9, the localization results of our proposed method have small errors and can satisfy the clinical requirements well.

Figure 14.14 provides visualization of predicted SPs for the fetal brain volumes. It shows that the detected planes are visually and spatially similar to the ground truth. It is interesting to see how the three SPs in fetal brain are closely adjacent. However, the proposed method is able to differentiate them and display related structures correctly. For example, the cerebellum (the "eye-glass" shape structure) can be seen clearly in the TC plane. Furthermore, in Figure 14.15, we show TC planes obtained by our proposed methods (*Ours*) and the competing methods, including S-Regression, SARL-WA, SARL, MARL, MARL-R, and G-MARL-R. From the pseudo-color images, it can be intuitively seen that planes predicted by our proposed method are closest to the ground truth.

Qualitative and Qualitative Results: PART 3

This section will present the results in ref. [26]. To demonstrate the efficacy of our proposed method, we performed the comparison with five SOTA approaches including regression-based, (i.e., RG_{Single}, RG_{ITN} [21]), registration-based (i.e., *Regist* [24]), and RL-based methods (i.e., RL_{AVP} [23], RL_{WSADT} [29]). Four criteria, including the spatial metrics (angle and distance between two planes, Ang & Dis) and content metrics (Structural Similarity Index and Normalized Cross-correlation, SSIM & NCC), were used to evaluate the performance.

As shown in Table 14.10, our proposed method outperforms all of the others on most of the metrics, indicating the superior ability of our method in SP localization tasks. Specifically, we can observe that RL gains large boosting in performance through pre-registration to ensure orientation consistency (RL_{AVP} vs. RL_{WSADT}). In comparison, our new formulation could enable the RL algorithm to achieve superiority even without pre-registration. Visual illustration of the results of our method in Figure 14.16 also shows the extent of the SP localization performance associated with the quantitative measures reported in Table 14.10.

TABLE 14.9
Scoring Results of Our Proposed Method

	TT	TV	TC
Score (Mean±STD)	0.803±0.130	0.759±0.124	0.783±0.133
Qualified rate (Score >0.6)	97.05%	96.08%	96.08%

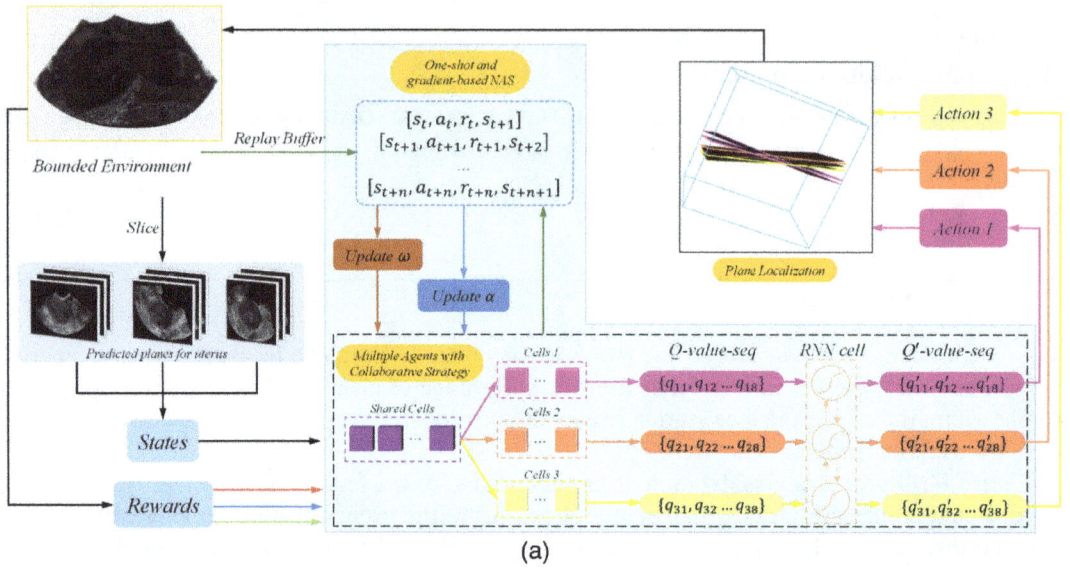

(a)

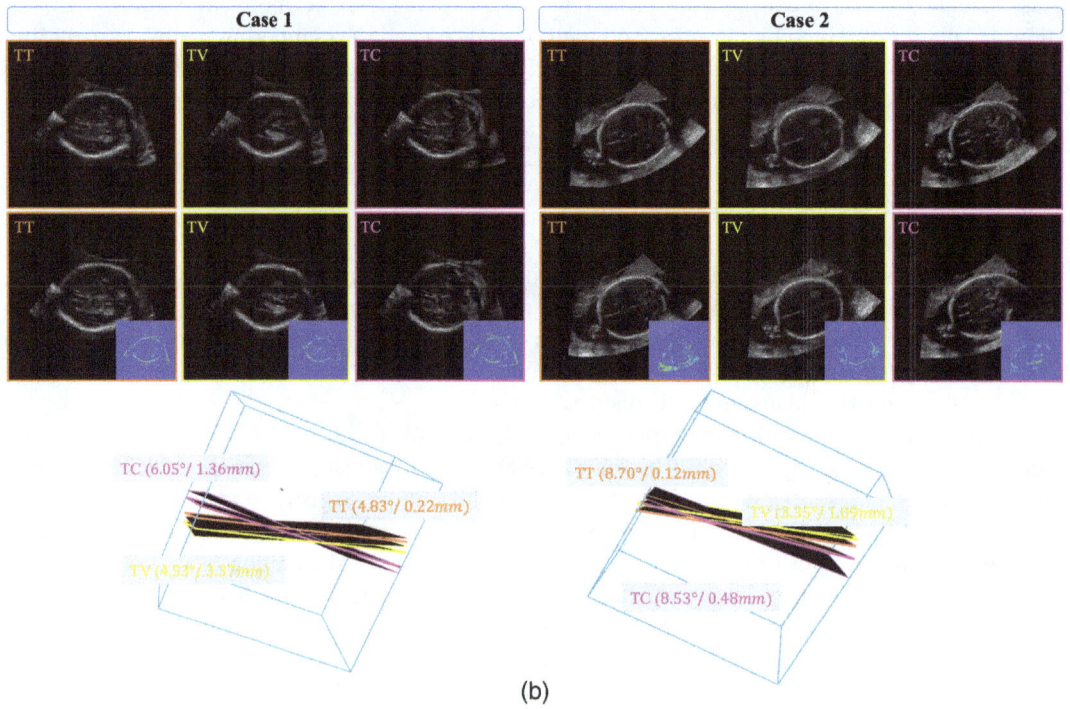

(b)

FIGURE 14.14 Visualization: Plane localization results for two fetal brain volumes (left and right). In both cases, the ground truth and prediction are shown in the first and second rows, respectively (the pseudo-color images in the lower right corner visualize the absolute intensity errors between the ground truth and prediction), and the third row illustrates their spatial difference.

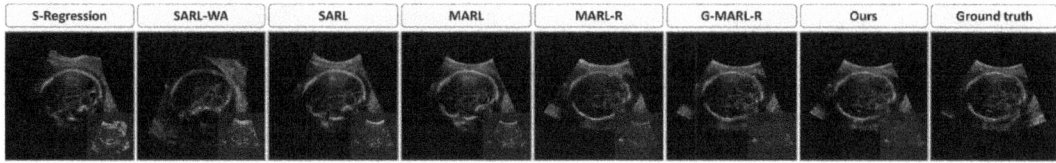

FIGURE 14.15 Visualization: TC planes obtained by different methods. The pseudo-color images in the lower right corner visualize the absolute intensity errors between the ground truth and prediction.

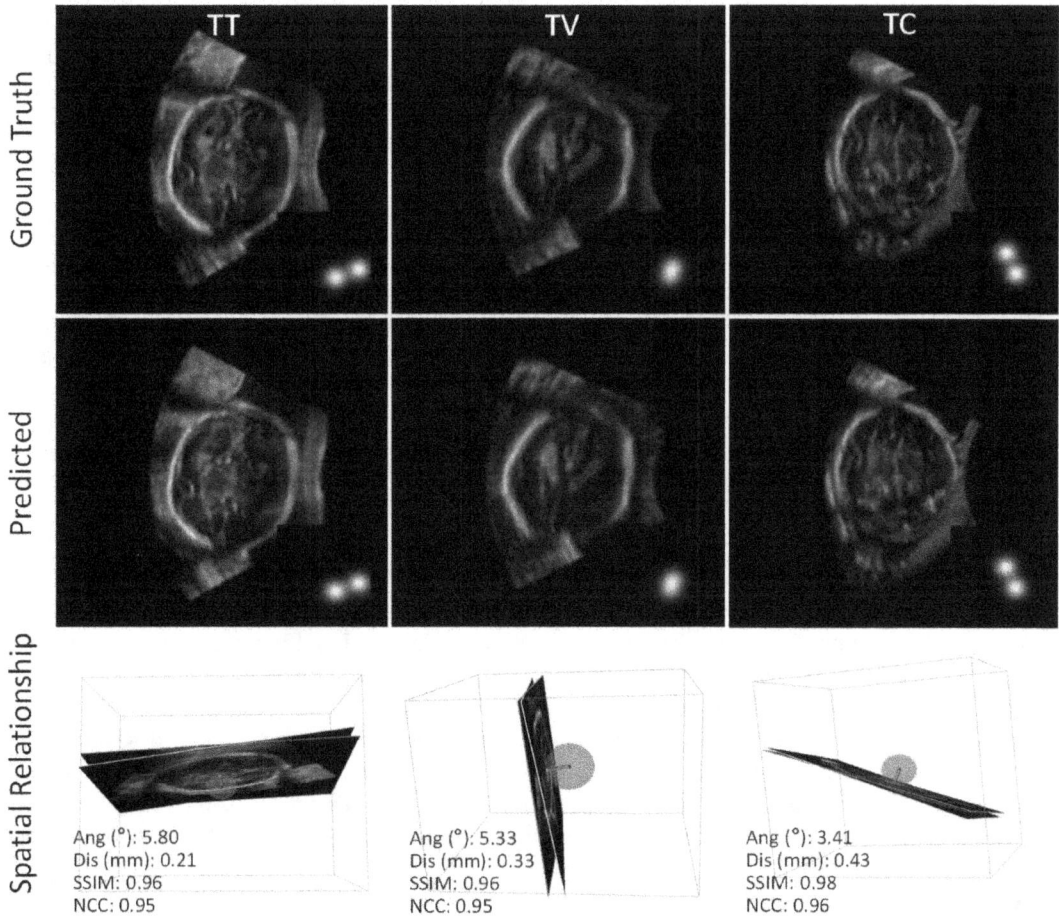

FIGURE 14.16 Visual example results of our methods. The first row shows the ground truth of three SPs; the second row shows the predicted plane with its landmark heatmap in the lower right corner; the third row shows the 3D spatial relationship between ground truth and prediction.

TABLE 14.10

Quantitative Comparison of Different Methods (Mean ± Std, Best Results Are Highlighted by Bold)

Methods	Metrics	RG_{single}	RG_{ITN}	$Regist$	RL_{AVP}	RL_{WSADT}	RL_{Ours}
* TT	Ang(°)	30.46±21.05	21.15±20.46	14.37±13.42	54.05±15.35	**10.48±5.80**	10.89±7.70
	Dis(mm)	3.53±2.19	0.94±0.75	2.12±1.42	4.34±2.97	2.02±1.33	**0.80±0.93**
	SSIM	0.58±0.14	0.85±0.05	0.83±0.09	0.64±0.06	0.78±0.06	**0.92±0.06**
	NCC	0.51±0.27	0.57±0.17	**0.83±0.13**	0.44±0.10	0.78±0.14	0.78±0.22
* TV	Ang(°)	38.78±25.40	26.80±25.55	13.40±4.68	53.77±14.58	10.39±4.03	**8.65±7.10**
	Dis(mm)	7.44±5.99	1.43±0.98	2.68±1.58	4.27±2.73	2.48±1.27	**1.16±2.45**
	SSIM	0.58±0.11	0.85±0.05	0.71±0.11	0.64±0.06	0.66±0.14	**0.92±0.06**
	NCC	0.53±0.18	0.56±0.16	0.56±0.27	0.43±0.11	0.57±0.28	**0.78±0.25**
* TC	Ang(°)	33.08±21.18	27.21±23.83	16.24±13.57	52.70±16.03	10.26±7.25	**9.75±8.45**
	Dis(mm)	3.82±3.30	1.26±1.06	3.47±2.39	4.20±2.65	2.52±2.13	**0.88±1.15**
	SSIM	0.59±0.13	0.84±0.05	0.68±0.18	0.64±0.07	0.64±0.14	**0.88±0.09**
	NCC	0.56±0.22	0.55±0.16	0.55±0.29	0.45±0.12	0.55±0.30	**0.69±0.25**

CONCLUSION AND DISSCUSSION

With this chapter, we present our investigation of RL-based framework for automatic plane localization in 3D US. Landmark-aware plane alignment module and imitation learning are vital for improving the learning efficiency of localization models. Constructing the network of agents, including the collaboration module, automatically designed architecture, and the branch for auxiliary task, plays an important role for improving the model performance. To reduce search space and decouple the action space, the plane localization task is proposed to be formulated as a tangent-point-based problem, which can facilitate agent learning. Spatial-anatomical reward is introduced to strengthen the obtained information and guide the learning trajectories of agents. An adaptive dynamic termination strategy is leveraged to stop the agent-environment interaction efficiently. All these designs are validated on fetal US volumes for plane localization (i.e., TT, TV, TC in fetal brain, and AM in fetal abdomen), and can be general for many other organs in 3D fetal US, or other types of US (e.g., pelvic US) and modalities (e.g., X-ray and MRI).

Although the RL-based automatic framework presented in this chapter is promising, there still exist many challenging problems for plane localization in 3D fetal US. First, the action space in the current RL system is discrete, it may further improve the model efficiency if extended to a continuous action space. Second, training or testing an RL framework is time-consuming. Thus, it is highly desired to design a time-saving RL training and testing strategy via some advanced RL updating methods or some adaptive learning-based tricks. Third, the experimental data used in the above-mentioned studies are all healthy cases. However, in clinical situation, deep models may fail in testing some abnormal cases. Improving the robustness of the model without abundant framework modification will be an interesting research direction.

ACKNOWLEDGMENTS

The work in this chapter was supported by the grant from National Key R&D Program of China (No. 2019YFC0118300), Shenzhen Peacock Plan (Nos. KQTD2016053112051497, KQJSCX20180328095606003), Natural Science Foundation of Shenzhen (JCYJ20190808115419619), National Natural Science Foundation of China (Nos. 62171290, 62101343), Shenzhen-Hong Kong Joint Research Program (No. SGDX20201103095613036), and Shenzhen Science and Technology Innovations Committee (No. 20200812143441001).

REFERENCES

1. L.J. Salomon, Z. Alfirevic, C.M. Bilardo, G.E. Chalouhi, T. Ghi, K.O. Kagan, T.K. Lau, A.T. Papageorghiou, N.J. Raine-Fenning, and J. Stirnemann, et al. Isuog practice guidelines: Performance of first-trimester fetal ultrasound scan. *Ultrasound in Obstetrics and Gynecology: The Official Journal of the International Society of Ultrasound in Obstetrics and Gynecology*, 41(1):102, 2013.

2. L.J. Salomon, Z. Alfirevic, V. Berghella, C. Bilardo, E. Hernandez-Andrade, S.L. Johnsen, K. Kalache, K.-Y. Leung, G. Malinger, and H. Munoz, et al. Practice guidelines for performance of the routine mid-trimester fetal ultrasound scan. *Ultrasound in Obstetrics and Gynecology*, 37(1):116–126, 2011.

3. D. Ni, X. Yang, X. Chen, C.-T. Chin, S. Chen, P. Ann Heng, S. Li, J. Qin, and T. Wang. Standard plane localization in ultrasound by radial component model and selective search. *Ultrasound in Medicine and Biology*, 40(11):2728–2742, 2014.

4. X. Yang, D. Ni, J. Qin, S. Li, T. Wang, S. Chen, and P. A. Heng. Standard plane localization in ultrasound by radial component. In *2014 IEEE 11th International Symposium on Biomedical Imaging (ISBI)*, pages 1180–1183, 2014.

5. B. Lei, L. Zhuo, S. Chen, S. Li, D. Ni, and T. Wang. Automatic recognition of fetal standard plane in ultrasound image. In *2014 IEEE 11th International Symposium on Biomedical Imaging (ISBI)*, pages 85–88, 2014.

6. L. Zhang, S. Chen, C.-T. Chin, T. Wang, and S. Li. Intelligent scanning: Automated standard plane selection and biometric measurement of early gestational sac in routine ultrasound examination. *Medical Physics*, 39(8):5015–5027, 2012.

7. H. Chen, D. Ni, J. Qin, S. Li, X. Yang, T. Wang, and P. A. Heng. Standard plane localization in fetal ultrasound via domain transferred deep neural networks. *IEEE Journal of Biomedical and Health Informatics*, 19(5):1627–1636, 2015.

8. H. Chen, D. Ni, X. Yang, S. Li, and P. A. Heng. Fetal abdominal standard plane localization through representation learning with knowledge transfer. In *Machine Learning in Medical Imaging*, pages 125–132, Cham, 2014. Springer International Publishing.

9. H. Chen, L. Wu, Q. Dou, J. Qin, S. Li, J. Cheng, D. Ni, and P. Heng. Ultrasound standard plane detection using a composite neural network framework. *IEEE Transactions on Cybernetics*, 47(6):1576–1586, 2017.

10 W. Huang, C. P. Bridge, J. A. Noble, and A. Zisserman. Temporal HeartNet: Towards human-level automatic analysis of fetal cardiac screening video. In MICCAI *2017*: Medical Image Computing and Computer-Assisted Intervention, pp. 341–349, 2017.

11 Y. Gao, and J. A. Noble. Detection and characterization of the fetal heartbeat in free-hand ultrasound sweeps with weakly-supervised two-streams convolutional networks. In MICCAI 2017: Medical Image Computing and Computer-Assisted Intervention, pp. 305–313, 2017.

12. C. F. Baumgartner, K. Kamnitsas, J. Matthew, T. P. Fletcher, S. Smith, L. M. Koch, B. Kainz, and D. Rueckert. Sononet: Real-time detection and localisation of fetal standard scan planes in freehand ultrasound. *IEEE Transactions on Medical Imaging*, 36(11):2204–2215, 2017.

13. J. Schlemper, O. Oktay, M. Schaap, M. Heinrich, B. Kainz, B. Glocker, and D. Rueckert. Attention gated networks: Learning to leverage salient regions in medical images. *Medical Image Analysis*, 53:197–207, 2019.

14. L. Wu, J. Cheng, S. Li, B. Lei, T. Wang, and D. Ni. FUIQA: Fetal ultrasound image quality assessment with deep convolutional networks. *IEEE Transactions on Cybernetics*, 47(5):1336–1349, 2017.

15. H. Luo, H. Liu, K.-J. Li, and B. Zhang. Automatic quality assessment for 2D fetal sonographic standard plane based on multi-task learning. *ArXiv*, abs/1912.05260, 2019.

16. Zehui Lin, Shengli Li, Dong Ni, Yimei Liao, Huaxuan Wen, Jie Du, Siping Chen, Tianfu Wang, and Baiying Lei. Multi-task learning for quality assessment of fetal head ultrasound images. *Medical Image Analysis*, 58:101548, 2019.

17. X. Lu, M.-P. Jolly, B. Georgescu, C. Hayes, P. Speier, M. Schmidt, X. Bi, R. Kroeker, D. Comaniciu, and P. Kellman, et al. Automatic view planning for cardiac mri acquisition. In *International Conference on Medical Image Computing and Computer-Assisted Intervention*, pp. 479–486. Springer, 2011.

18. K. Chykeyuk, M. Yaqub, and J. A. Noble. Class-specific regression random forest for accurate extraction of standard planes from 3D echocardiography. In *International MICCAI Workshop on Medical Computer Vision*, pp. 53–62. Springer, 2013.

19. H. Ryou, M. Yaqub, A. Cavallaro, F. Roseman, A. Papageorghiou, and J. A. Noble. Automated 3D ultrasound biometry planes extraction for first trimester fetal assessment. In *International Workshop on Machine Learning in Medical Imaging*, pp. 196–204. Springer, 2016.

20. C. Lorenz, T. Brosch, C. Ciofolo-Veit, K. Tobias, T. Lefevre, A. Cavallaro, I. Salim, A. Papageorghiou, C. Raynaud, and D. Roundhill, et al. Automated abdominal plane and circumference estimation in

3D us for fetal screening. In Medical Imaging 2018: Image Processing, volume 10574, pp. 105740I. International Society for Optics and Photonics, 2018.

21. Y. Li, B. Khanal, et al. Standard plane detection in 3D fetal ultrasound using an iterative transformation network. In MICCAI 2018: Medical Image Computing and Computer Assisted Intervention, pp. 392–400. Springer, 2018.

22. A. Schmidt-Richberg, N. Schadewaldt, T. Klinder, M. Lenga, R. Trahms, E. Canfield, D. Roundhill, and C. Lorenz. Offset regression networks for view plane estimation in 3D fetal ultrasound. In *Medical Imaging 2019: Image Processing*, volume 10949, pp. 109493K. International Society for Optics and Photonics, 2019.

23. A. Alansary, L. L. Folgoc, G. Vaillant, O. Oktay, Y. Li, W. Bai, J. Passerat-Palmbach, R. Guerrero, K. Kamnitsas, and B. Hou, et al. Automatic view planning with multi-scale deep reinforcement learning agents. In *International Conference on Medical Image Computing and Computer-Assisted Intervention*, pp. 277–285. Springer, 2018.

24. H. Dou, X. Yang, J. Qian, W. Xue, H. Qin, X. Wang, L. Yu, S. Wang, Y. Xiong, and P.-A. Heng, et al. Agent with warm start and active termination for plane localization in 3D ultrasound. In *International Conference on Medical Image Computing and Computer-Assisted Intervention*, pp. 290–298. Springer, 2019.

25. Y. Huang, X. Yang, R. Li, J. Qian, X. Huang, W. Shi, H. Dou, C. Chen, Y. Zhang, and H. Luo, et al. Searching collaborative agents for multi-plane localization in 3D ultrasound. In *International Conference on Medical Image Computing and Computer-Assisted Intervention*, pp. 553–562. Springer, 2020.

26. Y. Zou, H. Dou, Y. Huang, X. Yang, J. Qian, C. Zhen, X. Ji, N. Ravikumar, G. Chen, and W. Huang, et al. Agent with tangent-based formulation and anatomical perception for standard plane localization in 3D ultrasound. In *International Conference on Medical Image Computing and Computer-Assisted Intervention*, pp. 300–309. Springer, 2022.

27 J. Foerster, I. A. Assael, N. De Freitas, and S. Whiteson. Learning to communicate with deep multi-agent reinforcement learning. In Advances in Neural Information Processing Systems, pp. 2137–2145, 2016.

28. J. K. Gupta, M. Egorov, and M. Kochenderfer. Cooperative multi-agent control using deep reinforcement learning. In *International Conference on Autonomous Agents and Multiagent Systems*, pp. 66–83. Springer, 2017.

29. X. Yang, H. Dou, R. Huang, W. Xue, Y. Huang, J. Qian, Y. Zhang, H. Luo, H. Guo, and T. Wang, et al. Agent with warm start and adaptive dynamic termination for plane localization in 3d ultrasound. *IEEE Transactions on Medical Imaging*, 40(7):1950–1961, 2021.

30. X. Yang, Y. Huang, R. Huang, H. Dou, R. Li, J. Qian, X. Huang, W. Shi, C. Chen, and Y. Zhang, et al. Searching collaborative agents for multi-plane localization in 3D ultrasound. *Medical Image Analysis*, 72:102119, 2021.

31. V. Mnih, K. Kavukcuoglu, D. Silver, A. A. Rusu, J. Veness, M. G. Bellemare, A. Graves, M. A. Riedmiller, A. K. Fidjeland, G. Ostrovski, S. Petersen, C. Beattie, A. Sadik, I. Antonoglou, H. King, D. Kumaran, D. Wierstra, S. Legg, and D. Hassabis. Human-level control through deep reinforcement learning. *Nature*, 518:529–533, 2015.

32. F.-C. Ghesu, B. Georgescu, Y. Zheng, S. Grbic, A. K. Maier, J. Hornegger, and D. Comaniciu. Multi-scale deep reinforcement learning for real-time 3D-landmark detection in CT scans. *IEEE Transactions on Pattern Analysis and Machine Intelligence*, 41:176–189, 2019.

33. C. JCH Watkins, and P. Dayan. Q-learning. *Machine Learning*, 8(3-4):279–292, 1992.

34. H. V. Hasselt, A. Guez, and D. Silver. Deep reinforcement learning with double q-learning. In *Thirtieth AAAI Conference on Artificial Intelligence*, 2016.

35. Z. Wang, T. Schaul, M. Hessel, H. V. Hasselt, M. Lanctot, and N. de Freitas. Dueling network architectures for deep reinforcement learning. *ArXiv*, abs/1511.06581, 2016.

36. T. Schaul, J. Quan, I. Antonoglou, and D. Silver. Prioritized experience replay. *arXiv preprint arXiv:1511.05952*, 2015.

37. Ana IL Namburete, Richard V Stebbing, and J Alison Noble. Diagnostic plane extraction from 3D parametric surface of the fetal cranium. In *MIUA*, pages 27–32, 2014.

38. C. Lorenz, T. Brosch, C. Ciofolo-Veit, T. Klinder, T. Lefevre, A. Cavallaro, I. Salim, A. T. Papageorghiou, C. Raynaud, D. Roundhill, L. Rouet, N. Schadewaldt, and A. Schmidt-Richberg. Automated abdominal plane and circumference estimation in 3D US for fetal screening. In Elsa D. Angelini and Bennett A. Landman, editors, *Medical Imaging 2018: Image Processing*, volume 10574, pages 111–119. International Society for Optics and Photonics, SPIE, 2018.

39. R. Huang, W. Xie, and J. A. Noble. Vp-Nets: Efficient automatic localization of key brain structures in 3D fetal neurosonography. *Medical Image Analysis*, 47:127–139, 2018.

40. Ö. Çiçek, A. Abdulkadir, S. S. Lienkamp, T. Brox, and O. Ronneberger. 3D U-Net: Learning dense volumetric segmentation from sparse annotation. *ArXiv*, abs/1606.06650, 2016.

41. T. Hester, M. Vecerik, O. Pietquin, M. Lanctot, T. Schaul, B. Piot, D. Horgan, J. Quan, A. Sendonaris, and I. Osband, et al. Deep q-learning from demonstrations. In *Proceedings of the AAAI Conference on Artificial Intelligence*, volume 32, 2018.

42. L. BuSoniu, R. Babuška, and B. De Schutter. Multi-agent reinforcement learning: An overview. In *Innovations in Multi-Agent Systems and Applications-1*, pp. 183–221. Springer, 2010.

43. A. Vlontzos, A. Alansary, K. Kamnitsas, D. Rueckert, and B. Kainz. Multiple landmark detection using multi-agent reinforcement learning. In *International Conference on Medical Image Computing and Computer-Assisted Intervention*, pp. 262–270. Springer, 2019.

44. Y. Li. Deep reinforcement learning: An overview. *arXiv preprint arXiv:1701.07274*, 2017.

45. M. Jaderberg, V. Mnih, W. M. Czarnecki, T. Schaul, J. Z. Leibo, D. Silver, and K. Kavukcuoglu. Reinforcement learning with unsupervised auxiliary tasks. *arXiv preprint arXiv:1611.05397*, 2016.

46. P. Mirowski, R. Pascanu, F. Viola, H. Soyer, A. J. Ballard, A. Banino, M. Denil, R. Goroshin, L. Sifre, and K. Kavukcuoglu, et al. Learning to navigate in complex environments. *arXiv preprint arXiv:1611.03673*, 2016.

47. H. Liu, K. Simonyan, and Y. Yang. Darts: Differentiable architecture search. *arXiv preprint arXiv:1806.09055*, 2018.

48. C. J. Maddison, D. Tarlow, and T. Minka. A* sampling. In *Advances in Neural Information Processing Systems*, pp. 3086–3094, 2014.

49. X. Dong, and Y. Yang. Searching for a robust neural architecture in four GPU hours. In *Proceedings of the IEEE Conference on Computer Vision and Pattern Recognition*, pp. 1761–1770, 2019.

50. T. Elsken, J. H. Metzen, and F. Hutter. Neural architecture search: A survey. *arXiv preprint arXiv:1808.05377*, 2018.

51. J.-C. Yoo, and T. H. Han. Fast normalized cross-correlation. *Circuits, Systems and Signal Processing*, 28(6):819–843, 2009.

52. A. William Moore. Efficient memory-based learning for robot control. 1990.

53. A. G. Barto, R. S. Sutton, and C. W. Anderson. Neuronlike adaptive elements that can solve difficult learning control problems. *IEEE Transactions on Systems, Man, and Cybernetics*, 13(5):834–846, 1983.

54. J. C. Caicedo, and S. Lazebnik. Active object localization with deep reinforcement learning. In *Proceedings of the IEEE International Conference on Computer Vision*, pp. 2488–2496, 2015.

55. A. Paszke, S. Gross, F. Massa, A. Lerer, J. Bradbury, G. Chanan, T. Killeen, Z. Lin, N. Gimelshein, L. Antiga, A. Desmaison, A. Köpf, E. Yang, Z. DeVito, M. Raison, A. Tejani, S. Chilamkurthy, B. Steiner, L. Fang, J. Bai, and S. Chintala. PyTorch: An imperative style, high-performance deep learning library. In *33rd Conference on Neural Information Processing Systems*, Vancouver, 2019.

56. D. P. Kingma, and J. Ba. Adam: A method for stochastic optimization. *CoRR*, abs/1412.6980, 2015.

57. Z. Wang, A. C. Bovik, H. R. Sheikh, and E. P. Simoncelli. Image quality assessment: From error visibility to structural similarity. *IEEE Transactions on Image Processing*, 13:600–612, 2004.

58. A. Alansary, L. L. Folgoc, G. Vaillant, O. Oktay, Y. Li, W. Bai, J. Passerat-Palmbach, R. Guerrero, K. Kamnitsas, B. Hou, S. McDonagh, B. Glocker, B. Kainz, and D. Rueckert. Automatic view planning with multi-scale deep reinforcement learning agents. In *Medical Image Computing and Computer Assisted Intervention – MICCAI 2018*, pp. 277–285, Cham, 2018. Springer International Publishing.

59. H. Dou, X. Yang, J. Qian, W. Xue, H. Qin, X. Wang, L. Yu, S. Wang, Y. Xiong, P.-A. Heng, and D. Ni. Agent with warm start and active termination for plane localization in 3D ultrasound. In *Medical Image Computing and Computer Assisted Intervention – MICCAI 2019*, pp. 290–298, Cham, 2019. Springer International Publishing.

60. A. Graves, and J. Schmidhuber. Framewise phoneme classification with bidirectional LSTM and other neural network architectures. *Neural Networks*, 18(5-6):602–610, 2005.

Index

For Product Safety Concerns and Information please contact our EU
representative GPSR@taylorandfrancis.com
Taylor & Francis Verlag GmbH, Kaufingerstraße 24, 80331 München, Germany